Practical Statistics for the Analytical Scientist
A Bench Guide
2nd Edition

Practical Statistics for the Analytical Scientist
A Bench Guide
2nd Edition

Stephen L R Ellison
LGC, Queens Road, Teddington, Middlesex, UK

Vicki J Barwick
LGC, Queens Road, Teddington, Middlesex, UK

Trevor J Duguid Farrant
Cadbury (RSSL), Global Science Centre, University of Reading Campus, Reading, Berkshire, UK

RSCPublishing

ISBN: 978-0-85404-131-2

A catalogue record for this book is available from the British Library

© LGC Limited 2009

All rights reserved

Apart from fair dealing for the purposes of research for non-commercial purposes or for private study, criticism or review, as permitted under the Copyright, Designs and Patents Act 1988 and the Copyright and Related Rights Regulations 2003, this publication may not be reproduced, stored or transmitted, in any form or by any means, without the prior permission in writing of The Royal Society of Chemistry or the copyright owner, or in the case of reproduction in accordance with the terms of licences issued by the Copyright Licensing Agency in the UK, or in accordance with the terms of the licences issued by the appropriate Reproduction Rights Organization outside the UK. Enquiries concerning reproduction outside the terms stated here should be sent to The Royal Society of Chemistry at the address printed on this page

Published by The Royal Society of Chemistry,
Thomas Graham House, Science Park, Milton Road,
Cambridge CB4 0WF, UK

Registered Charity Number 207890

For further information see our website at www.rsc.org

Preface

An analyst working in an analytical laboratory today is not only expected to understand their instrumental methods and the chemistry of their test materials and analytes. They are also expected to understand and to follow basic principles of good practice in analytical measurement. One summary of these is the set of six principles currently disseminated by the Chemical and Biological Metrology Programme (see http://www.nmschembio.org.uk for details):

- *Principle 1:* Analytical measurements should be made to satisfy an agreed requirement.
- *Principle 2:* Analytical measurements should be made using methods and equipment which have been tested to ensure they are fit for purpose.
- *Principle 3:* Staff making analytical measurements should be both qualified and competent to undertake the task.
- *Principle 4:* There should be a regular independent assessment of the technical performance of a laboratory.
- *Principle 5:* Analytical measurements made in one location should be consistent with those elsewhere.
- *Principle 6:* Organisations making analytical measurements should have well-defined quality control and quality assurance procedures.

Of these, Principle 1 is usually identified with the procedures of contract review, planning and selection of methods and sampling protocols, Principle 2 with method validation, Principle 3 with proper education, training and qualification of analytical staff, Principle 4 with regular proficiency testing (PT), Principle 5 with the establishment of traceability through calibration of equipment and Principle 6 with the implementation of appropriate analytical quality assurance (QA) and quality control (QC) procedures, typically backed up by formal accreditation or other third-party assessment of quality systems.

Although there is much more to this than statistics, effective use of statistical methods certainly has a role in many of these processes. Method development and validation obviously depend very heavily on statistical methods, both for the design of suitable experiments and for the statistical tests used to identify problems and improvements and to confirm acceptability. Calibration of instruments almost invariably uses statistical methods, particularly least-squares regression. Proficiency testing organisers use statistical methods to summarise participant data and calculate scores and the interpretation of PT scores within a laboratory relies on basic statistical techniques and concepts. In sampling, the idea of a 'representative sample' is itself a statistical concept and proper sampling is fundamental; no analytical measurement is better than the sample analysed, and deciding on the appropriate number and type of test items is a key part of planning in many analytical projects. Day-to-day analytical quality control is also heavily dependent on statistical methods. It is no surprise, then, that third-party assessors invariably look for a sound statistical basis for all of these processes. This book is therefore intended to help analytical scientists use statistics effectively in their day-to-day work.

The first edition was written as a bench manual, covering the basic statistics needed in an analytical laboratory and providing detailed calculations for many of the more common operations. This second edition is a very substantial revision of the first. The changes have three principal causes. First, we wished to take advantage of the revision to improve the clarity and authority of the text, essentially all of which has been revised. In the process, we have added references to the principal literature sources for many tests and reviewed and amended many of the statistical procedures and the reference data. The Grubbs and Dixon tables in particular have been updated to follow recent published implementations and we have reverted to Grubbs' original implementation of Grubbs test 3 for greater consistency with other texts. The interpretation of two-way ANOVA has also been amended significantly, largely to allow valid interpretation irrespective of whether the effects under study are random or fixed.

Second, there have been significant changes in the availability of statistical software. Essentially every analytical scientist now has access to software for common statistical operations, including t-tests, regression and analysis of variance, even if only the implementations in common spreadsheet software. Because of this, the text has been revised to concentrate on the interpretation of output from the calculations, rather than on the calculations themselves. The calculations are included for reference, but are now to be found in appendices rather than in the body text.

Third, we wished to add new material. A new chapter on graphical assessment of data reflects the importance of visual review and interpretation, an essential step in data analysis. Experimental design is covered in another entirely new chapter, covering the basics of experimental design and planning and introducing some of the more advanced experimental designs to be found in specialist software. Analytical QA practices now routinely include the topic of measurement uncertainty; another new chapter therefore provides a summary of uncertainty estimation procedures, key references and a comparatively simple methodology for practical use. A new chapter on method validation explains the application of basic statistical methods and experimental design to method validation studies. These join existing chapters on proficiency testing and sampling, also important parts of analytical QA.

The text has been restructured with the aim of making sections easier to find and to provide additional explanation and guidance, both in the body of the text and, for example, in the extended section on distributions and types of data. Examples in the body text have also been rewritten using experimental data wherever possible and the exercises revised and expanded to cover the new topics.

As before, the book is intended primarily as a reference to useful procedures, rather than a theoretical textbook. The Introduction is intended as the starting point, providing cross-references to the procedures necessary for common analytical tasks. Wherever possible in the main text, topics describe a step-by-step procedure for implementing the necessary procedures. Finally, the question section provides exercises and fully worked answers to cover all the applications, so that a newcomer can check their understanding before applying the procedures to their own data.

S L R Ellison
V J Barwick

Acknowledgements

This comprehensive revision would not have been possible without many contributions. We would like to thank all those who provided detailed comments and suggestions for improvement on the first edition, particularly Professor M Thompson, of Birkbeck College. Our thanks also go to Professor Thompson, to Professor J Miller, of Loughborough University, to Alex Williams, former Government Chemist and Chair of the Eurachem Measurement Uncertainty Working Group, and to Dr C Burgess (Burgess Consultancy) for their help in checking the revised manuscript.

The preparation of this book was supported under contract with the National Measurement Office as part of the Chemical and Biological Metrology Programme.

Contents

Chapter 1	**Introduction – Choosing the Correct Statistics**	**1**
	1.1 Introduction	1
	1.2 Choosing the Right Statistical Procedures	2
	1.2.1 Planning Experiments	2
	1.2.2 Representative Test Portions for Large Samples	2
	1.2.3 Reviewing and Checking Data	2
	1.2.4 Reporting Results – Summarising and Describing Data	3
	1.2.5 Decisions About Differences and Limits	3
	1.2.6 Calibrating Instruments	4
	1.2.7 Describing Analytical Method Performance	4
	1.2.8 Analytical Method Validation	5
	1.2.9 Analytical Quality Control	5
	1.2.10 Testing Laboratory Performance – Proficiency Testing	6
	1.2.11 Measurement Uncertainty	6
Chapter 2	**Graphical Methods**	**7**
	2.1 Some Example Data	7
	2.2 Dot Plots	7
	2.3 Stem-and-Leaf Plots	8
	2.4 Tally Charts	9
	2.5 Histograms	9
	2.6 Frequency Polygon	10
	2.7 Cumulative Distribution	10
	2.8 Box Plots	11
	2.9 Scatter Plots	12
	2.10 Normal Probability Plots	13
Chapter 3	**Distributions and Types of Data**	**16**
	3.1 Introduction	16
	3.2 Describing Distributions	16

Practical Statistics for the Analytical Scientist: A Bench Guide, 2nd Edition
Stephen L R Ellison, Vicki J Barwick, Trevor J Duguid Farrant
© LGC Limited 2009
Published by the Royal Society of Chemistry, www.rsc.org

	3.3	Distributions of Analytical Data	17
		3.3.1 The Normal Distribution	17
		3.3.2 The Lognormal Distribution	20
		3.3.3 Poisson and Binomial Distributions	20
	3.4	Distributions Derived from the Normal Distribution	20
		3.4.1 Distribution of Student's t	20
		3.4.2 The Chi-squared Distribution	21
		3.4.3 The F Distribution	22
	3.5	Other Distributions	22
		3.5.1 Rectangular Distribution (or Uniform Distribution)	22
		3.5.2 Triangular Distribution	22
	3.6	Populations and Samples	23
	3.7	Checking Normality	23
	3.8	Types of Data	24
	References		24

Chapter 4 Basic Statistical Techniques — 25

	4.1	Summarising Data: Descriptive Statistics	25
		4.1.1 Introduction	25
		4.1.2 Counts, Frequencies and Degrees of Freedom	25
		4.1.3 Measures of Location	26
		4.1.4 Measures of Dispersion	27
		4.1.5 Skewness	29
		4.1.6 Kurtosis	29
	4.2	Significance Testing	29
		4.2.1 Introduction	29
		4.2.2 A Procedure for Significance Testing	30
		4.2.3 Tests on One or Two Mean Values – Student's t-Test	34
		4.2.4 Comparing Two Observed Standard Deviations or Variances – the F-Test	42
		4.2.5 Comparing Observed Standard Deviation or Variance with an Expected or Required Standard Deviation Using Tables for the F Distribution	45
	4.3	Confidence Intervals for Mean Values	46

Chapter 5 Outliers in Analytical Data — 48

	5.1	Introduction	48
	5.2	Outlier Tests	48
		5.2.1 The Purpose of Outlier Tests	48
		5.2.2 Action on Detecting Outliers	49
		5.2.3 The Dixon Tests	49
		5.2.4 The Grubbs Tests	51
		5.2.5 The Cochran Test	53
	5.3	Robust Statistics	53
		5.3.1 Introduction	53
		5.3.2 Robust Estimators for Population Means	54
		5.3.3 Robust Estimates of Standard Deviation	55

Contents   xi

5.4	When to Use Robust Estimators	57
References		58

Chapter 6 Analysis of Variance 59

6.1	Introduction	59
6.2	Interpretation of ANOVA Tables	60
	6.2.1 Anatomy of an ANOVA Table	60
	6.2.2 Interpretation of ANOVA Results	62
6.3	One-way ANOVA	63
	6.3.1 Data for One-way ANOVA	63
	6.3.2 Calculations for One-way ANOVA	63
6.4	Two-factor ANOVA	65
	6.4.1 Applications of Two-factor ANOVA	65
6.5	Two-factor ANOVA With Cross-classification	66
	6.5.1 Two-factor ANOVA for Cross-classification Without Replication	66
	6.5.2 Two-factor ANOVA for Cross-classification With Replication	69
6.6	Two-factor ANOVA for Nested Designs (Hierarchical Classification)	76
	6.6.1 Data for Two-factor ANOVA for Nested Designs	76
	6.6.2 Results Table for Two-factor ANOVA for Nested Designs	76
	6.6.3 Variance Components	77
	6.6.4 F-Tests for Two-factor ANOVA on Nested Designs	77
6.7	Checking Assumptions for ANOVA	79
	6.7.1 Checking Normality	79
	6.7.2 Checking Homogeneity of Variance – Levene's Test	80
6.8	Missing Data in ANOVA	81
Appendix: Manual Calculations for ANOVA		82

Chapter 7 Regression 92

7.1	Linear Regression	92
	7.1.1 Introduction to Linear Regression	92
	7.1.2 Assumptions in Linear Regression	92
	7.1.3 Visual Examination of Regression Data	93
	7.1.4 Calculating the Gradient and Intercept	94
	7.1.5 Inspecting the Residuals	97
	7.1.6 The Correlation Coefficient	98
	7.1.7 Uncertainty in Predicted Values of x	99
	7.1.8 Interpreting Regression Statistics from Software	100
	7.1.9 Testing for Non-linearity	102
	7.1.10 Designing Linear Calibration Experiments	103
	7.1.11 Two Common Mistakes	107
7.2	Polynomial Regression	108
	7.2.1 Polynomial Curves and Non-linearity	108
	7.2.2 Fitting a Quadratic (Second-order Polynomial)	109
	7.2.3 Using Polynomial Regression for Checking Linearity	109
Appendix: Calculations for Polynomial Regression		109
References		113

Chapter 8	**Designing Effective Experiments**		**114**
	8.1 Some New Terminology		114
	8.2 Planning for Statistical Analysis		115
	8.2.1 Measuring the Right Effect		115
	8.2.2 Single- *Versus* Multi-factor Experiments		115
	8.3 General Principles		115
	8.4 Basic Experimental Designs for Analytical Science		116
	8.4.1 Simple Replication		116
	8.4.2 Linear Calibration Designs		116
	8.4.3 Nested Designs		118
	8.4.4 Factorial Designs		118
	8.5 Number of Samples		119
	8.5.1 Number of Samples for a Desired Standard Deviation of the Mean		119
	8.5.2 Number of Samples for a Given Confidence Interval Width		120
	8.5.3 Number of Samples for a Desired *t*-Test Power		122
	8.5.4 Number of Observations for Other Applications and Tests		124
	8.6 Controlling Nuisance Effects		124
	8.6.1 Randomisation		125
	8.6.2 Pairing		128
	8.6.3 Blocked Designs		129
	8.6.4 Latin Square and Related Designs		131
	8.6.5 Validating Experimental Designs		133
	8.7 Advanced Experimental Designs		133
	8.7.1 Fractional Factorial Designs		133
	8.7.2 Optimisation Designs		135
	8.7.3 Mixture Designs		137
	8.7.4 D-optimal Designs		138
	8.7.5 Advanced Blocking Strategies		139
	Appendix: Calculations for a Simple Blocked Experiment		140
	References		143
Chapter 9	**Validation and Method Performance**		**144**
	9.1 Introduction		144
	9.2 Assessing Precision		145
	9.2.1 Types of Precision Estimate		146
	9.2.2 Experimental Designs for Evaluating Precision		146
	9.2.3 Precision Limits		149
	9.2.4 Statistical Evaluation of Precision Estimates		149
	9.3 Assessing Bias		150
	9.3.1 Statistical Evaluation of Bias Data		151
	9.4 Accuracy		152
	9.5 Capability of Detection		153
	9.5.1 Limit of Detection		153
	9.5.2 Limit of Quantitation		155
	9.6 Linearity and Working Range		156

	9.7	Ruggedness	157
		9.7.1 Planning a Ruggedness Study	158
		9.7.2 Evaluating Data from a Ruggedness Study	158
	References		159

Chapter 10 Measurement Uncertainty — 161

	10.1	Definitions and Terminology	161
	10.2	Principles of the ISO Guide to the Expression of Uncertainty in Measurement	162
		10.2.1 Steps in Uncertainty Assessment	162
		10.2.2 Specifying the Measurand	162
		10.2.3 Identifying Sources of Uncertainty – the Measurement Equation	162
		10.2.4 Obtaining Standard Uncertainties for Each Source of Uncertainty	164
		10.2.5 Converting Uncertainties in Influence Quantities to Uncertainties in the Analytical Result	165
		10.2.6 Combining Standard Uncertainties and 'Propagation of Uncertainty'	166
		10.2.7 Reporting Measurement Uncertainty	168
	10.3	Practical Implementation	168
		10.3.1 Using a Spreadsheet to Calculate Combined Uncertainty	168
		10.3.2 Alternative Approaches to Uncertainty Evaluation – Using Reproducibility Data	170
	10.4	A Basic Methodology for Uncertainty Estimation in Analytical Science	170
	References		171

Chapter 11 Analytical Quality Control — 173

	11.1	Introduction	173
	11.2	Shewhart Charts	173
		11.2.1 Constructing a Shewhart Chart	173
		11.2.2 Shewhart Decision Rules	175
	11.3	CuSum Charts	175
		11.3.1 Constructing a CuSum Chart	175
		11.3.2 CuSum Decision Rules	177
	References		178

Chapter 12 Proficiency Testing — 180

	12.1	Introduction	180
	12.2	Calculation of Common Proficiency Testing Scores	180
		12.2.1 Setting the Assigned Value and the Standard Deviation for Proficiency Assessment	180
		12.2.2 Scoring PT Results	184
	12.3	Interpreting and Acting on Proficiency Test Results	185
	12.4	Monitoring Laboratory Performance – Cumulative Scores	187
	12.5	Ranking Laboratories in Proficiency Tests	188
	References		188

Chapter 13 Simple Sampling Strategies — 189

    13.1 Introduction — 189
    13.2 Nomenclature — 189
    13.3 Principles of Sampling — 190
        13.3.1 Randomisation — 190
        13.3.2 Representative Samples — 190
        13.3.3 Composite Samples — 190
    13.4 Sampling Strategies — 191
        13.4.1 Simple Random Sampling — 191
        13.4.2 Stratified Random Sampling — 192
        13.4.3 Systematic Sampling — 194
        13.4.4 Cluster and Multi-stage Sampling — 195
        13.4.5 Quota Sampling — 197
        13.4.6 Sequential Sampling — 197
        13.4.7 Judgement Sampling — 198
        13.4.8 Convenience Sampling — 198
        13.4.9 Sampling in Two Dimensions — 199
    13.5 Uncertainties Associated with Sampling — 201
    13.6 Conclusion — 201
    References — 201

Appendices — 203

Appendix A Statistical Tables — 205

Appendix B Symbols, Abbreviations and Notation — 216

Appendix C Questions and Solutions — 220

    Questions — 220
    Solutions — 234

Subject Index — 263

CHAPTER 1

Introduction – Choosing the Correct Statistics

1.1 Introduction

Analytical scientists answer customers' questions. The variety of these is enormous and so, therefore, is the range of measurements that have to be made. Perhaps meaningful information can be provided only if a number of test samples are analysed. This might be because there is doubt about the homogeneity of the material or because more confidence is required in the result presented. To answer the customers' questions adequately, the analyst will need to decide on issues such as the number of samples to analyse and on what experiments and numerical comparisons must be done. These decisions require the use of statistics.

Using statistics effectively is an important part of any analytical scientist's job. For any meaningful decision, an analyst must plan the right experiments using properly validated methods, calibrate instruments, acquire the data under appropriately controlled conditions, review and check the results, present the data concisely and make objective decisions based on the data. Every step in this sequence of events is supported by some statistical procedure.

This book describes a range of the most useful statistical procedures for analytical scientists. It is intended to focus on procedure rather than theory and gives detailed instructions on how to carry out particular statistical operations, such as taking averages, evaluating standard deviations and constructing calibration curves. In general, each Chapter builds on topics from previous Chapters. Chapter 2 describes useful graphical methods for examining and presenting data. Chapter 3 provides some essential background information on data and the more common distributions relevant for assessing analytical data. Chapter 4 provides an array of basic statistical techniques, including the most common 'summary statistics' and the main statistical tests used in day-to-day analytical science. Outliers – a common feature of experimental data – are discussed in Chapter 5, which provides methods for detecting outliers and also some appropriate methods for handling data with some outlying values present. Chapters 6 and 7 describe analysis of variance and linear regression. Experimental design – a crucial preliminary activity – is discussed in Chapter 8, as it relies heavily on earlier concepts. Chapters 9–12 consider the statistical aspects of analytical method validation, measurement uncertainty estimation, quality control and proficiency testing, and Chapter 13 provides an introduction to sampling strategies for obtaining representative sets of test samples.

Practical Statistics for the Analytical Scientist: A Bench Guide, 2nd Edition
Stephen L R Ellison, Vicki J Barwick, Trevor J Duguid Farrant
© LGC Limited 2009
Published by the Royal Society of Chemistry, www.rsc.org

1.2 Choosing the Right Statistical Procedures

Statistics provides a large set of tools for analysing data. Many of the most useful for routine analysis, analytical method development and quality assurance are described in this book. But which tools are appropriate for different tasks?

The following sections form a short guide to the parts of this book which are best suited to solving particular problems. They are arranged as a series of questions that analysts might ask themselves, grouped according to the type of problem.

Throughout this section, questions are denoted by ▷ and answers by ▶.

1.2.1 Planning Experiments

▷ 'How many test samples and replicates do I need?'
▶ Chapter 8, Section 8.5, discusses number of samples or observations for several different situations.
▷ 'In what order do I run my test materials for a single run?'
▶ For a simple experiment in a single run, randomised order (Chapter 8, Section 8.6.1) is simple and effective.
▷ 'How do I minimise operator or run effects in a comparative experiment?'
▶ Either randomise or (better) treat Operator or Run as a 'blocking factor' in a randomised block design (Chapter 8, Section 8.6.3).
▷ 'How do I test for several effects at a time without confusing them?'
▶ Factorial designs (Chapter 8, Section 8.4.4) or, more economically, fractional factorial designs (Chapter 8, Section 8.7.1) can test several effects simultaneously without compromising test power.
▷ 'What experiments help most to improve analytical methods or processes?'
▶ Using fractional factorial designs (Chapter 8, Section 8.7.1) to identify important effects, followed up with optimisation designs (Chapter 8, Section 8.7.2) to locate optimal conditions is often considered the most effective and economical approach.

1.2.2 Representative Test Portions for Large Samples

▷ 'The sample sent to the laboratory is far too big to analyse every item and all the items look the same. Which items should be taken for analysis?'
▶ With a collection of apparently identical objects, simple random sampling is usually appropriate. This is described in Chapter 13, Section 13.4.1. A systematic sampling scheme (Chapter 13, Section 13.4.3) may also be appropriate if there is no reason to suspect regular variation in the laboratory sample.
▷ 'The sample sent to the laboratory is large and there are clearly two or more different types of item present. Which items should be taken for analysis to obtain an average value?'
▶ With two or more identifiable groups, stratified random sampling is often the most effective. This is described in Chapter 13, Section 13.4.2.

1.2.3 Reviewing and Checking Data

▷ 'How do I locate outliers in a set of experimental results?'

Introduction – Choosing the Correct Statistics

▶ Inspection using a dot plot (Chapter 2, Section 2.2) or box plot (Chapter 2, Section 2.8) is often the quickest way to find possible outliers. Follow up suspect values with *outlier tests* (see below).
▷ 'How do I decide whether an apparent outlier is a chance occurrence or requires closer examination?'
▶ Perform either Dixon's test or Grubbs' test (G') for a single outlier (Chapter 5, Sections 5.2.3 and 5.2.4, respectively).
▷ 'How do I check two possible outliers simultaneously?'
▶ Perform the Grubbs' tests G'' and G''' for outlying pairs (Chapter 5, Section 5.2.4).
▷ 'If I have data from several different sources, how do I find out if one of the sources has an abnormally high (or low) spread of data?'
▶ If you have a set of standard deviations, s, or variances, s^2, you can carry out the Cochran test following the procedure in Chapter 5, Section 5.2.5, to check whether one of the variances is abnormal.
▷ 'If I have outliers among my data, what should I do?'
▶ The appropriate action to take on finding outliers when using outlier tests is discussed in Chapter 5, Section 5.2.2. Alternatively, if outliers are likely, consider using robust statistics as described in Chapter 5, Section 5.3.

1.2.4 Reporting Results – Summarising and Describing Data

▷ 'How can I best summarise my data?'
▶ (a) If you have a set of analytical results from a single source (such as one method, location or time), it is normally sufficient to give the number of results, the mean and the standard deviation. Confidence limits or range may be included. All these statistics are described in Chapter 4, Section 4.1.
(b) If you have a set of results from several different sources and it is not appropriate to treat all the data as a single set, each set can be summarised separately as in (a). For multiple groups of data on the same material, it may also be useful to calculate within- and between-group standard deviations as described in Chapter 6, Section 6.3.2.2. For data on precision of methods, estimation of repeatability and reproducibility as described in Chapter 9, Section 9.2.2, is also useful.
▷ 'It looks like the results spread out more at higher concentrations. Is there a simple way of describing this?'
▶ Often, analytical precision is approximately proportional to concentration when far from the detection limit. This can be conveniently summarised across modest concentration ranges using the relative standard deviation or coefficient of variation described in Chapter 4, Section 4.1.4.4.

1.2.5 Decisions About Differences and Limits

1.2.5.1 Comparing Results with Limits

▷ 'I have a single measurement; how do I decide whether my result indicates that the limit has been exceeded?'
▶ This question is now most commonly addressed by considering the measurement uncertainty associated with the observation. If the result exceeds the limit by more than the expanded uncertainty (Chapter 10, Section 10.2.7), the limit is considered exceeded.

- ▷ 'I have made a number of measurements on a material; how do I decide if my results indicate that the limit has been exceeded?'
- ▶ Refer to Chapter 4, Section 4.2.3.1 (Comparing the mean with a stated value) and carry out a one-tailed test to see if the mean significantly exceeds the limit.
- ▷ 'The limit is zero. What do I do?'
- ▶ *Either* carry out a significance test (Chapter 4, Section 4.2.3.1) to see if the mean result significantly exceeds zero, *or* compare the result with the *critical value* described in Chapter 9, Section 9.5.1.

1.2.5.2 Differences Between Sets of Data

- ▷ 'Is there a significant difference between my two sets of results?'
- ▶ Use the procedure in Chapter 4, Section 4.2.3.2 (Comparing the means of two independent sets of data).
- ▷ 'Is my new method performing to the same (or better) precision than the method previously used?'
- ▶ If you have a set of data on the same material for each method, then compare the variances using the *F*-test (Chapter 4, Section 4.2.4).

1.2.6 Calibrating Instruments

- ▷ 'I have calibrated an instrument by measuring the instrument response for a set of solutions of differing concentrations of an analytical standard. What do I do next?'
- ▶ Refer to Chapter 7, Section 7.1, on linear regression. Calculate the regression line as described in Chapter 7, Section 7.1.4. Examine the residuals (Chapter 7, Section 7.1.5) to see whether the data are adequately represented by the chosen line. You should also examine the regression statistics (Chapter 7, Sections 7.1.6 and 7.1.8) to make sure the calibration is adequate.
- ▷ 'How do I calculate the concentration of a given test material from the calibration line?'
- ▶ Measure the instrument response (or responses, if replicated) for the given sample solution, then use the gradient and intercept of the calibration line as described in Chapter 7, Section 7.1.7.
- ▷ 'How do I estimate the statistical uncertainty of a concentration calculated from a regression line?'
- ▶ Refer to Chapter 7, Section 7.1.7, to calculate the standard error on the predicted concentration.
- ▷ 'How do I fit data to a curved line?'
- ▶ Some moderately curved calibration lines can be effectively fitted using polynomial regression as shown in Chapter 7, Section 7.2. Special forms (such as immunoassay response) are better treated with specialist software to fit the correct type of curve.

1.2.7 Describing Analytical Method Performance

Method performance is normally described using a range of parameters, such as precision (including repeatability and reproducibility), bias, recovery and detection and quantitation limits.

Introduction – Choosing the Correct Statistics

These are usually determined during method validation, which is discussed further below and in more detail in Chapter 9. Simple summary statistics for describing bias and dispersion are also covered in Chapter 4.

1.2.8 Analytical Method Validation

▷ 'How do I test for significant method bias using a certified reference material?'
▶ Experiments for bias are described in Chapter 9, Section 9.3. Statistical tests are usually based on the t-tests described in Chapter 4, Section 4.2.3.
▷ 'How do I test for bias compared to a reference method?'
▶ This question is answered in Chapter 9, Section 9.3.1.2. It is worth remembering that paired experiments (Chapter 8, Section 8.6.2) and the corresponding paired test in Chapter 4, Section 4.2.3.3, provide sensitive tests for differences between methods.
▷ 'How do I estimate the repeatability standard deviation?'
▶ The repeatability standard deviation (see Chapter 9, Sections 9.2.1 and 9.2.2) can be estimated from repeated measurements on identical test materials using the same method, the same operator and within short intervals of time. It can also be estimated in conjunction with reproducibility (see below).
▷ 'How do I estimate the reproducibility standard deviation?'
▶ The reproducibility standard deviation (Chapter 9, Sections 9.2.1 and 9.2.2) can be estimated from the results of repeated measurements on identical test materials using the same method, but in different laboratories. Usually, reproducibility is determined using simple nested experiments (Chapter 8, Section 8.4.3) and calculated from ANOVA tables as described in Chapter 9, Section 9.2.2.
▷ 'How do I determine the detection limit?'
▶ Experiments and calculations for characterising detection capability are discussed in detail in Chapter 9, Section 9.5.
▷ 'How do I assess linearity?'
▶ Linearity is assessed visually using scatter plots (Chapter 2, Section 2.9) of both raw data and residuals from linear regression (Chapter 7, Section 7.1.5). Objective tests for linearity are described in Chapter 7, Section 7.1.9.
▷ 'How do I carry out a ruggedness study?'
▶ Experiments and calculations for ruggedness studies usually use fractional factorial designs (Chapter 8, Section 8.7.1); their application to method validation is discussed in Chapter 9, Section 9.7.
▷ 'What do I do if I need to compare more than two methods?'
▶ Refer to Chapter 6, which describes analysis of variance (ANOVA). The type of ANOVA test depends on the information available. For results on the same material for each method, one-way ANOVA as described in Section 6.3 is appropriate. When several materials have been tested, two-factor ANOVA (Section 6.5) is appropriate.

1.2.9 Analytical Quality Control

▷ 'How do I set up a quality control (QC) chart?'
▶ Obtain data from the analysis of a stable check sample repeated over time (ideally several weeks or months) to set up either a Shewhart chart (Section 11.2) or a CuSum chart (Section 11.3).

▷ 'What is the best chart to detect unusual individual results?'
▶ A Shewhart chart (Chapter 11, Section 11.2) is often the most useful chart for identifying short-term deviations such as individual analytical run problems.
▷ 'How do I set up a chart to detect small changes over time?'
▶ A CuSum chart (Chapter 11, Section 11.3) is particularly good for identifying small, sustained shifts in a process or for retrospective examination to locate the time of a change.

1.2.10 Testing Laboratory Performance – Proficiency Testing

▷ 'How did we do?'
▶ Review the scores from your last proficiency test (PT) following Chapter 12, Section 12.3. Large z-scores indicate poorer performance.
▷ 'How do PT scores work?'
▶ PT scores compare the difference between a laboratory's result and an assigned value with some measure of the expected spread of results for competent laboratories. The most common are z-scores and Q-scores described in Chapter 12, Section 12.2.2.
▷ 'What does a z-score outside ± 2 indicate?'
▶ Usually, a z-score over 2 is an indication that the laboratory result was unusually distant from the assigned value. If above 3, investigative and corrective actions are nearly always required (see Chapter 12, Section 12.3).
▷ 'Is there a way of keeping track of our laboratory performance over time?'
▶ Chapter 12, Section 12.4, gives some useful pointers.
▷ 'We had the best score in our last PT round. Does that make us the best laboratory?'
▶ Sadly, no. Ranking in PT schemes is not a good indication of performance. Chapter 12, Section 12.5 explains why.

1.2.11 Measurement Uncertainty

▷ 'What is measurement uncertainty?'
▶ Measurement uncertainty is a single number that summarises the reliability of individual measurement results. It is defined and discussed in detail in Chapter 10.
▷ 'Is there a quicker way of calculating measurement uncertainty?'
▶ A particularly useful method using an ordinary spreadsheet is described in Chapter 10, Section 10.3.1.
▷ 'I already have the published repeatability and reproducibility data for my analytical method. Can I use this for uncertainty estimation?'
▶ Repeatability and reproducibility data can be used as the basis for uncertainty estimation in testing laboratories. The general principle is described briefly in Chapter 10, Section 10.3.2 and references to some relevant international documents (some freely available) are provided.

CHAPTER 2
Graphical Methods

2.1 Some Example Data

Graphical methods greatly simplify the assessment of analytical data. This chapter covers a range of graphical tools appropriate for reviewing analytical results. Most of the plots will use part or all, of the small data set shown in Table 2.1. The data are experimental data from an in-house validation study of a routine analytical method for antibiotic residues in foodstuffs.

2.2 Dot Plots

Small data sets are best reviewed as *dot plots*, sometimes also known as *strip charts*. For example, the day 2 data for thiamphenicol produces the dot plot in Figure 2.1a. The value at 66.58 stands out immediately as a possible suspect value. For larger data sets, dot plots begin to suffer from overlap between data points. This is usually treated by stacking points vertically to avoid overlap (Figure 2.1b).

Dot plots can also be plotted separately for individual groups in a larger data set. The complete thiamphenicol data set from Table 2.1 is shown grouped by day in Figure 2.2.

Dot plots and strip charts are routinely provided by statistics packages, but rarely by spreadsheets. However, a dot plot can be produced relatively easily in a spreadsheet using an 'X–Y' or scatter plot, with the *y*-axis values set to some arbitrary constant value for all the observations. For grouped data, as in Figure 2.2, the *y*-axis values are simply set to a different value for each group.

Table 2.1 Thiamphenicol in food (spiked at $50 \, \mu g \, kg^{-1}$).

Day 1:	54.72	56.78	53.33	48.59	53.80	52.06
Day 2:	44.31	45.48	49.69	52.17	54.69	66.58
Day 3:	52.55	47.34	64.89	45.76	63.52	59.13

Practical Statistics for the Analytical Scientist: A Bench Guide, 2nd Edition
Stephen L R Ellison, Vicki J Barwick, Trevor J Duguid Farrant
© LGC Limited 2009
Published by the Royal Society of Chemistry, www.rsc.org

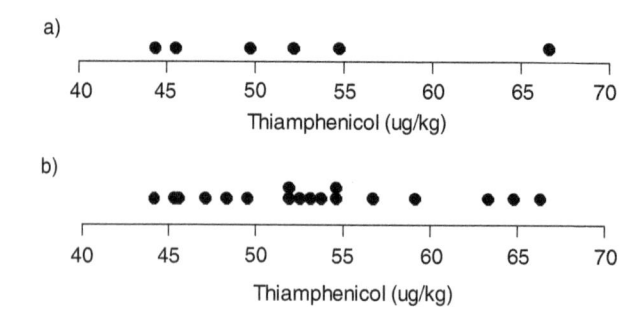

Figure 2.1 Dot plots. (a) Thiamphenicol data (Table 2.1) for day 2 only; (b) data for all three days.

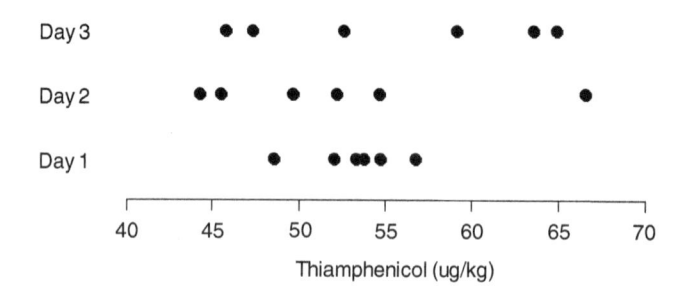

Figure 2.2 Dot plot for grouped data. The figure shows the thiamphenicol data from Table 2.1, grouped by day.

2.3 Stem-and-Leaf Plots

Stem-and-leaf plots are a traditional graphical method, still occasionally useful for quick paper summaries and suitable for viewing medium-sized data sets. Stem-and-leaf plots use the digits in the rounded raw data to form natural intervals. Rounding to two figures, the whole data set becomes, in order of appearance:

55, 57, 53, 49, 54, 52, 44, 45, 50, 52, 55, 67, 53, 47, 65, 46, 64, 59

Writing the first digit(s) in the first column, and the next digit as a list to the right, gives a 'plot' of the following form:

```
4    9, 4, 5, 7, 6
5    5, 7, 3, 4, 2, 0, 2, 5, 3, 9
6    7, 5, 4
```

In this very simple variant, the first line lists the values 4 9 = 49, 4 4 = 44, *etc*; these are just the values from 40 to 49 found in the data table, read in order. Spacing the subsequent digits approximately equally (or, traditionally, not spacing them at all) gives a quick visual impression of the distribution.

More sophisticated stem-and-leaf plots may further divide the rows and additionally sort the 'leaf' digits into order as shown in Figure 2.3. Again, the first line lists the value 44, the second 45, 46, *etc*. Note that in this particular listing, the 'leaves' are between 0 and 4 for the first instance of a value in the first column and 5–9 for the second instance. This is most obvious for the two '5' rows

Graphical Methods

```
4 | 4
4 | 5679
5 | 022334
5 | 5579
6 | 1
6 | 57
```

Figure 2.3 Stem-and-leaf plot for thiamphenicol results (all data). The decimal point is one digit to the right of the | symbol.

in the layout shown in Figure 2.3; the first includes all the rounded data from 50 to 54 (that is, 50, 52, 52, 53, 53 and 54); the second includes all the data from 55 to 59.

Note that it is important to make sure that the intervals for each row are the same size, so stem-and-leaf plots will use intervals either from 0 to 9 per row or (as in Figure 2.3) intervals of 0–4 and 5–9. Unequal intervals will give uneven frequencies and make interpretation much more difficult.

2.4 Tally Charts

Another 'back of an envelope' method of examining the distribution of larger data sets is to divide the set into a number of intervals, count the number of observations in each group and write these in tabular form. The 'tally marks' provide a quick visual indication of the distribution of data, as shown in Table 2.2. Here, the data visibly cluster around approximately 53 µg kg^{-1}. (The frequency data are not necessary for this simple plotting method, but can be used in histogram drawing and other methods.)

2.5 Histograms

A histogram is usually presented as a vertical bar chart showing the number of observations in each of a series of (usually equal) intervals, exactly as in Table 2.2. The horizontal axis is divided into segments corresponding to the intervals. On each segment a rectangle is constructed whose *area* is proportional to the frequency in the group. Figure 2.4a shows a typical example.

Stem-and-leaf plots, tally charts and histograms are all somewhat sensitive to the choice of intervals. Figure 2.4b shows the same data as Figure 2.4a with a different start point. Even

Table 2.2 Tally chart for thiamphenicol results (all data).

Range (µg kg^{-1})	Midpoint	Tally marks	f^a	C^b	C(%)
<40	–		0	0	0.0
40–44.99	42.5	|	1	1	5.6
45–49.99	47.5	|||||	5	6	33.3
50–54.99	52.5	|||||||	7	13	72.2
55–59.99	57.5	||	2	15	83.3
60–64.99	62.5	||	2	17	94.4
65–69.99	67.5	|	1	18	100.0
>70	–		0	18	100.0

[a] f=frequency or count.
[b] C=cumulative frequency; sum of counts up to and including the current group. C (%) is the cumulative frequency as a percentage of the total number.

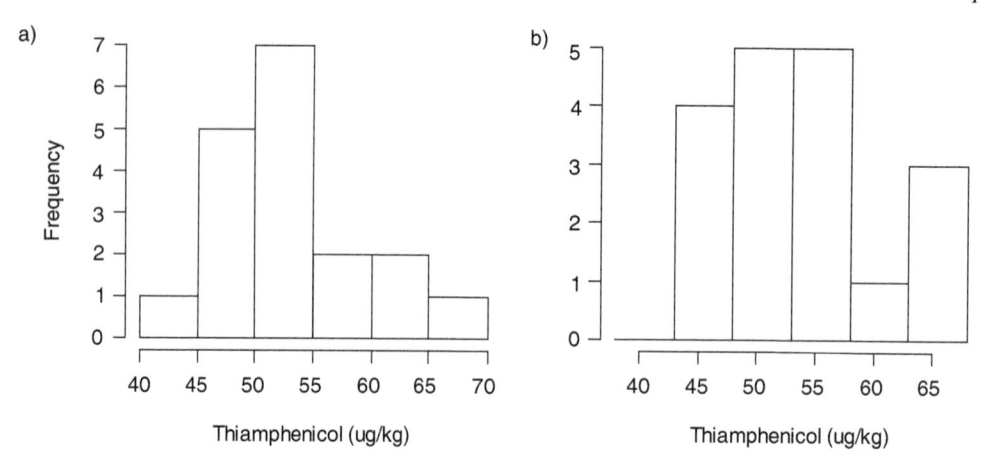

Figure 2.4 Histograms for thiamphenicol data (all data). (a) Intervals of width 5 from 40; (b) intervals of width 5 from 38.

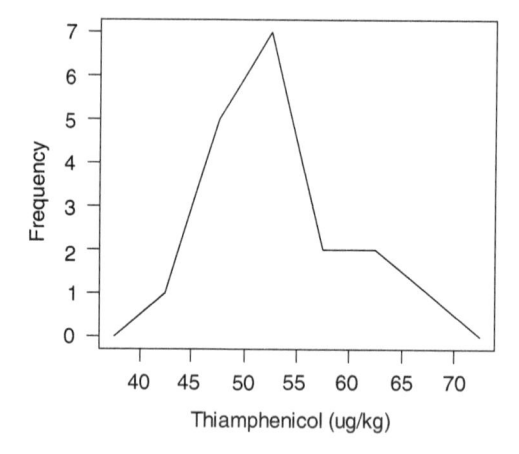

Figure 2.5 Frequency polygon for the thiamphenicol data in Table 2.1.

relatively small changes in start point or interval width can noticeably change the appearance of the plot, especially for smaller data sets.

2.6 Frequency Polygon

If the variable is discrete (*i.e.* restricted to particular values, such as integers), an appropriate method of plotting such data is to erect lines or bars at a height proportional to the frequencies. However, it is also possible to plot the points and join them with straight lines, forming a polygon. The same format can also be used with continuous data by plotting frequencies at the mid-points in each group (Figure 2.5). Like the histogram, the shape is sensitive to the choice of intervals.

2.7 Cumulative Distribution

The cumulative distribution curve (or, more correctly for observed data, the 'empirical cumulative distribution function') is a plot of the proportion of the data (on the vertical axis) included up to the

Graphical Methods

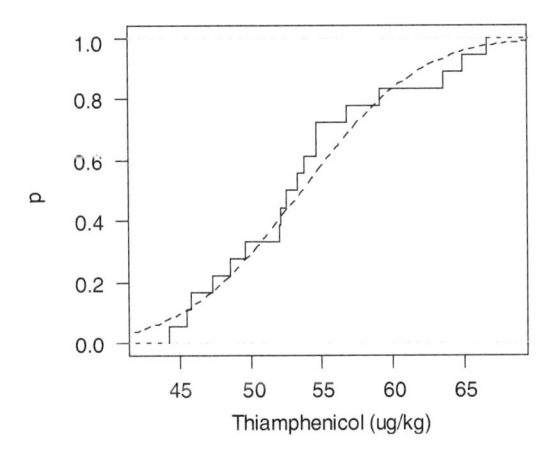

Figure 2.6 Cumulative distribution (all thiamphenicol data).

current position on the *x*-axis. For observed data, the plot is a step function (Figure 2.6). One way of visualising the meaning is to imagine walking, from the left, through the sorted data. Each time you reach an observation, the curve takes one step *upwards*. Each step upwards is equal to $1/n$, where n is the number of observations, so the vertical axis always runs from 0 to 1 (or from 0 to 100%). A cumulative frequency plot can be used to find the proportion of results falling above or below a certain value or to estimate *quantiles* such as the median or quartiles (see Chapter 4, Section 4.1).

A cumulative distribution curve is more commonly seen by analytical scientists for smooth distribution functions; the dashed curve in Figure 2.6 shows the cumulative probability for the normal distribution with the same mean and standard deviation as the thiamphenicol data. The characteristic S-shape (ogive curve) arises when the frequency has a maximum well inside the range of the variable.

2.8 Box Plots

A box plot – often also called a 'box-and-whisker' plot – is a useful method of summarising larger data sets, particularly where the data fall into groups. A box plot for the three days' thiamphenicol data is shown in Figure 2.7. The different features can represent different statistics, but the most common choice is as follows:

- The central solid line is the *median* for each group (see Chapter 4, Section 4.1.3.2).
- The bottom and top of the rectangular 'box' show the lower and upper quartiles, respectively. In effect, the box shows the range of the central 50% of the data points. The length of the box is the *interquartile range*, a useful indicator of the dispersion of the data (see Chapter 4, Section 4.1.4.6).
- The lines extending upwards and downwards from each box – the 'whiskers' – are drawn from the end of the box to the last data point within 1.5 times the interquartile range of the box end. The lines are often terminated with a horizontal line, or 'fence', at this value.
- Individual observations outside the 'fences' are drawn as separate points on the plot (there are none in the thiamphenicol data).

For a normal distribution, observations outside the 'fences' are expected about 0.7% of the time, so individual points outside the fences are generally considered to be outliers (see Chapter 5).

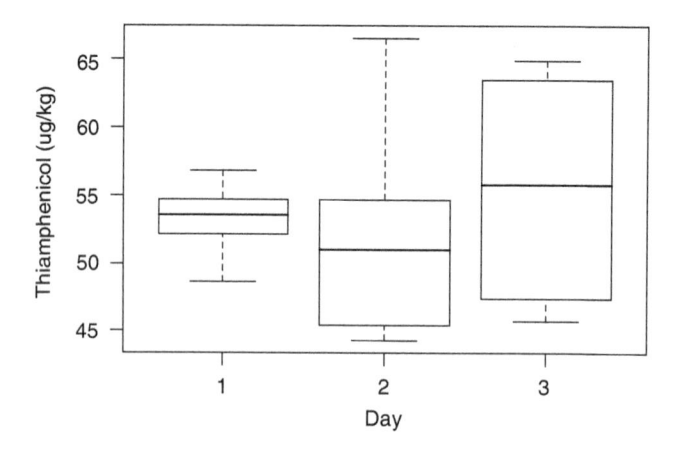

Figure 2.7 Box plot: thiamphenicol data by day.

The plot therefore gives a good deal of information. The central line shows the central value for each group of data; the box indicates the dispersion; the whiskers and fences show the range of the bulk of the data and single data points outside that range are outliers meriting checking and investigation. Since these data are summarised for several groups simultaneously, we can also compare groups very easily. For example, in Figure 2.7, we can see immediately that there are no seriously suspect points; that the data for day 1 are clustered more tightly than for the other two days; that the medians seem broadly consistent (at least by comparison with the ranges); and that on day 2, there is a suggestion of asymmetry, although the most extreme high value is not marked as an outlier (compare the box plot with the dot plots in Figure 2.1a and Figure 2.2).

Although it is not shown here, it is worth noting that one additional refinement, the *notched* box plot, adds a notch in each box, centred on the median. The notch size is chosen so that non-overlapping notches indicate a statistically significant difference between the medians, allowing a visual assessment of the significance of differences between the medians.

Box plots work best for data sets with more than about five data points per group and are recommended for examining grouped data.

2.9 Scatter Plots

Scatter plots, such as that in Figure 2.8, are the most common method of reviewing or presenting bivariate data, that is, data relating to two variables (conventionally x and y). The plot shows a data point at each pair of (x,y) coordinates. A graph of a calibration curve is probably the most familiar form to analytical scientists. Figure 2.8 shows a different example; a plot of observed value against expected value, with a line fitted through the data. It is easy to see that the observed value is (as expected) proportional to the spiking level and the increasing scatter of data around the line also provides a quick indication that the dispersion increases with concentration. The figure uses the conventional arrangement of known ('predictor' or 'independent') variable – the added amount, in this case – on the horizontal x-axis and the measured or 'dependent' variable on the vertical y-axis. It is also perfectly permissible to plot two measured values on a scatter plot, for example to compare responses for two different analytical methods on a series of test samples or simply to see whether there seems to be any relation between two measured variables. (Note that simple

Graphical Methods 13

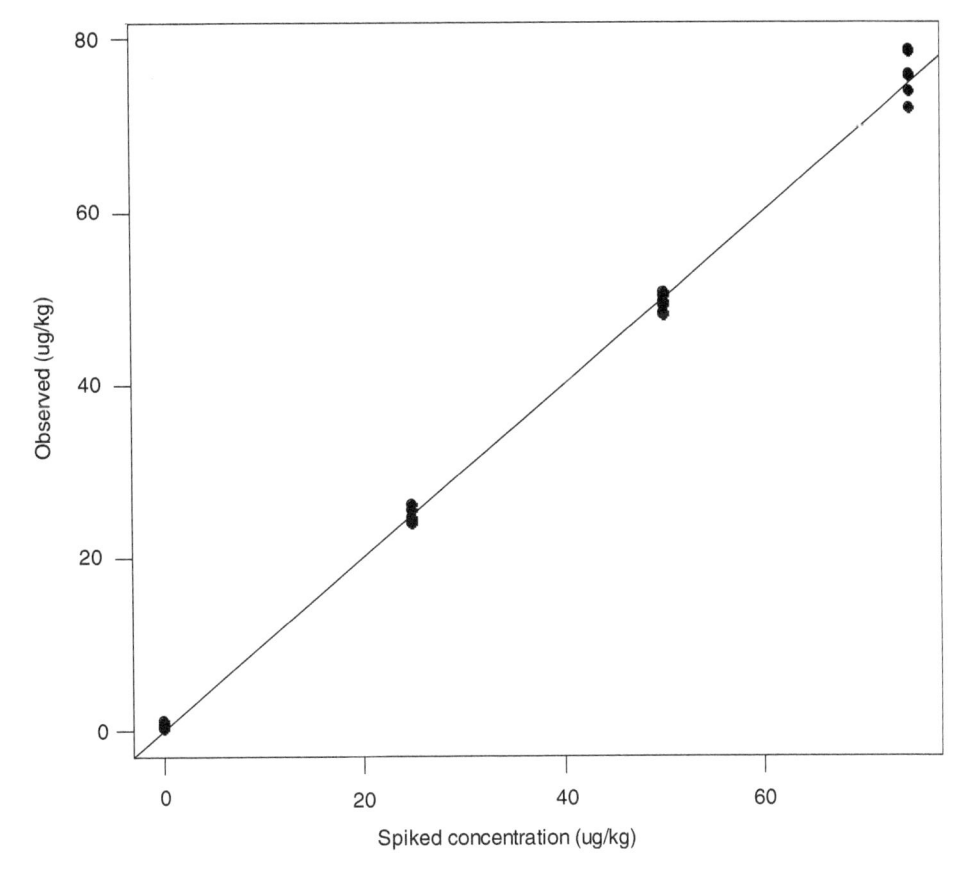

Figure 2.8 Scatter plot showing validation data for chlorpromazine spiked into bovine liver at three concentrations (and also the untreated liver with no added material). The line is the simple least-squares linear regression line through the data (see Chapter 7 for more details on regression).

least-squares regression between two variables may give misleading results when both have significant errors, as the resulting gradient and intercept are then biased.)

2.10 Normal Probability Plots

A special example of a scatter plot is a normal probability plot, sometimes also called a quantile–quantile plot or 'Q–Q plot' (Q–Q plots can also be drawn for other distributions). It is used as a rapid check to see whether a particular data set is approximately normally distributed. A normal probability plot of the thiamphenicol data in Table 2.1 is shown in Figure 2.9. If the points fall on or close to a straight line, the data are close to normally distributed. Deviations from the line (usually at either end) indicate some deviation from normality.

Although most statistical software constructs normal probability plots very easily, the plot can also be constructed manually. The plot is constructed by plotting 'Normal scores' (which we shall call z_i) against the original data. One way of forming approximate normal scores for n data points $x_1, \ldots, x_n$ is as follows:

1. Obtain the *ranks* r_i for the data set (the rank is the position of the data point when the data are placed in ascending order).

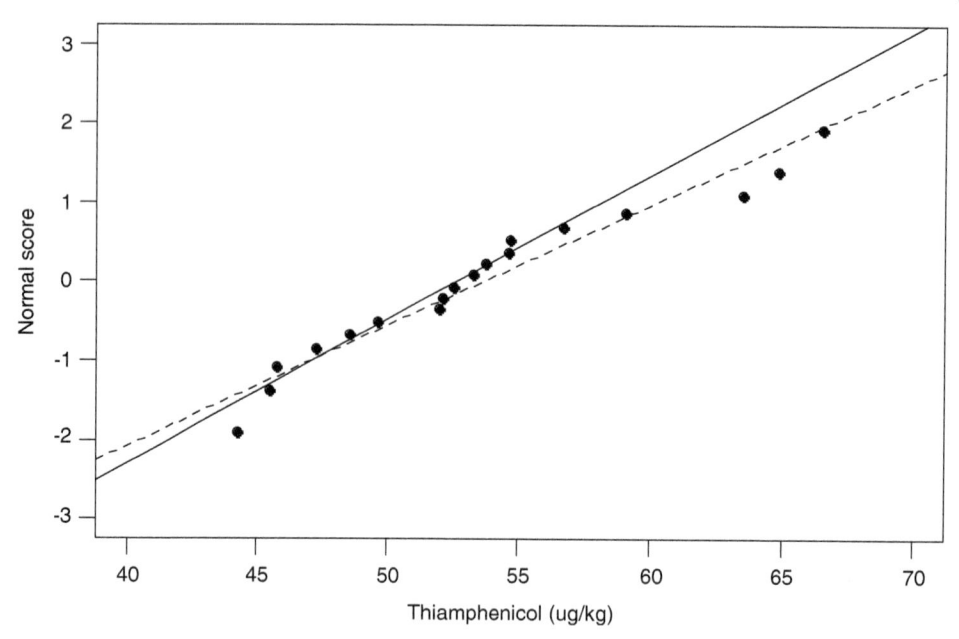

Figure 2.9 Normal probability plot. The figure shows a normal probability plot, or Q–Q plot, for the thiamphenicol data in Table 2.1. The solid line is drawn through the quartiles of the data and their scores [equation (2.3)]; the dashed line is based on the mean and standard deviation of the data [equation (2.4)].

2. Calculate normal probabilities p_i for each data point. These are usually estimated from

$$p_i = \frac{i - a}{n - 1 + 2a} \qquad (2.1)$$

where a is set to 3/8 (0.375) if n is less than or equal to 10 and 0.5 otherwise.

3. Calculate estimated scores z_i from

$$z_i = 4.91\left[p_i^{0.14} - (1 - p_i)^{0.14}\right] \qquad (2.2)$$

Plotting z_i against x_i gives the normal probability plot.

The line on the plot is drawn only to assist visual comparison to a straight line. One simple option is to draw the line through the upper and lower quartiles of x and z (Chapter 4, Section 4.1.4.6 describes quartiles and their calculation). If the lower and upper quartiles Q_1 and Q_3 for z and x are $[Q_1(z), Q_3(z)]$ and $[Q_1(x), Q_3(x)]$ respectively, this gives a line with the slope and intercept shown in the following equation:

$$\begin{aligned}\text{Slope:} \quad & [Q_1(z) - Q_3(z)]/[Q_1(x) - Q_3(x)] \\ \text{Intercept:} \quad & Q_1(z) - Q_1(x)[Q_1(z) - Q_3(z)]/[Q_1(x) - Q_3(x)]\end{aligned} \qquad (2.3)$$

This line, shown as a solid line in Figure 2.9, is fairly insensitive to extreme values, so outliers and extreme values show up well. Note that the last three points in the figure are relatively distant from the solid line, reflecting the slight positive skew visible in the histogram of the same data in Figure 2.4a.

Graphical Methods

Another option for a guide line uses the mean $\bar{x}$ and standard deviation s of the data; the slope and intercept are then

$$\begin{aligned} \text{Slope:} &\quad 1/s \\ \text{Intercept:} &\quad -\bar{x}/s \end{aligned} \quad (2.4)$$

This line is shown as the dashed line in Figure 2.9. In this case, most of the points fall fairly close to the line; only the slight curvature in the data suggests any non-normality.

CHAPTER 3
Distributions and Types of Data

3.1 Introduction

Figure 3.1 shows a typical measurement data set from a method validation exercise. The tabulated data show a range of values. Plotting the data in histogram form shows that observations tend to cluster near to the centre of the data set. The histogram is one possible graphical representation of the *distribution* of the data.

If the experiment were repeated, a visibly different data distribution is usually observed. However, as the number of observations in an experiment increases, the distribution becomes more consistent from experiment to experiment, tending towards some underlying form. This underlying form is sometimes called the *parent* distribution. In Figure 3.1, the smooth curve is a plot of a possible parent distribution; in this case, a *normal distribution* with a mean and standard deviation estimated from the data (the terms 'mean', 'variance' and 'standard deviation' are described in detail in Chapter 4).

There are several important features of the parent distribution shown in Figure 3.1. First, it can be represented by a mathematical equation – a distribution function – with a relatively small number of parameters. For the normal distribution, the parameters are the mean and population standard deviation. Knowing that the parent distribution is normal, it is possible to summarise a large number of observations simply by giving the mean and standard deviation. This allows large sets of observations to be summarised in terms of the distribution type and the relevant parameters. Second, the distribution can be used predictively to make statements about further observations; in Figure 3.1, for example, the curve indicates that observations in the region of 2750–2760 mg kg^{-1} will occur only rarely. The distribution is accordingly important both in describing data and in drawing inferences from the data.

3.2 Describing Distributions

Distributions are usually pictured using a *distribution function* or a *density function*, either of which is an equation describing the probability of obtaining a particular value. For a discrete variable (for example, a number of items, which can only take integer values), the relevant function is called a *probability distribution function* because it shows the probability of obtaining each discrete value. For continuous variables, the function is called a *probability density function* (often abbreviated

Distributions and Types of Data

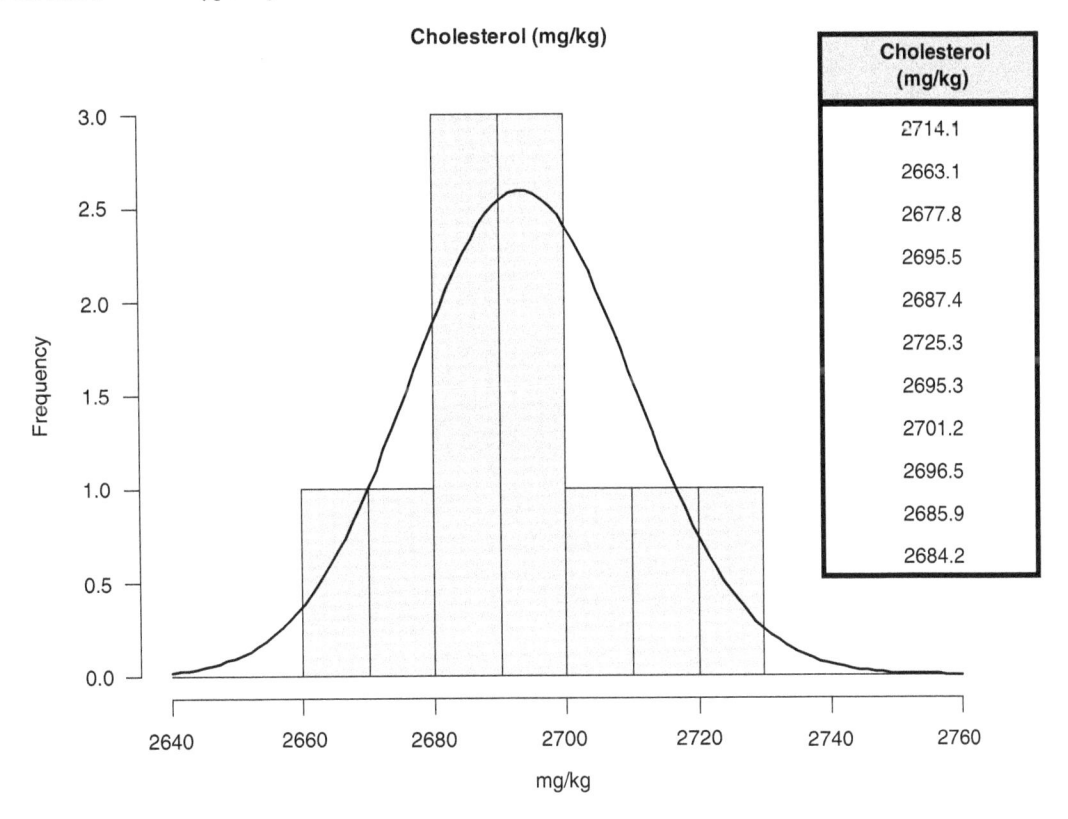

Cholesterol (mg/kg)
2714.1
2663.1
2677.8
2695.5
2687.4
2725.3
2695.3
2701.2
2696.5
2685.9
2684.2

Figure 3.1 Typical measurement data. The figure shows a histogram of the tabulated data from 11 replicate analyses of a certified reference material with a certified value of 2747 ± 90 mg kg^{-1} cholesterol. The curve is a normal distribution with mean and standard deviation calculated from the data and with vertical scaling adjusted for easy comparison with the histogram.

PDF). The 'probability density' is the probability per unit change in the variable and not the probability of obtaining the particular value. Unlike probability, the probability density at a point x can be higher than one. For most practical purposes, however, the height of the plotted curve can be thought of as representing the likelihood of obtaining the value.

Distributions can also be represented by their cumulative distribution (see Chapter 2, Section 2.7), which shows the probability of finding a value *up to* x, instead of the probability (or probability density) associated with finding x itself.

3.3 Distributions of Analytical Data

Replicated measurement results can often be expected to follow a normal distribution and, in considering statistical tests for ordinary cases in this book, this will be the assumed distribution. However, some other distributions are important in particular circumstances. Table 3.1 lists some common distributions, whose general shapes are shown in Figure 3.2. The most important features of each are described in the following subsections.

3.3.1 The Normal Distribution

The normal distribution was first studied in the 18th century, when it was found that observed distributions of errors could be closely approximated by a curve that was called the 'normal curve

Table 3.1 Common distributions in analytical data.

Distribution	Density function	Mean	Variance	Remarks
Normal	$\frac{1}{\sigma\sqrt{2\pi}}\exp\left[-\frac{(x-\mu)^2}{2\sigma^2}\right]$	μ	σ^2	Arises naturally from summation of many small random errors from any distribution.
Lognormal	$\frac{1}{x\sigma\sqrt{2\pi}}\exp\left\{-\frac{[\ln(x)-\mu]^2}{2\sigma^2}\right\}$	$\exp\left(\mu+\frac{\sigma^2}{2}\right)$	$\exp(2\mu+\sigma^2)[\exp(\sigma^2)-1]$	Arises naturally from the product of many terms with random errors. Approximates to normal for small standard deviation.
Poisson	$\lambda^x \exp(-\lambda)/x!$	λ	λ	Distribution of events occurring in an interval; important for radiation counting. Approximates to normality for large λ.
Binomial	$\binom{n}{x}p^x(1-p)^{(n-x)}$	np	$np(1-p)$	Distribution of x, the number of successes in n trials with probability of success p. Common in counting at low to moderate levels, such as microbial counts; also relevant in situations dominated by particulate sampling.
Contaminated normal	Various			'Contaminated normal' is the most common assumption given the presence of a small proportion of aberrant results. The 'correct' data follow a normal distribution; aberrant results follow a different, usually much broader, distribution.

Distributions and Types of Data

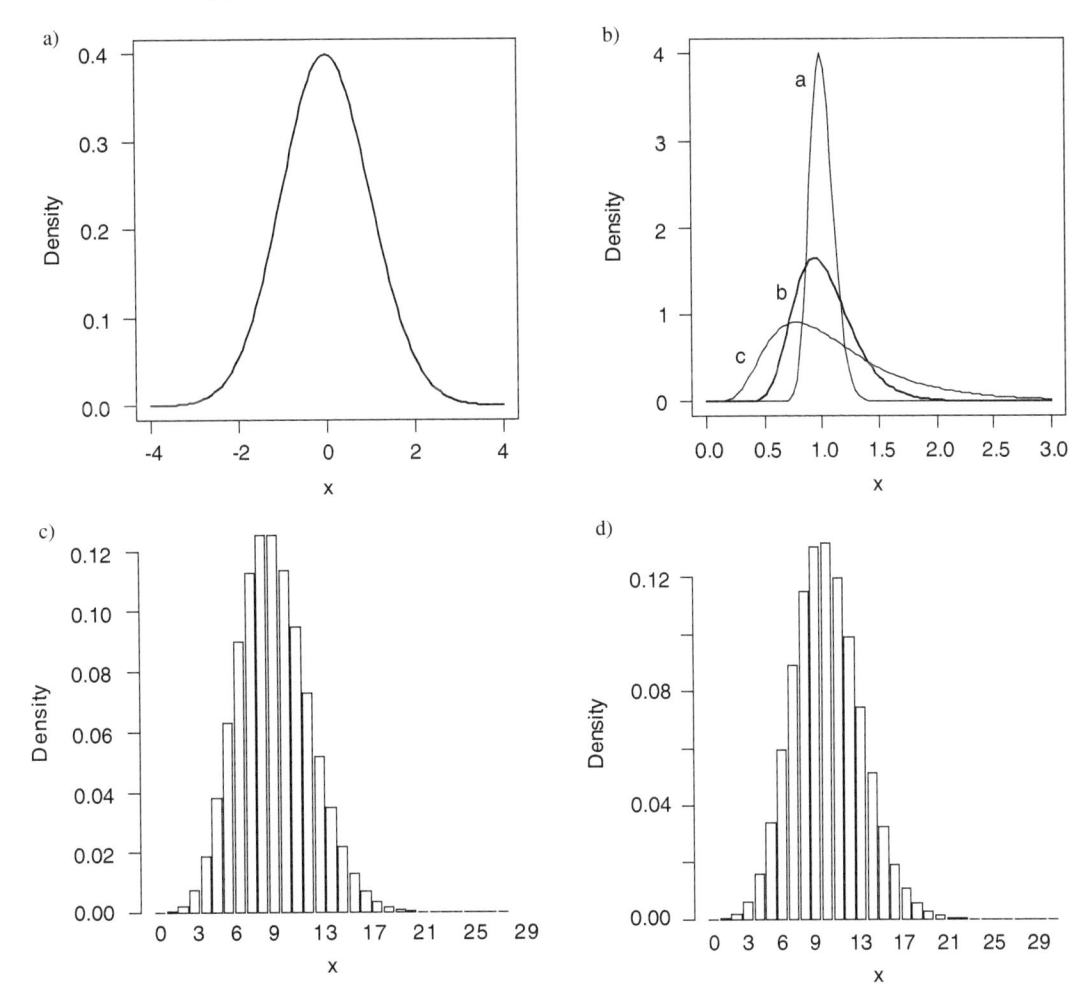

Figure 3.2 Measurement data distributions. (a) The standard normal distribution (mean = 0, standard deviation = 1.0). (b) Lognormal distributions. Mean on log scale: 0; standard deviation on log scale: a, 0.1; b, 0.25; c, 0.5. (c) Poisson distribution, $\lambda = 10$. (d) Binomial distribution; 100 trials, p(success) = 0.1. Note that this provides the same mean as in (c).

of errors'. The probability density function for a normal distribution of mean μ and variance σ^2 is shown in Table 3.1. The related cumulative distribution function is the integral of the density function [equation (3.1)]; this gives the probability of a value *up to* x occurring.

$$\int_{-\infty}^{\infty} \frac{1}{\sigma\sqrt{2\pi}} \exp\left[-(x-\mu)^2/2\sigma^2\right] dx = 1 \tag{3.1}$$

A normal distribution is completely determined by the parameters μ and σ. The mean can take any value and the standard deviation any non-negative value. The distribution is symmetric about the mean and, although the density falls off sharply, is actually infinite in extent. The higher probability of finding measurements close to μ is emphasised by the fact that 68.27% of the observations are expected to lie in the interval $\mu \pm \sigma$, 95.45% between $\mu \pm 2\sigma$ and only 0.27% beyond $\mu \pm 3\sigma$.

The smooth curve in Figure 3.1 shows a particular normal distribution. Figure 3.2a shows the *standard normal distribution* (or, more accurately, its probability density function). The standard normal distribution is a normal distribution of mean 0 and standard deviation 1.0.

Many distributions found in observational and experimental work are approximately normal. This is because the normal distribution arises naturally from the additive combination of many small effects, even, according to the central limit theorem, when those effects do not themselves arise from a normal distribution. This has an important consequence for means; the mean of as few as three or four observations can often be taken to be normally distributed even where the parent distribution of individual observations is not. Further, since small effects generally behave approximately additively, a very wide range of measurement systems show approximately normally distributed error.

3.3.2 The Lognormal Distribution

The lognormal distribution, illustrated in Figure 3.2b, is closely related to the normal distribution; the logarithms of values from a lognormal distribution are normally distributed. It most commonly arises when errors combine multiplicatively, instead of additively. The lognormal distribution itself is generally asymmetric, with positive skew. However, as shown in the figure, the shape depends on the ratio of standard deviation to mean and approaches that of a normal distribution as the standard deviation becomes small compared with the mean. The simplest method of handling lognormally distributed data is to take logarithms and treat the logged data as arising from a normal distribution.

3.3.3 Poisson and Binomial Distributions

The Poisson and binomial distributions (Figure 3.2c and d, respectively) describe counts and accordingly are *discrete* distributions; they have non-zero probability only for integer values of the variable. The Poisson distribution is applicable to cases such as radiation or microbiological counting; the binomial most appropriate for systems dominated by sampling, such as number of defective parts in a batch, number of microbes in a fixed volume or the number of contaminated particles in a sample from an inhomogeneous mixture. In the limit of large counts, the binomial distribution tends to the normal distribution; for small probability, it tends to the Poisson distribution. Similarly, the Poisson distribution tends towards normality for small probability and large counts. Thus, the Poisson distribution is often a convenient approximation to the binomial distribution and, as counts increase, the normal distribution can be used to approximate either.

3.4 Distributions Derived from the Normal Distribution

It is important to be aware that other distributions are important in analysing measurement data with normally distributed error. The most important for this book are described below.

3.4.1 Distribution of Student's *t*

The *t*-distribution is routinely used in *t*-tests (see Chapter 4, Section 4.2.3) for checking results for significant bias or for comparing observations with limits, and is also important for calculating confidence intervals (see Chapter 4, Section 4.3). The distribution is symmetric and resembles the normal distribution except for rather stronger 'tails'. It is described by the number of degrees of

Distributions and Types of Data 21

freedom v (the Greek letter v is pronounced 'nu' or 'new'). As the number of degrees of freedom increases, the distribution becomes closer to the normal distribution (Figure 3.3a). Chapter 4, Section 4.1.2, explains the concept of degrees of freedom in more detail.

3.4.2 The Chi-squared Distribution

The chi-squared (sometimes written χ^2) distribution describes, among other things, the distribution of estimates of variance. Specifically, for a standard deviation s taken from n data points, the variable $(n-1)s^2/\sigma^2$ has a chi-squared distribution with $v = n-1$ degrees of freedom. The chi-squared distribution is asymmetric with mean v and variance $2v$. Some examples are shown in Figure 3.3b.

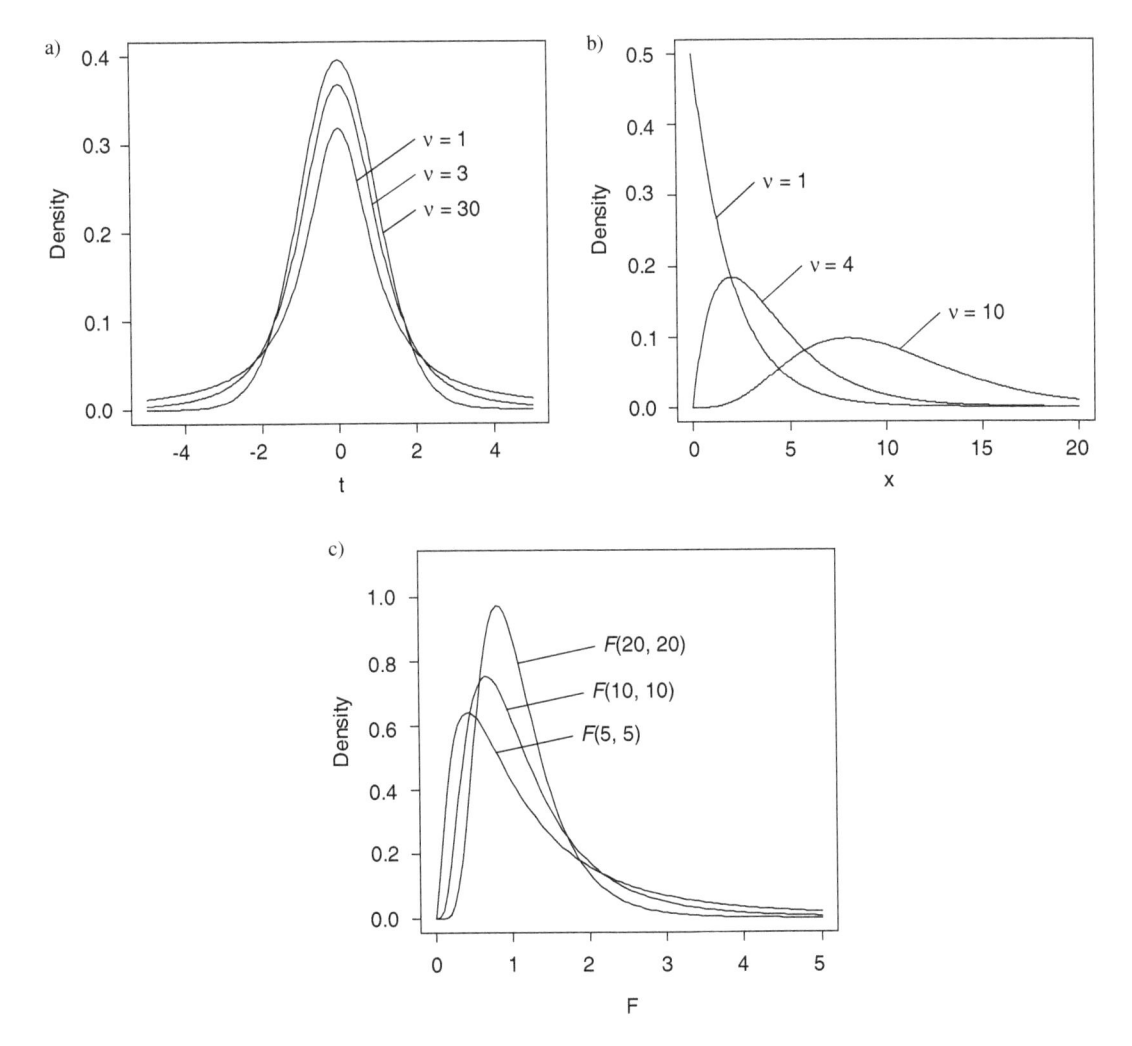

Figure 3.3 Distributions related to the normal distribution. (a) Examples of the *t*-distribution with different degrees of freedom; (b) chi-squared distributions with different degrees of freedom; (c) examples of the *F* distribution.

3.4.3 The *F* Distribution

The *F* distribution describes the distribution of ratios of variance estimates. This is important in comparing the spread of two different data sets and is used extensively in analysis of variance (ANOVA) in addition to being useful for comparing the precision of alternative methods of measurement. Because the *F* distribution is a ratio of two variances, it is characterised by the number of degrees of freedom for each. The *F* distribution is asymmetric and increases in width as the respective degrees of freedom decrease (Figure 3.3c).

3.5 Other Distributions

Two additional distributions will be referred to in this book in discussing uncertainty of measurement (Chapter 10): the rectangular, or uniform, distribution and the triangular distribution (Figure 3.4).

3.5.1 Rectangular Distribution (or Uniform Distribution)

The rectangular, or uniform, distribution is shown in Figure 3.4a. The distribution describes a variable which can take any value within a particular range with equal probability, but cannot take values outside the range. It is characterised by its width ($2a$ in the figure) and location or by its upper and lower limits. With width $2a$, the distribution has standard deviation $a/\sqrt{3}$ or approximately $0.577a$.

Many random number generators [including the RAND( ) function in many spreadsheets] produce approximately random numbers from a uniform distribution.

3.5.2 Triangular Distribution

The triangular distribution (Figure 3.4b) is used to describe a variable that is limited to a particular range but is most likely to have values in the middle of the range. Like the rectangular distribution, it is chiefly characterised by its width. A triangular distribution of width $2a$ has standard deviation $a/\sqrt{6}$ or approximately $0.408a$.

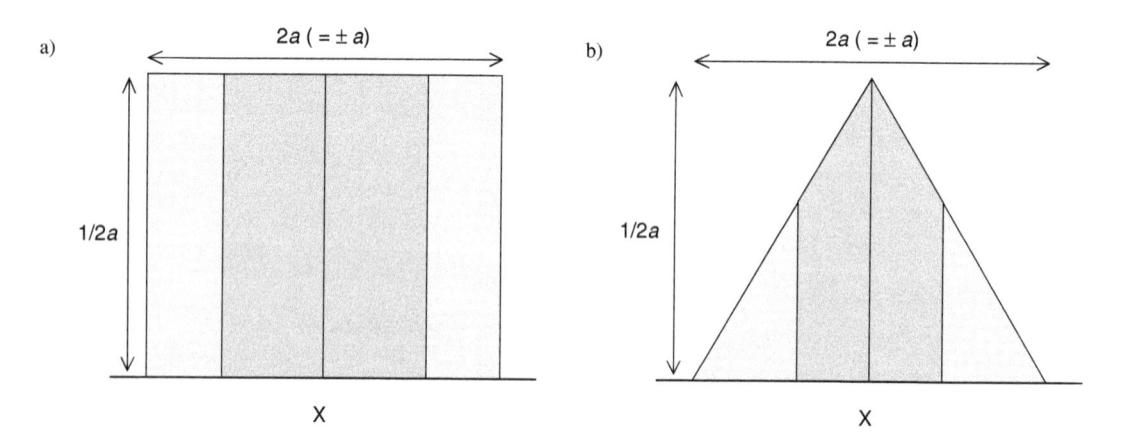

Figure 3.4 Other distributions. (a) Rectangular distribution; (b) triangular distribution. The shaded areas in both figures show the region between $\pm\sigma$, where σ is the standard deviation calculated as in Section 3.5.

Distributions and Types of Data 23

3.6 Populations and Samples

The theory of statistics as it is usually applied to analytical science generally assumes that a set of observations is a random sample from some potentially infinite population of possible results. This has two practical implications. First, analysts almost invariably use, and report, sample statistics (see Chapter 4). Second, statisticians – and analytical scientists – need to be careful to distinguish between observations and true values. For an analyst, observations are generally considered to be estimates of some unknown true value. In statistical terminology, the *population* parameters (such as the population mean) are the unknown true values of interest. If we had the complete population, we could calculate the exact population parameter. In practice this is rare, so 'samples' – sets of data taken from the population – are used to calculate *sample estimates* of the population parameters. This gives rise to a consistent distinction between population statistics – the population mean, population standard deviation *etc.* – and their respective sample estimates. The latter are usually abbreviated to *sample statistics* such as *sample mean* and *sample standard deviation*. Conventionally, unknown population parameters are represented in Greek script, whereas sample parameters, including summary statistics of analytical results, are represented in Roman script; this convention is retained in this book.

A random sample is a set of observations drawn from a population so that every possible observation has an equal chance of being drawn. Random sampling is important because for random samples, some important sample statistics (for example, the sample mean $\bar{x}$ and sample variance s^2) are unbiased estimates of the corresponding population parameters. If the sampling is not random, bias may be introduced and the calculated parameters may show systematic deviations from their true values.

3.7 Checking Normality

Since our knowledge of a population is generally incomplete, we generally assume that a particular distribution will adequately model the distribution for a particular variable. This is often justified by experience or prior knowledge. As with any other assumption, however, this should be checked.

Unfortunately, although there are many statistical tests for departures from normality, few are sufficiently powerful to be useful on the relatively small data sets common in analytical science. For that reason, no general tests for normality of distributions are described in detail in this book. The only statistical tests that relate to normality are the various tests for outliers described in Chapter 5, because outliers represent a particularly common departure from normal distributions in analytical data. Instead, it is recommended that analysts rely chiefly on prior knowledge of the system under study and use graphical tools, including the scatter plots and normal probability plots described in Chapter 2, to review data sets and identify significant anomalies.

If specific tests for normality are considered essential, perhaps the most common are the Kolmogorov–Smirnov test and the Shapiro–Wilk test,[1] both found in most statistical software. A more powerful modification of the Kolmogorov–Smirnov test is the Anderson–Darling test.[2] Others include the chi-squared goodness-of-fit test and tests for significant skewness and kurtosis (see Chapter 4, Sections 4.1.5 and 4.1.6). Many of these tests can be used to compare data against other distributions. Reference 3 provides a useful web-based reference describing most of these tests.

Prior knowledge is important in determining the distribution because the underlying physical processes often determine the expected distribution of errors. Processes in which errors tend to be additive will usually show approximately normally distributed error. Processes in which effects tend to multiply (including dilution and any multiplicative calculations) will tend to generate lognormal errors, although in practice, with relative standard deviations below about 20%, the distribution is usually sufficiently close to the normal distribution for the assumption of normality to be used, at least for tests at up to 95% confidence. Counting and sampling will generally produce approximately Poisson and binomial error, respectively.

3.8 Types of Data

There are four broad types of data that are important in general statistics. Although this book does not provide methods for all of them, it is useful to be aware of the differences. The different types of data are:

- *Categorical data:* Data which simply describe a category or grouping; for example, laboratory name, sample number or analytical method or analytical method identifier. For categorical data, there is no 'ordering relation'; ordering results by sample identifier, for example, or by analyst name, has no special meaning. In analytical science, categorical data are most important as an indicator of grouping of analytical results.
- *Ordinal data:* Ordinal data are categorical data in which there is some meaningful ordering but in which differences between values cannot be directly compared. For example, a test kit providing semi-quantitative responses 'Low', 'Medium' or 'High' generates ordinal data; the values can be meaningfully ordered, but it is not meaningful to take a difference between 'Medium' and 'Low'.
- *Data from an interval scale ('interval data'):* Interval data are ordered data in which differences are meaningful, but ratios are not. Temperatures on the Celsius scale are 'interval data'; the difference of 2 °C between 25 and 27 °C means the same as that between 19 and 21 °C, but the ratio of 25 and 2.5 °C does *not* mean that one is '10 times hotter' than another. Interval scale data can be continuous-valued or discrete.
- *Data from a ratio scale:* A ratio scale is a scale in which both differences and ratios have meaning. Analyte concentrations are a good example; a concentration of $2\,\mathrm{mg\,L^{-1}}$ implies that there is indeed twice as much analyte present as a concentration of $1\,\mathrm{mg\,L^{-1}}$. Ratio scale data can be continuous-valued or discrete.

This book is largely concerned with continuous-valued data on interval or ratio scales. Statistical methods for categorical or ordinal data are not discussed except in connection with counts or frequencies (see Chapter 4, Section 4.1.2).

References

1. S. S. Shapiro and M. B. Wilk, *Biometrika*, 1965, **52**, 591.
2. M. A. Stephens, *J. Am. Stat. Assoc.*, 1974, **69**, 730.
3. NIST/SEMATECH, *e-Handbook of Statistical Methods*, http://www.itl.nist.gov/div898/handbook/ (accessed 22 April 2008).

CHAPTER 4
Basic Statistical Techniques

4.1 Summarising Data: Descriptive Statistics

4.1.1 Introduction

For many purposes it is necessary to provide a concise summary of the information contained in a collection of data. The statistics most commonly used to represent the properties of a distribution, or a data set, fall into the following categories:

1. Counts, such as the number of data points.
2. Measures of location – giving the location of the central or typical value.
3. Measures of dispersion – showing the degree of spread of the data round the central value.

Two additional terms, 'skewness' and 'kurtosis', are occasionally used to describe the general shape of distributions of observations and are consequently included here for completeness.

4.1.2 Counts, Frequencies and Degrees of Freedom

When providing a summary of a data set (for example, by quoting the mean and standard deviation), the number of values in the data set and/or the degrees of freedom should also be reported. The term 'degrees of freedom' refers to a measure of the number of independent pieces of data that have been used to evaluate a particular parameter. In general, the degrees of freedom will be the number of data points (n) less the number of parameters estimated from the data. In the case of the sample standard deviation, for example, the degrees of freedom is equal to $n-1$ as the mean, $\bar{x}$, has to be calculated from the data as part of the calculation of the standard deviation [see equation (4.4b)]. The number of degrees of freedom is denoted by the Greek letter v and is frequently abbreviated to 'df' or 'DOF'.

Data can also be summarised by dividing the data set into a number of intervals and counting the number of values that fall into each interval. This frequency information is usually presented in the form of a histogram (see Chapter 2, Section 2.5).

4.1.3 Measures of Location

4.1.3.1 Arithmetic Mean

Usually simply called the mean, this is the sum of all observations (x_i) divided by the number of observations (n):

$$\bar{x} = \frac{\sum_{i=1}^{n} x_i}{n} \qquad (4.1)$$

If the sample is representative of the population, $\bar{x}$ is the best estimate of μ (the population mean).

4.1.3.2 Median

When observations are arranged in ascending order, the median is the central member of the series; there are equal numbers of observations smaller and greater than the median. For a symmetrical distribution, the mean and median have the same expected value. The median is more robust, in that it is less affected by extreme values, but (in the absence of outliers) varies more than the mean from one sample to the next. The median of n ordered values $x_1, \ldots, x_n$ is given by

$$\tilde{x} = \begin{cases} n \text{ odd}: & x_{(n+1)/2} \\ n \text{ even}: & \dfrac{x_{n/2} + x_{(n+2)/2}}{2} \end{cases} \qquad (4.2)$$

Note that there is no generally agreed or widely used symbol for the median. In this book, the symbol $\tilde{x}$ is usually used, as it is the most common in current use. However, the notations median(x) or med(x) and, occasionally, $\mu_{1/2}(x)$ may also be found.

4.1.3.3 Mode

The mode is the value of the variable that occurs most frequently. It is most often used with discrete distributions; it is also sometimes used with categorical data.

Examples
1. For the set [1, 3, 3, 3, 3, 4, 5, 5, 6, 7, 8, 9, 9, 9], the mode is 3.
2. In a microbiological inspection, the incidence (number of items containing the microbe) might be: *E. coli* 0157, 3; *Salmonella*, 123; *Listeria*, 14; *Campylobacter*, 26. The modal value is *Salmonella*, as the most frequently occurring species.

It is possible to have more than one mode; for small- to medium-sized discrete data sets or in histograms of continuous-valued data, this is a fairly common chance occurrence. The presence of two or more reproducible modes, however, may indicate a non-homogeneous data set; for example, apparently multimodal data may be observed in proficiency testing when different participants use two or three markedly different analytical methods.

For a normal distribution, the mode coincides with the mean and the median.

4.1.4 Measures of Dispersion

4.1.4.1 Variance

The population variance is the mean squared deviation of the individual values from the population mean μ and is denoted σ^2 or sometimes σ_n^2. s^2 (and, particularly on calculator keypads, the strictly incorrect form σ_{n-1}^2) is used to denote the sample variance ('sample variance' can be considered as shorthand for 'sample estimate of the population variance'). The variance and the standard deviation both indicate the extent to which the results differ from each other; the larger the variance (or standard deviation), the greater the spread of data.

$$\text{Population variance:} \quad \sigma^2 = \frac{\sum_{i=1}^{n}(x_i - \mu)^2}{n} \tag{4.3a}$$

$$\text{Sample variance:} \quad s^2 = \frac{\sum_{i=1}^{n}(x_i - \bar{x})^2}{n-1} \tag{4.3b}$$

Remember that population parameters are very rarely relevant in analytical science; the sample statistics should be used unless there is a very good reason to do otherwise.

4.1.4.2 Standard Deviation

The standard deviation is the positive square root of the variance:

$$\text{Population standard deviation:} \quad \sigma = \sqrt{\frac{\sum_{i=1}^{n}(x_i - \mu)^2}{n}} \tag{4.4a}$$

$$\text{Sample standard deviation:} \quad s = \sqrt{\frac{\sum_{i=1}^{n}(x_i - \bar{x})^2}{n-1}} \tag{4.4b}$$

The standard deviation is usually more useful than the variance for analytical work because it has the same units as the raw data. For example, for data quoted in milligrams, the standard deviation will also be in milligrams, whereas the variance would be in mg^2. The variance is more commonly used in theoretical statistics, because the sample variance s^2 is an unbiased estimator of the population variance σ^2, whereas the sample standard deviation s is – perhaps surprisingly at first sight – a slightly biased estimate of the population standard deviation σ. This has important consequences for 'pooling' data; we shall see later that where we wish to obtain a 'pooled' standard deviation from several separate estimates, the standard deviations are always squared before summation.

4.1.4.3 Standard Deviation of the Mean

The standard deviation of the mean, $s(\bar{x})$, also known as the standard error, is calculated using the equation

$$s(\bar{x}) = \frac{s}{\sqrt{n}} \tag{4.5}$$

The standard deviation of the mean represents the variation associated with a mean value. For most practical purposes, it represents the part of the uncertainty which arises from the random variation observable in a particular experiment. The standard deviation of the mean is less than the sample standard deviation because it estimates the variation of averages, which are more precise than single observations.

4.1.4.4 Relative Standard Deviation and Coefficient of Variation

The relative standard deviation (RSD), also called the coefficient of variation (CV), is a measure of the spread of data in comparison with the mean of the data. It may be expressed as a fraction:

$$\text{RSD} = \text{CV} = \frac{s}{\bar{x}} \tag{4.6}$$

or (like any other fraction) as a percentage:

$$\%\text{RSD} = \%\text{CV} = \frac{s}{\bar{x}} \times 100 \tag{4.7}$$

The relative standard deviation is a particularly useful measure of dispersion when the dispersion is approximately proportional to concentration. This is fairly common in analytical measurements when far from the limit of detection; the RSD (or CV) is often constant over a wide concentration range.

The RSD and/or CV should only be used for ratio scale data (see Chapter 3, Section 3.8).

4.1.4.5 Range

The range is the difference between the highest and lowest values in a set of results. It is a useful quick summary statistic, particularly for small data sets where the range can be determined by inspection. It is possible to derive an estimate of standard deviation from the range using appropriate statistical tables, and some older quality control charting standards use this method for convenience of calculation. However, the range is very variable from one data set to the next and very badly affected by errors which cause outliers. It is therefore a poor method for estimating population dispersion (except for only two data points, where all dispersion measures are equally affected by outliers and variability).

4.1.4.6 Quartiles and the Interquartile Range

The median [see equation (4.2)] divides a data set in half; the three quartiles – lower quartile, median and upper quartile – divide it into quarters. Like the median, if there is no data point exactly at these positions, the upper and lower quartiles are found by interpolation. Interpolation methods differ in different software; one of the most common is as follows:

1. Rank the data set in ascending order, to give n ordered values $x_1, \ldots, x_n$.
2. Calculate the values $l_1 = 1 + (n-1)/4$, $l_3 = 1 + 3(n-1)/4$. If l_i (either l_1 or l_3) is not an integer, calculate the next lower and higher integers l_{i-} and l_{i+}. Then,

$$Q_i = \begin{cases} l_i \text{ integer}: & x_{l_i} \\ l_i \text{ fractional}: & (l_{i+} - l_i)x_{l_{i-}} + (l_i - l_{i-})x_{l_{i+}} \end{cases} \tag{4.8}$$

The interquartile range (IQR) is the difference between the lower (or first) quartile, Q_1, and the upper (or third) quartile, Q_3 (Q_2 would be the median). The interquartile range is used in

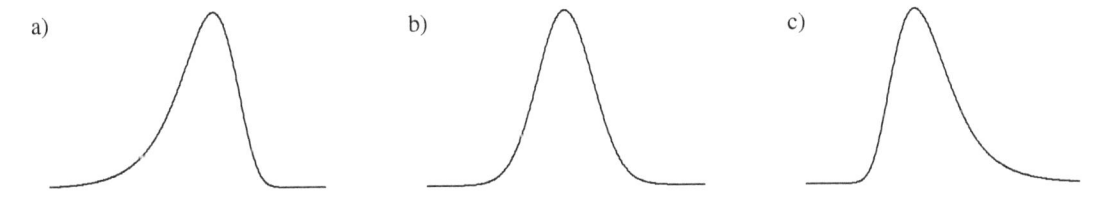

Figure 4.1 Skewness. (a) Negative skew; (b) symmetric (no skew); (c) positive skew.

the construction of box plots (see Chapter 2, Section 2.8). It can also be used as a robust estimate of standard deviation if the underlying distribution is believed to be normal. Since the standard normal distribution (see Chapter 3, Section 3.3.1) has an interquartile range of 1.35, the standard deviation of an underlying normal distribution can be estimated from

$$s \approx IQR/1.35 \qquad (4.9)$$

The median absolute deviation is, however, a slightly more stable estimate (see Chapter 5, Section 5.3.3.1).

Example
For the ordered data set [9.6, 10.0, 11.3, 11.8, 12.0, 12.0, 12.2, 12.3, 12.5, 12.6, 13.1, 13.2]:
There are 12 data points. $l_1 = 1 + (12-1)/4 = 3.75$; $l_3 = 1 + 3(12-1)/4 = 9.25$. These are not integers, so for Q_1 we calculate $l_{1-} = 3$, $l_{1+} = 4$. Then, $Q_1 = (4 - 3.75) \times 11.3 + (3.75 - 3) \times 11.8 = 11.675$. For Q_3, $l_{3-} = 9$ and $l_{3+} = 10$, so $Q_3 = (10 - 9.25) \times 12.5 + (9.25 - 9) \times 12.6 = 12.525$.
The interquartile range is $Q_3 - Q_1 = 12.525 - 11.675 = 0.85$.

4.1.5 Skewness

Skewness is a measure of the degree of asymmetry in a distribution. It can be calculated where necessary using equation (4.10). The term is usually used in a qualitative sense in analytical measurement; since a symmetrical distribution has skewness equal to zero and a distribution with a longer 'tail' towards higher values has a positive value of skewness, the general descriptions in Figure 4.1 are usually used.

$$\text{skewness} = \frac{\sum_{i=1}^{n}(x_i - \bar{x})^3}{(n-1) \times s^3} \qquad (4.10)$$

4.1.6 Kurtosis

Kurtosis is descriptive of the degree of peakedness of a distribution. Figure 4.2 shows some distributions with different kurtosis. The terms 'mesokurtic', 'leptokurtic' and 'platykurtic' are rarely used in analytical science.

4.2 Significance Testing

4.2.1 Introduction

Figure 4.3 shows two sets of data with means of $\bar{x}_A$ and $\bar{x}_B$. Although the observed means are different, it is possible that both data sets come from the same population (with a mean equal to μ) and that the difference in the observed means is due simply to random variation in the data. This is the

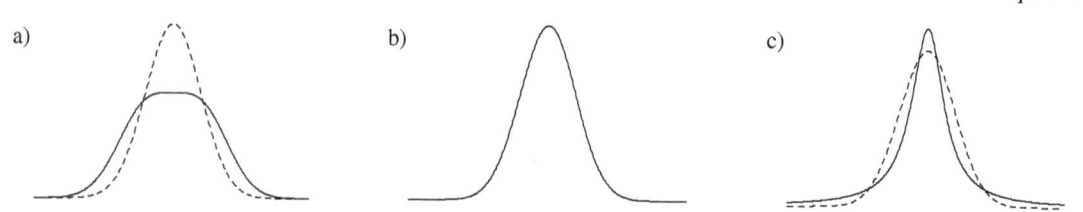

Figure 4.2 Kurtosis. The figure shows three distributions (solid lines) with different kurtosis. (a) Platykurtic (kurtosis<0); (b) mesokurtic (kurtosis=0); (c) leptokurtic (kurtosis>0). The dashed lines in (a) and (c) show the normal distribution, which is mesokurtic, for comparison.

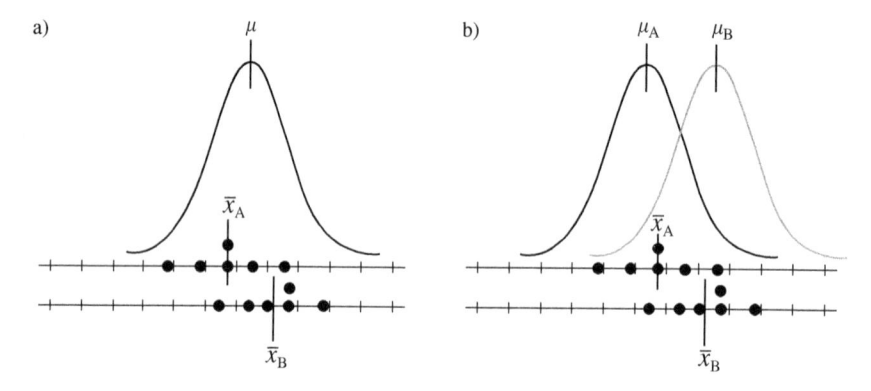

Figure 4.3 Comparing data sets. The figure shows two possible interpretations of the same data sets: (a) the data sets might be drawn from the same population with mean μ, or (b) the data sets might be evidence of different populations with means μ_A and μ_B.

situation shown in Figure 4.3a. However, it is also possible that the data sets are drawn from two different populations of data (with means equal to μ_A and μ_B). This is the situation illustrated in Figure 4.3b. Significance tests provide an objective method of deciding which is the more likely alternative.

4.2.2 A Procedure for Significance Testing

4.2.2.1 Steps in Significance Testing

Although the calculations differ, most significance tests rely on the same general steps. In the significance tests described in this section, these steps are organised into the following general procedure:

1. State the null hypothesis – a statement about the data that you want to test.
2. State the alternative hypothesis – an alternative statement that will be taken as true if the null hypothesis is rejected.
3. Check the distribution of the data. In this section, all the tests assume that the data are approximately normally distributed.
4. Select the appropriate test.
5. Choose the level of significance for the test.
6. Choose the number of 'tails' for the test. This usually follows directly from the exact form of the alternative hypothesis.
7. Calculate the test statistic.

Basic Statistical Techniques

8. Obtain the critical value for the test.
9. Compare the test statistic with the critical value.

The separate steps are all important and are described below before turning to individual tests.

4.2.2.2 Stating the Hypotheses

The null and alternative hypotheses define the question to be answered by the significance test. The null hypothesis, which is often denoted H_0, is usually that there is no difference between the values being compared. For example, when comparing the means of two data sets (A and B), the null hypothesis is usually that the population means are equal (*i.e.* that the data sets are from the same population):

$$H_0: \mu_A = \mu_B$$

The natural alternative hypothesis (usually denoted H_1 or H_A) is that the population means are not equal (*i.e.* the data sets are from different populations):

$$H_1: \mu_A \neq \mu_B$$

This pair of hypotheses simply asks 'is there a difference?'; it makes no prior assumptions about which mean might be greater than the other. Sometimes, the sign of the difference is important and this changes the probability of seeing a particular outcome. If a method (say, method A) was intended or expected to generate lower values than method B, a lower mean value for method A ought to be more likely. So if we found a *higher* mean for method A, we ought to regard this as more significant than if we had no prior expectation. To build this difference into the test, it is necessary to start with a different set of hypotheses, usually written as

$$H_0: \mu_B = \mu_A$$
$$H_1: \mu_B > \mu_A$$

It is important to decide on the hypotheses independently of the data. It will be seen, later, that the hypothesis changes the critical value and can therefore affect the outcome of the test. The critical values are based on probabilities. If we allow the data to dictate the choice of hypothesis, we implicitly bias the test against the null hypothesis by increasing the probability of a significant test result. For example, we are generally interested in whether the recovery of analyte from a test sample differs significantly from 100%; this implies a hypothesis of the form $\mu \neq \mu_0$ (where μ is the population mean and μ_0 is the hypothesised true value, in this example equal to 100%). But if we see a mean recovery above 100%, we might ask, 'Is it really greater than 100%?' and, because of the wording, erroneously choose a one-tailed test (see Section 4.2.2.6). This is a mistake because we have allowed the data to dictate the hypothesis; we are inevitably more likely to see a 'significantly higher' test result if we choose to test for a higher mean only when we see a higher mean value and *vice versa*. In fact, choosing the test from the direction indicated by the data exactly doubles the probability of incorrectly rejecting the null hypothesis.

4.2.2.3 Checking the Distribution

All statistical significance tests rely to some extent on assumed properties of the parent distribution from which the data are assumed to arise. Significance tests can give very badly misleading results if the assumptions are inappropriate for the data. Here, we assume that the data are approximately normally distributed. Inspection of the data using graphical methods (Chapter 2) is among the most effective ways of detecting departures from normality; severe outliers and visible skew will generally compromise the significance tests described in this book. (Alternative tests should be used if the data show clear departures from normality. The so-called 'non-parametric' tests are often appropriate when the data do not follow a normal distribution but are approximately symmetrically distributed. Positively skewed distributions can often be returned to approximate normality by taking logarithms. Detailed descriptions are beyond the scope of the present text; advice from a qualified statistician is strongly recommended.)

4.2.2.4 Selecting the Appropriate Test

There are many different types of significance test which can be used to examine different statistical parameters. Two of the most common statistical tests are the *t*-test, which is used to compare mean values, and the *F*-test, which is used to compare variances. These tests are appropriate for approximately normally distributed error. They are discussed in more detail in Sections 4.2.3 and 4.2.4, respectively. Each test has an equation for calculating the test statistic, which is then compared with the relevant critical value (see Section 4.2.2.8).

4.2.2.5 Choosing the Level of Significance

The aim of significance testing is to identify outcomes that are unlikely to have happened by chance if the null hypothesis is true. If the difference between the means of two data sets is so great that it is unlikely to have happened by chance if the null hypothesis (H_0: $\mu_A = \mu_B$) is true, the null hypothesis is rejected and the difference between the mean values is described as 'statistically significant'. When carrying out a significance test, we therefore need to define what we would consider to be 'unlikely' in terms of a probability. This probability is the significance level for the test, usually denoted α. The majority of tests are carried out using a significance level of 0.05. If the probability of observing a particular outcome *if the null hypothesis is true* is less than 0.05, then the null hypothesis is rejected. Note that if the significance level is $\alpha = 0.05$, there will be a 5% chance of rejecting the null hypothesis even if it is correct (this is known as a Type I error; different types of error are discussed in more detail in Chapter 8, Section 8.1). If a 5% chance of incorrectly rejecting the null hypothesis is considered too high, then a smaller value for α should be used, such as $\alpha = 0.01$.

In some texts and statistical packages, the significance level is expressed in terms of a confidence level (CL). The confidence level, often expressed as a percentage, is related to α as follows:

$$\text{CL} = 100 \times (1 - \alpha) \,\% \qquad (4.11)$$

A significance level of $\alpha = 0.05$ is therefore equivalent to a confidence level of 95%.

4.2.2.6 Choosing the Number of 'Tails'

Two alternative hypotheses are given in Section 4.2.2.2. The first, H_1: $\mu_A \neq \mu_B$, describes a 'two-tailed' test. This is because we want to know if there is a significant difference between the means *in either direction*. However, we may only be interested in determining whether the mean of data set B is significantly *greater than* the mean of data set A. This is the second alternative hypothesis shown, H_1: $\mu_B > \mu_A$. Similarly, we may only be interested in knowing whether the mean of data set B is significantly *less than* the mean of data set A. In this case, the alternative hypothesis is H_1: $\mu_B < \mu_A$.

In both of these cases, the significance test is 'one-tailed' (or 'one-sided') as we are only interested in whether there is a difference between the means *in one direction*.

Basic Statistical Techniques

The distinction between one- and two-tailed tests is important as the number of tails influences the critical value for the test (see Section 4.2.2.8).

4.2.2.7 Calculating the Test Statistic

The *test statistic* is a measure of the difference under study and is calculated from the data. Each different statistical test is associated with a particular test statistic. The relevant equations for calculating the test statistics for *t*- and *F*-tests are given later in this chapter.

4.2.2.8 Obtaining the Critical Value

Each significance test has an associated set of critical values which are used to determine whether an observed result can be considered significant. Critical values are related to the distribution for the particular test statistic. For example, distributions of values of Student's *t* for three different values of v are illustrated in Figure 3.3a in Chapter 3. The critical values for the test are the extremes of the range containing a fraction $(1 - \alpha)$ of the area under the probability density curve (remember that the area under a part of the density curve is the probability of a value occurring in that range; see Chapter 3, Section 3.2). For example, in the case of the *t*-distribution with three degrees of freedom shown in Figure 4.4, the light region marked '0.95' in each case includes 95% of the area; the shaded 'tail areas' include the remaining 5%, corresponding to $\alpha = 0.05$. The light region in Figure 4.4a includes the 'most likely 95%' of values for *t* given the null hypothesis in a one-tailed test, so a value for *t* in the shaded area is unlikely if the null hypothesis is true. Similarly, for a two-tailed test, the 'most likely 95%' of values for *t* are centred on $t = 0$, leaving the least likely 5% equally distributed in the two tails as in Figure 4.4b. The ends of the 'likely' region for the null hypothesis are, as shown, the relevant critical values for 95% confidence in either case. (Figure 4.4b only shows the upper critical value, because the test statistic is arranged to give a positive value in all cases; see Section 4.2.3 for details.)

In practice, the critical values for a particular test are obtained from statistical tables (see Appendix A) or from software. To obtain the appropriate critical value for a particular significance test, one will usually need to know the number of degrees of freedom associated with the calculated test statistic (see Section 4.1.2), the level of significance for the test and whether the test is one- or two-tailed.

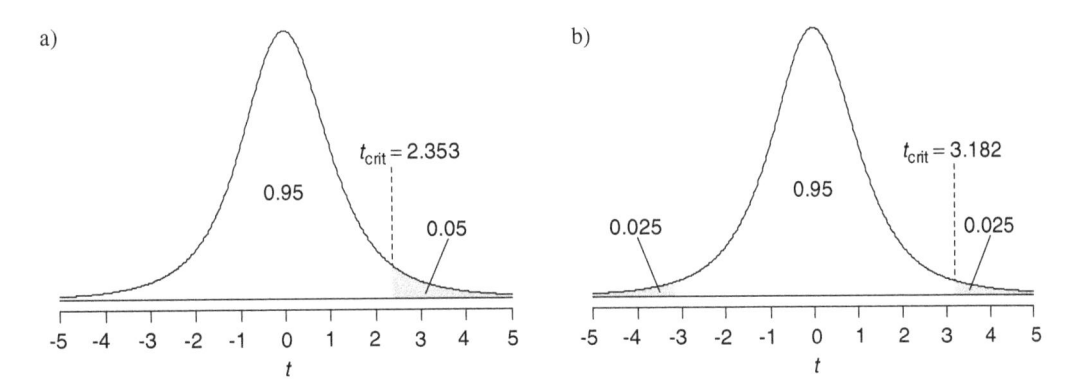

Figure 4.4 Distribution of Student's *t* values for $v = 3$. (a) One-tailed critical value, $\alpha = 0.05$; (b) two-tailed value, $\alpha = 0.05$. The shaded areas have total area 0.05 in both cases. Note that tables generally quote only the upper critical value, since the test statistic for a two-tailed *t*-test is the absolute value $|t|$ and the test statistic for a one-tailed *t*-test is always arranged so that a high positive value of *t* is significant.

4.2.2.9 *Comparing the Test Statistic with the Critical Value*

To decide whether the null hypothesis should be accepted or rejected, the calculated test statistic is compared with the appropriate critical value for the test. Most tests arrange the test statistic so that high values are unlikely given the null hypothesis. Usually, therefore, if the calculated test statistic exceeds the critical value at the chosen level of confidence, the null hypothesis is rejected and the result of the test is described as 'significant'.

4.2.2.10 *Using p-Values*

If statistical software is used to carry out significance tests, the software will return a *p*-value in addition to the calculated test statistic and the critical value. The *p*-value represents the probability of observing a value of the test statistic greater than or equal to the critical value if the null hypothesis is true. If this probability is low, the observed result is unlikely to arise by chance. Consequently, a *p*-value *less than* the significance level used for the significance test indicates that it is unlikely that the results would have been obtained if the null hypothesis was true. The null hypothesis should therefore be rejected. Simple rules for interpretation of *p*-values can therefore be summarised as follows:

- Low *p*-value: test result statistically significant; reject the null hypothesis.
- High *p*-value: test result *not* statistically significant; accept the null hypothesis.

The *p*-value has the advantage that the same critical value (of *p*) has the same interpretation for any statistical test, so it is well worth remembering the additional rules for interpreting *p*-values.

4.2.3 Tests on One or Two Mean Values – Student's *t*-Test

The *t*-test can be used to compare the following:

- the mean of a data set with a stated value;
- the means of two independent data sets;
- the effect of two treatments applied once each to a range of different test objects (this is called a paired comparison).

The different tests for each of these cases are described below. The descriptions follow the general procedure indicated in Section 4.2.2.1, except that (i) the choice of test (step 4) is assumed to be the *t*-test and (ii) it will be assumed that the level of significance (step 5) is chosen following Section 4.2.2.5.

4.2.3.1 *Comparing the Mean with a Stated Value (One-sample Test)*

This test is used to determine whether the sample mean $\bar{x}$ of a data set differs significantly from a target value or limit, μ_0. μ_0 may be zero, in a test for the presence of a material, or some other value such as a limit or reference value. One common use of this test in analytical chemistry is to determine whether the mean of a set of results obtained from the analysis of a certified reference material is significantly different from the certified value for the material; another is to decide whether there is evidence that a statutory or contractual upper limit has been breached.

Basic Statistical Techniques

Stating the hypotheses

The null hypothesis is always $\mu = \mu_0$. For a test for significant difference, irrespective of the sign of the difference, the alternative hypothesis is $\mu \neq \mu_0$. To determine whether there is evidence that a stated value is exceeded, the alternative hypothesis is $\mu > \mu_0$; for testing whether a result is significantly below a stated value, the alternative hypothesis is $\mu < \mu_0$.

Checking the distribution

The *t*-test assumes approximate normality; check for this using the graphical methods in Chapter 2, paying particular attention to outliers and asymmetry.

Number of tails

The number of tails for each hypothesis is shown in Table 4.1.

Calculating the test statistic

The calculations for the test statistic are summarised for the different hypotheses in Table 4.1. All the test statistics involve the difference divided by the standard deviation of the mean. Note, however, that for the first case ($\mu = \mu_0$ vs $\mu \neq \mu_0$) the absolute value of the difference is used, so the calculated value of *t* is always positive. For the other two cases, the sign of the comparison is important and the calculation in Table 4.1 has been arranged so that the expected difference is positive if the alternative hypothesis is true. In all cases, therefore, larger positive values of *t* will be evidence against the null hypothesis and smaller positive values or *any* negative value are taken to be less than the critical value and therefore do not indicate significance.

The critical value

The critical value is found from the relevant tables (Appendix A, Table A.4) using $n - 1$ degrees of freedom and the appropriate number of tails as shown in Table 4.1. The level of significance is chosen following Section 4.2.2.5. Note that in Table A.4 in Appendix A the level of significance is arranged by the number of tails for the test; for a one-tailed test, select the level of confidence from the row marked '1T' and for the two-tailed test, use the confidence levels in row '2T'.

Critical values can also be found from software. For example, in MS Excel and OpenOffice spreadsheets, the critical value for a two-tailed test with a significance level of 0.05 (confidence level of 95%) and five degrees of freedom would be

$$= \text{TINV}(0.05, 5).$$

Table 4.1 Hypotheses and test statistics for the one-sample *t*-test (degrees of freedom: $n - 1$ for all tests).

Null and alternative hypotheses	Abbreviated hypotheses	Test statistic	Tails
(a) 'The population mean μ is equal to the given value μ_0' versus 'μ is *not equal to* the given value'	$H_0: \mu = \mu_0$ $H_1: \mu \neq \mu_0$	$t = \dfrac{\lvert \bar{x} - \mu_0 \rvert}{s/\sqrt{n}}$	2
(b) 'The population mean μ is equal to the given value μ_0' versus 'μ is *greater than* the given value'	$H_0: \mu = \mu_0$ $H_1: \mu > \mu_0$	$t = \dfrac{\bar{x} - \mu_0}{s/\sqrt{n}}$	1
(c) 'The population mean μ is equal to the given value μ_0' versus 'μ is *less than* the given value'	$H_0: \mu = \mu_0$ $H_1: \mu < \mu_0$	$t = \dfrac{\mu_0 - \bar{x}}{s/\sqrt{n}}$	1

Only the two-tailed value is available directly from this function; to obtain a one-tailed critical value for significance level α and v degrees of freedom, use

$$= \mathrm{TINV}(2\alpha,\ \nu).$$

For example, to obtain the one-tailed critical value for 95% confidence with five degrees of freedom, use

$$= \mathrm{TINV}(0.10,\ 5).$$

Comparing with the critical value
In all cases, the calculated t statistic is compared directly with the critical value. If the calculated value exceeds the critical value, the null hypothesis is rejected. The comparison takes the sign into account if necessary (cases b and c in Table 4.1). For example, -3.2 should be considered less than 2.353.

Example
An analyst is validating a method for the determination of cholesterol in milk fat. As part of the validation study, 10 portions of a certified reference material (CRM) are analysed. The results are shown in Table 4.2. The concentration of cholesterol in the CRM is certified as 274.7 mg per 100 g. The analyst wants to know whether the mean of the results obtained differs significantly from the certified value of the reference material.

The data are approximately symmetrically distributed (Figure 4.5), with the majority of the data near the centre and no strong outliers. The distribution can therefore be taken as approximately normal and a t-test is appropriate.

The relevant hypotheses are:

$$H_0: \mu = 274.7$$
$$H_1: \mu \neq 274.7$$

Table 4.2 Results from the analysis of an anhydrous milk fat CRM (results expressed as mg of cholesterol per 100 g of sample).

271.4	266.3	267.8	269.6	268.7
272.5	269.5	270.1	269.7	268.6

Sample mean	269.42
Sample standard deviation	1.75

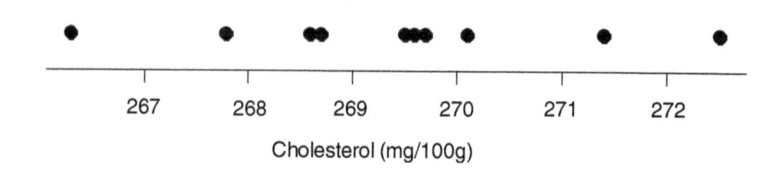

Figure 4.5 Dot plot of cholesterol data from Table 4.2.

Basic Statistical Techniques

The t statistic is calculated using the equation for case (a) in Table 4.1:

$$t = \frac{|269.42 - 274.7|}{1.75/\sqrt{10}} = 9.54$$

The two-tailed critical value for t with significance level $\alpha = 0.05$ and degrees of freedom $v = 9$ (i.e. $v = n - 1$) is 2.262 (see Appendix A, Table A.4).

The calculated value exceeds the critical value so the null hypothesis is rejected. The analyst therefore concludes that the test method gives results that are significantly different from the true value, i.e. the results are *biased* (see Chapter 9, Section 9.3).

4.2.3.2 Comparing the Means of Two Independent Sets of Data (Two-sample Test)

The two-sample test is typically used to decide whether two test items or treatments are different, by comparing the means of a number of observations on each. For example, comparing the concentrations of an active ingredient in ostensibly similar products, or examining the effect of a change of extraction solvent on analytical recovery, might both use two-sample t-tests. The principle of the two-sample test is similar to that of the one-sample test; the test statistic is a difference divided by the standard error of the difference. The calculations are altered, however, because instead of one observed standard deviation, there are now two observed standard deviations – one for each data set – and (often) two different numbers of observations. We therefore start with two data sets (1 and 2) with hypothesised true means μ_1 and μ_2 and observed mean, standard deviation and number of observations $\bar{x}_1$, s_1, n_1 and $\bar{x}_2$, s_2, n_2, respectively.

Stating the hypotheses
The null hypothesis is usually written as $\mu_1 = \mu_2$. To test for a difference irrespective of sign, the alternative hypothesis is $\mu_1 \neq \mu_2$. To test whether data set 1 implies a significantly higher mean than data set 2, the alternative hypothesis becomes $\mu_1 > \mu_2$ and *vice versa*.

Checking the distribution
As before, graphical inspection to confirm approximate normality is recommended.

Number of tails
The alternative hypothesis $\mu_1 \neq \mu_2$ implies a two-tailed test; the alternative hypotheses $\mu_1 > \mu_2$ and $\mu_1 < \mu_2$ imply one-tailed tests.

Table 4.3 Hypotheses and test statistics for two-sample t-tests.

Null and alternative hypotheses	Abbreviated hypotheses	Test statistic[a]	Tails
(a) 'Population mean μ_1 is equal to population mean μ_2' versus 'μ_1 and μ_2 are *not equal*'	H$_0$: $\mu_1 = \mu_2$ H$_1$: $\mu_1 \neq \mu_2$	$t = \dfrac{\lvert \bar{x}_1 - \bar{x}_2 \rvert}{s_{\text{diff}}}$	2
(b) 'Population mean μ_1 is equal to population mean μ_2' versus 'μ_1 is *greater* than μ_2'	H$_0$: $\mu_1 = \mu_2$ H$_1$: $\mu_1 > \mu_2$	$t = \dfrac{\bar{x}_1 - \bar{x}_2}{s_{\text{diff}}}$	1
(c) 'Population mean μ_1 is equal to population mean μ_2' versus 'μ_1 is *less* than μ_2'	H$_0$: $\mu_1 = \mu_2$ H$_1$: $\mu_1 < \mu_2$	$t = \dfrac{\bar{x}_2 - \bar{x}_1}{s_{\text{diff}}}$	1

[a] See equations (4.12)–(4.16) for calculation of s_{diff} and degrees of freedom.

Calculating the test statistic

The general forms of the test statistics corresponding to the various possible hypotheses are listed in Table 4.3. The calculation of the term s_diff in the table, however, depends on whether the standard deviations for each group can be treated as arising from the same population of errors. If they can, the test is known as an *equal variance t-test*. If not – that is, if the standard deviations for each set are significantly different – the test becomes an *unequal variance t-test*. The appropriate degrees of freedom v for the critical value are also affected.

The different calculations are as follows:

1. For the equal variance *t*-test:

$$s_\text{diff} = \sqrt{\left(\frac{1}{n_1} + \frac{1}{n_2}\right)\left[\frac{s_1^2(n_1 - 1) + s_2^2(n_2 - 1)}{n_1 + n_2 - 2}\right]} \tag{4.12}$$

(this is the 'pooled standard deviation').

$$v = n_1 + n_2 - 2 \tag{4.13}$$

Note that when n_1 and n_2 are equal, the equation for calculating s_diff can be simplified to

$$s_\text{diff} = \sqrt{\frac{s_1^2 + s_2^2}{n}} \tag{4.14}$$

where n is the number of observations in each group.

2. For the unequal variance *t*-test:

$$s_\text{diff} = \sqrt{\frac{s_1^2}{n_1} + \frac{s_2^2}{n_2}} \tag{4.15}$$

$$v = \frac{\left(\frac{s_1^2}{n_1} + \frac{s_2^2}{n_2}\right)^2}{\frac{(s_1^2/n_1)^2}{n_1 - 1} + \frac{(s_2^2/n_2)^2}{n_2 - 1}} \tag{4.16}$$

Because the choice of test depends on whether the two standard deviations are significantly different, it is good practice to compare the two standard deviations using an *F*-test (see Section 4.2.4) before carrying out the *t*-test. However, visual inspection or prior knowledge can also help. It may also be worth noting that the unequal variance *t*-test will give very nearly identical results to the equal variance *t*-test when the standard deviations are very similar. If in doubt, therefore, carry out an unequal variance test.

The critical value

The critical value is obtained from Appendix A, Table A.4 or from similar tables or software, using the chosen confidence level and the degrees of freedom calculated using equation (4.13) for the equal variance *t*-test or equation (4.16) for the unequal variance *t*-test. Equation (4.16), however, usually generates a non-integer number of degrees of freedom. It is more conservative to round the calculated degrees of freedom down to the next lower integer, and this is usually the recommended approach. (Here, 'conservative' indicates that the test is less likely to generate a spurious significant test result.) However, if rounding to the nearest integer instead would make a

Basic Statistical Techniques

difference to the test result, it is clearly prudent to consider the test result to be marginal and act accordingly. Note that some software implementations – including the spreadsheet functions mentioned in Section 4.2.3.1 – round down automatically, whereas others, particularly statistical software, may calculate an exact critical value and p-value for non-integer v.

Comparison with the critical value
As in Section 4.2.3.1, a calculated test statistic greater than the critical value should be interpreted as indicating a significant result at the chosen level of confidence and the null hypothesis should be rejected in favour of the alternative.

Example

An analyst wants to compare the performances of two test methods for the determination of selenium in vegetables. The analyst selects a suitable sample and analyses eight portions of the sample using Method 1 and eight portions using Method 2. The results are shown in Table 4.4. The analyst wishes to check whether the mean values of the results imply a significant difference.

The relevant hypotheses are:

$$H_0: \mu_1 = \mu_2$$
$$H_1: \mu_1 \neq \mu_2$$

The plotted data (Figure 4.6) show a possible low outlier for Method 1, but in such a small data set the observation could arise by chance from a normal distribution (in fact, an outlier test of the type considered in Chapter 5 returns a p-value of about 0.02 – enough to warrant close inspection, but insufficient to rule out chance). A *t*-test therefore remains appropriate. The standard deviations in Table 4.4 are fairly similar for the number of observations and the plot shows a similar spread, so an equal variance test is chosen.

Table 4.4 Results for the determination of selenium in cabbage (results expressed as mg per 100 g).

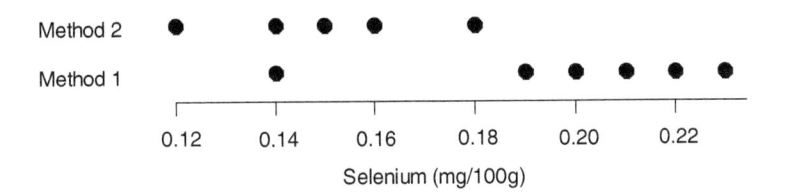

									$\bar{x}$	s
Method 1	0.20	0.19	0.14	0.19	0.23	0.21	0.22	0.21	0.1988	0.0275
Method 2	0.18	0.14	0.14	0.16	0.18	0.15	0.14	0.12	0.1513	0.0210

Figure 4.6 Dot plots for selenium data from Table 4.4.

The t statistic is calculated using case (a) in Table 4.3 with s_{diff} from equation (4.12):

$$t = \frac{|0.1988 - 0.1513|}{\sqrt{\left(\frac{1}{8} + \frac{1}{8}\right)\left[\frac{0.0275^2 \times (8-1) + 0.0210^2 \times (8-1)}{8+8-2}\right]}}$$

$$= \frac{0.0475}{\sqrt{0.25 \times \left(\frac{0.00838}{14}\right)}} = 3.88$$

Note that as the data sets contain the same number of observations, it would also be possible to use equation (4.14) to calculate s_{diff}.

The two-tailed critical value for t with significance level $\alpha = 0.05$ and degrees of freedom $\nu = 14$ (i.e. $\nu = n_1 + n_2 - 2$) is 2.145 (see Appendix A, Table A.4).

The calculated value (3.88) exceeds the critical value (2.145), so the null hypothesis is rejected. The analyst therefore concludes that there is a significant difference between the means of the results produced by the two methods.

For comparison, an unequal variance t-test on the same data gives a test statistic of 3.88, degrees of freedom 13.1 (rounded to 13) and critical value of 2.160 – an almost identical result.

4.2.3.3 Paired Comparisons

When comparing the performances of two methods, it may not be possible to generate two replicated data sets and apply the t-test described in Section 4.2.3.2. For example, it may not be practical to obtain more than one result from each method on any one test item. In such cases, the paired comparison test is very useful. This requires pairs of results obtained from the analysis of different test materials as illustrated in Figure 4.7.

In Figure 4.7a, the points represent pairs of results for six different test materials. In each pair, the open and filled circles represent results obtained from different treatments A and B; for example, different methods or different analysts. Since the sets of results from each treatment vary

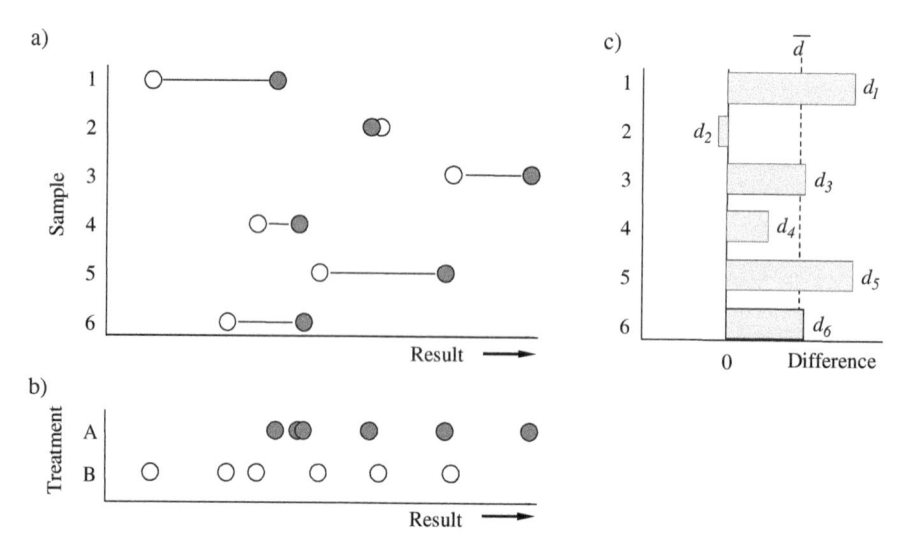

Figure 4.7 Comparison using paired samples.

Basic Statistical Techniques

due to genuine differences between the test materials and also because of random measurement variation (Figure 4.7b), it would be misleading to evaluate the data using the *t*-test given in Table 4.3; any difference between the treatments could be masked by differences between the test materials. However, the difference $d_i = a_i - b_i$ between results a_i and b_i in each pair is due only to the effects of the treatments and to random variation. Taking the differences gives a new set of data, the differences shown in Figure 4.7c. If there is no difference between the results obtained using the two treatments, the average of the differences is expected to be zero. To test for a significant difference between treatments without interference from differences between test materials, therefore, we test whether the mean difference μ_d is zero. Following the usual procedure:

Stating the hypotheses
The null hypothesis is generally taken as $\mu_d = 0$. Taking the different treatments as treatment A and treatment B and assuming that $d_i = a_i - b_i$, if the test is for a difference irrespective of sign, the alternative hypothesis H_1 is $\mu_d \neq 0$. If it is important to test for a particular sign of the difference – such as in a test to check for improved analytical recovery – the alternative hypothesis becomes $\mu_d > 0$ if the alternative relates to A generating higher responses than B, or $\mu_d < 0$ if the alternative relates to A generating lower responses than B.

Checking the distribution
The paired *t*-test does not require that the raw data be normally distributed, only that the *differences* be approximately normally distributed. Checking the differences using graphical methods is usually sufficient. It is important, however, to check that the size of the difference does not depend very strongly on the concentration in the different test materials. This can be done either by checking that the concentrations in the different materials are broadly similar (for example, with a range less than about 20% of the mean value) or by plotting absolute difference against mean value for each material and checking for a visible trend.

Number of tails
As usual, the alternative hypothesis $\mu_d \neq 0$ implies a two-tailed test, whereas the alternative hypotheses $\mu_d > 0$ and $\mu_d < 0$ imply one-tailed tests.

Calculating the test statistic
The test statistics associated with the various hypotheses are summarised in Table 4.5. Note that in the one-tailed tests, the test statistic is given in terms of the differences to emphasise the change in sign required in changing from a test for A > B (case b) to A < B (case c).

Table 4.5 Hypotheses and test statistics for paired *t*-tests.

Null and alternative hypotheses	Abbreviated hypotheses[a]	Test statistic[b]	Tails
(a) 'The difference in treatment effects is zero' versus 'There is a difference between treatments'	$H_0: \mu_d = 0$ $H_1: \mu_d \neq 0$	$t = \dfrac{\lvert \bar{d} \rvert}{s_d/\sqrt{n}}$	2
(b) 'Treatment A gives the same results as treatment B' versus 'Treatment A gives *higher* results than treatment B'	$H_0: \mu_d = 0$ $H_1: \mu_d > 0$	$t = \dfrac{\sum(a_i - b_i)/n}{s_d/\sqrt{n}}$	1
(c) 'Treatment A gives the same results as treatment B' versus 'Treatment A gives *lower* results than treatment B'	$H_0: \mu_d = 0$ $H_1: \mu_d < 0$	$t = \dfrac{\sum(b_i - a_i)/n}{s_d/\sqrt{n}}$	1

[a] The mean difference $\bar{d}$ is defined such that $d_i = a_i - b_i$.
[b] $\bar{d}$ and s_d are the mean and the standard deviations of the differences; a_i and b_i are the individual observations for methods A and B, respectively.

The critical value
The critical value is obtained from tables such as Table A.4 in Appendix A or from software, using the appropriate level of significance and degrees of freedom exactly as for the one-sample test in Section 4.2.3.1. For n pairs, the number of degrees of freedom is $v = n - 1$.

Comparing with the critical value
If the calculated test statistic is greater than the critical value, the test is significant at the chosen level of confidence and the null hypothesis rejected.

Example
Two methods are available for determining the concentration of vitamin C in vegetables. An analyst wants to know whether there is any significant difference in the results produced by the two methods. Eight different test materials are analysed using the two methods. The results are shown in Table 4.6 and Figure 4.8.

The relevant hypotheses are:

$$H_0: \mu_d = 0$$
$$H_1: \mu_d \neq 0$$

The plotted differences show no strong departure from normality and the absolute differences show no relationship with the mean for each test material. A paired t-test is therefore appropriate. The hypotheses correspond to case (a) in Table 4.5; the t statistic is calculated accordingly:

$$t = \frac{1.1}{10.8/\sqrt{8}} = 0.288$$

There are eight pairs, so $8 - 1 = 7$ degrees of freedom. The two-tailed critical value for t with significance level of $\alpha = 0.05$ and degrees of freedom $v = 7$ is 2.365 (see Appendix A, Table A.4). The calculated value is less than the critical value so the null hypothesis is accepted. The analyst therefore concludes that there is no significant difference between the results produced by the different methods.

4.2.4 Comparing Two Observed Standard Deviations or Variances – the *F*-Test

The *F*-test compares two experimentally observed variances, s_a^2 and s_b^2. Since variances are simply squared standard deviations, it may also be used to compare standard deviations. This is useful for comparing the precision of analytical methods to see if one is significantly better than the other; it is also fundamental to another very useful statistical tool, the analysis of variance described in Chapter 6.

Table 4.6 Results from the determination of vitamin C in vegetables (results expressed as mg per 100 g).

	Test material							
	1	2	3	4	5	6	7	8
Method A	291	397	379	233	365	291	289	189
Method B	272	403	389	224	368	282	286	201
Mean	281.5	400	384	228.5	366.5	286.5	287.5	195
Difference ($d = a - b$)	19	−6	−10	9	−3	9	3	−12
Mean difference	1.1							
Standard deviation of differences	10.8							

Basic Statistical Techniques

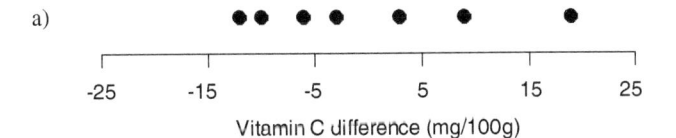

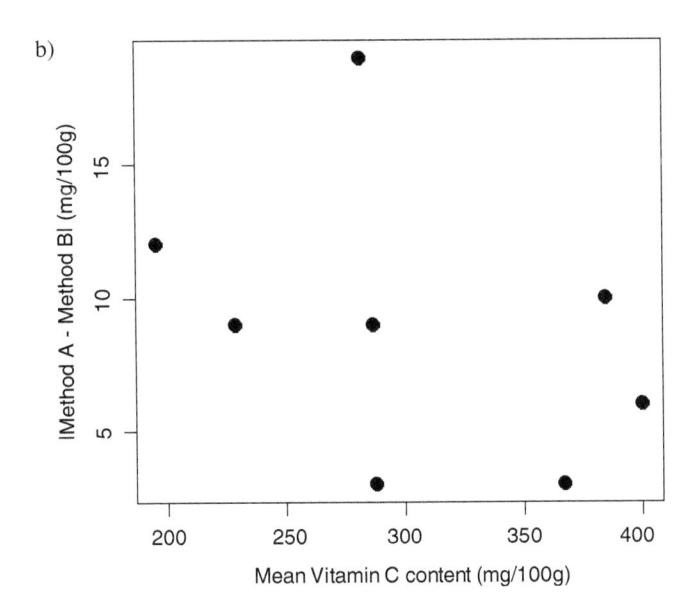

Figure 4.8 Vitamin C data from Table 4.6. (a) Dot plot of the differences in Table 4.6; (b) absolute differences ($|d|$) plotted against mean for each test material.

Stating the hypotheses
Hypotheses for the *F*-test are all about comparing two variances, σ_A^2 and σ_B^2. The usual null hypothesis is $\sigma_A^2 = \sigma_B^2$. For a two-tailed test, for example when testing one method against another to see whether the performance is similar, the alternative hypothesis is $\sigma_A^2 \neq \sigma_B^2$. When testing whether variance A is greater, the alternative hypothesis is $\sigma_A^2 > \sigma_B^2$ and *vice versa*. The one-sided hypothesis is certainly appropriate for situations when one variance is known to include more sources of variation (as it will in analysis of variance). It might also be appropriate if a change has been made that is specifically intended to improve precision.

Checking the distribution
Like the *t*-test, the *F*-test depends on an assumption of normality. For small data sets, the distributions are most usefully checked by inspection using dot plots and, where necessary, normal probability plots, as described in Chapter 2. As usual, outliers and excessive skew will compromise the test.

Calculating the test statistic
The test statistics are summarised in Table 4.7, which also indicates the number of tails appropriate to the hypotheses. Note that for the two-tailed test, the calculation is arranged so that the value of *F* is always greater than 1.0, whereas for the one-tailed tests, the numerator and denominator follow the order implied by the hypotheses. Another feature is that *F* is always a ratio of *variances*; standard deviations must be squared to obtain the correct value of *F*.

Table 4.7 Hypotheses and test statistics for the *F*-test.

Null and alternative hypotheses	Abbreviated hypotheses	Test statistic	Tails	DOF
(a) 'Population variance σ_a^2 is equal to population variance σ_b^2' versus 'σ_a^2 is not equal to σ_b^2'	$H_0: \sigma_a^2 = \sigma_b^2$ $H_1: \sigma_a^2 \neq \sigma_b^2$	$^aF = \dfrac{s_{max}^2}{s_{min}^2}$	2	ν_{max}, ν_{min}
(b) 'Population variance σ_a^2 is equal to population variance σ_b^2' versus 'σ_a^2 is greater than σ_b^2'	$H_0: \sigma_a^2 = \sigma_b^2$ $H_1: \sigma_a^2 > \sigma_b^2$	$F = \dfrac{s_a^2}{s_b^2}$	1	ν_a, ν_b
(c) 'Population variance σ_a^2 is equal to population variance σ_b^2' versus 'σ_a^2 is less than σ_b^2'	$H_0: \sigma_a^2 = \sigma_b^2$ $H_1: \sigma_a^2 < \sigma_b^2$	$F = \dfrac{s_b^2}{s_a^2}$	1	ν_b, ν_a

$^a s_{max}^2$ is the larger and s_{min}^2 the smaller of s_a^2 and s_b^2.

The critical value

The *F*-test compares two variances, so the degrees of freedom for each variance are both important. Software will always require entry of both degrees of freedom and also the appropriate probability. Tables of critical values for *F* usually show upper one-tailed critical values for a single probability, with degrees of freedom for the numerator (top row) of the test statistic in the top row of the table and the degrees of freedom for the denominator down the left side (see Appendix A, Table A.5).

The fact that the tables for *F* show only one-tailed probabilities requires some care in selecting the correct table. Like the *t*-test, critical values for a two-tailed test with significance level α are the same as critical values of a one-tailed test with significance $\alpha/2$ (check Appendix A, Table A.4). It is vital to use the following two rules for selecting the correct probability to look up:

1. For a one-tailed *F*-test with level of significance α, use a table of critical values for significance level α or confidence level $100(1-\alpha)\%$.
2. For a two-tailed test with level of significance α, use a table of critical values for significance level $\alpha/2$ or confidence level $100(1-\alpha/2)\%$.

These rules also apply to most – but not all – software functions for providing critical values for the *F*-distribution. It is therefore important to check the software documentation.

As a check on correct use of the tables and rules, for a two-tailed test with significance level 0.05, $\nu_{max} = 7$ and $\nu_{min} = 4$, the critical value is 9.074.

These rules, and the selection of the correct test statistic, are summarised in Figure 4.9.

Example

In the example in Section 4.2.3.2, the means of two sets of data were compared. Prior to carrying out the *t*-test, it is good practice to check whether the variances of the data sets are significantly different, to determine which form of the *t*-test should be used. The relevant hypotheses for comparing the variances are:

$$H_0: \sigma_1^2 = \sigma_2^2$$
$$H_1: \sigma_1^2 \neq \sigma_2^2$$

The data were approximately normally distributed (Figure 4.6), so the *F*-test may be used. Using the data in Table 4.4 and the calculation shown in case (a) in Table 4.7, the test statistic *F* is

Basic Statistical Techniques

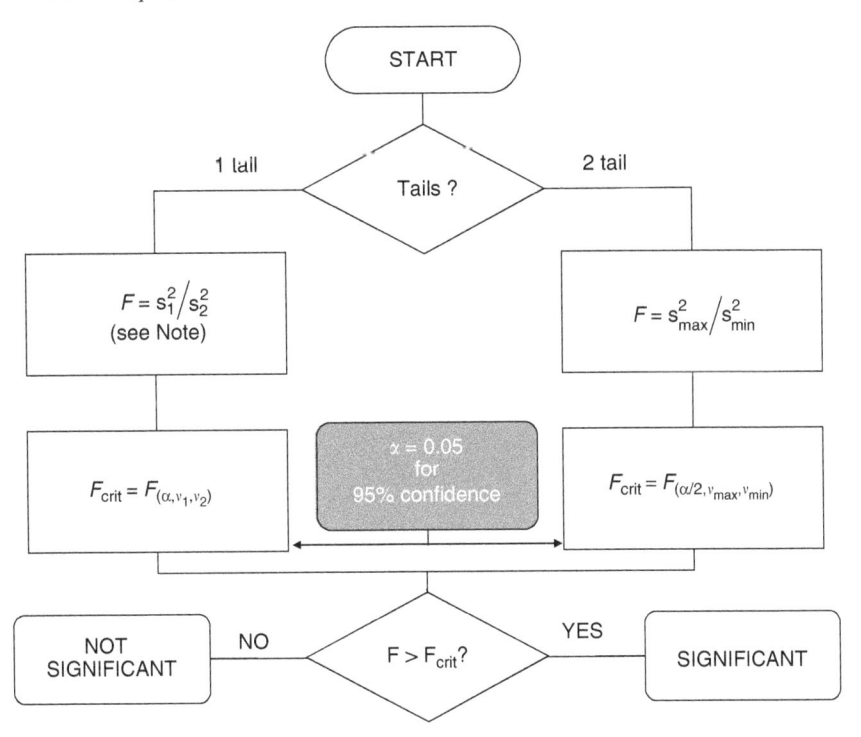

Figure 4.9 The *F*-test procedure. Note: consult Table 4.7 to select s_1 and s_2 correctly for the one-tailed test. For example, for the alternative hypothesis H_1: $\sigma_a > \sigma_b$, s_1 would be the standard deviation of data set a and s_2 would be the standard deviation of data set b.

calculated as follows:

$$F = \frac{s_{max}^2}{s_{min}^2} = \frac{0.027^2}{0.021^2} = 1.65$$

Note that the larger variance s_{max}^2 has been placed in the numerator, as is appropriate for an alternative hypothesis of the form $\sigma_1^2 \neq \sigma_2^2$. The significance level for the test is $\alpha = 0.05$. However, as this is a two-tailed test, the correct critical value is obtained from the tables for $\alpha/2 = 0.025$. The critical value for $v_{max} = v_{min} = 7$ (i.e. $n_{max} - 1$ and $n_{min} - 1$) is 4.995 (see Appendix A, Table A.5b). The calculated value of *F* is less than the critical value. The null hypothesis is therefore accepted and the conclusion is that there is no significant difference between the variances of the sets of data.

4.2.5 Comparing Observed Standard Deviation or Variance with an Expected or Required Standard Deviation Using Tables for the *F* Distribution

The *F*-test is usually used to compare two observed variances. Sometimes, however, it is useful to compare an observed variance (or standard deviation) with an expected or required value. For example, if a validation requirement specifies a maximum permitted repeatability standard deviation σ_0, it becomes necessary to compare the observed variance s^2 with the required standard deviation to see if the true variance σ^2 for the method is greater than the required variance σ_0^2. This differs from the *F*-test in that the variance σ_0^2 is known exactly.

Observed variances follow a scaled chi-squared distribution (see Chapter 3, Section 3.4.2), so a test for a single variance against a fixed value can use critical values based on the chi-squared distribution. However, the relevant scaled chi-squared distribution is just the same as the F-distribution with one observed variance with $n-1$ degrees of freedom and one exactly known variance which can be treated as having infinite degrees of freedom. In practice, therefore, the relevant test is simply a special case of the F-test.

To compare an experimental standard deviation s having $n-1$ degrees of freedom with a required or fixed value, therefore, carry out an F-test as in Section 4.2.4, replacing s_a with the observed standard deviation s and s_b with the fixed value σ_0. The critical value is read from the F tables as in Section 4.2.4, using $v_a = n-1$ and $v_b = \infty$.

Example
Do the data in the cholesterol example (Table 4.2) indicate that the method is providing precision significantly better than 1% RSD?

The mean is 269.42 mg per 100 g, so 1% RSD corresponds to a standard deviation of $0.01 \times 269.42 \approx 2.69$ mg per 100 g. We take this as our fixed standard deviation, σ_0. The observed standard deviation s is 1.75 mg per 100 g, with $10 - 1 = 9$ degrees of freedom. The question 'is the observed precision significantly *better* than 1% ...' implies – because better precision is a lower standard deviation – an alternative hypothesis of $\sigma < \sigma_0$ or, expressed in terms of variances, $H_1: \sigma^2 < \sigma_0^2$. This is case (c) in Table 4.7. Replacing s_a with s and s_b with σ_0 in the calculation of the test statistic F, we obtain

$$F = \frac{\sigma_0^2}{s^2} = \frac{2.69^2}{1.75^2} = 2.363$$

The test is one-tailed and we choose the usual significance level of $\alpha = 0.05$. The correct probability to look up is, following Figure 4.9, $\alpha = 0.05$, corresponding to $100(1-0.05)\% = 95\%$ confidence, so the correct table is Appendix A, Table A.5a. The critical value for F with $v_1 = \infty$ and $v_2 = 9$ is 2.707. The test value, 2.363, is less than the critical value, so the null hypothesis is retained; there is insufficient evidence at the 95% level of confidence to conclude that the method provides better than 1% RSD.

4.3 Confidence Intervals for Mean Values

The confidence interval gives a range of values for the true mean μ which would be considered to be consistent with an observed mean $\bar{x}$ with a given level of confidence. For a population with a mean μ and a standard deviation σ, 95% of the sample means, $\bar{x}$, will lie within the range given by the equation

$$\mu - 1.96(\sigma/\sqrt{n}) < \bar{x} < \mu + 1.96(\sigma/\sqrt{n}) \tag{4.17}$$

The value of 1.96 is the two-tailed Student's t value for $\alpha = 0.05$ and $v = \infty$.

In general, we will have an estimate of $\bar{x}$ (from experimental data) and need to know the confidence interval for the population mean. Equation (4.17) is therefore rearranged:

$$\bar{x} - 1.96(\sigma/\sqrt{n}) < \mu < \bar{x} + 1.96(\sigma/\sqrt{n}) \tag{4.18}$$

This is the 95% confidence interval for the mean if we know the population standard deviation σ exactly.

Basic Statistical Techniques

Typically, only a relatively small number of data points are used to estimate $\bar{x}$. We therefore need to replace σ and the value 1.96 in equation (4.18) with the sample standard deviation s and the two-tailed Student's t value for $n-1$ degrees of freedom. The confidence interval for the mean is then:

$$x \pm t(s/\sqrt{n}) \tag{4.19}$$

CHAPTER 5
Outliers in Analytical Data

5.1 Introduction

An outlier may be defined as an observation in a set of data that appears to be inconsistent with the remainder of that set. Usually, this means a value that is visibly distant from the remainder of the data. Outlying values generally have an appreciable influence on calculated mean values and even more influence on calculated standard deviations. Random variation generates occasional extreme values by chance; these are part of the valid data and should generally be included in any calculations. Unfortunately, a common cause of outliers – particularly very extreme values – is human error or other aberration in the analytical process, such as instrument failure. Obviously, values arising from a faulty procedure should not be allowed to influence any conclusions drawn from the data, making it important to be able to minimise their impact on the statistics.

There are two general strategies for minimising the effect of outliers. The first is outlier testing, which is intended to identify outliers and distinguish them from chance variation, allowing the analyst to inspect suspect data and if necessary correct or remove erroneous values. Outlier testing methods are described in Section 5.2. The second strategy is to use *robust statistics*: statistical procedures which are not greatly affected by the presence of occasional extreme values, but which still perform well when no outliers are present. Some robust methods are described in Section 5.3.

5.2 Outlier Tests

5.2.1 The Purpose of Outlier Tests

Visual inspection using dot plots, box plots, *etc.* (see Chapter 2) will often show one or two outlying values in a data set. Outlier tests show whether such values could reasonably arise from chance variation or are so extreme as to indicate some other cause, for example, an unusual test sample, a mistake or an instrument fault. This both protects the analyst from taking unnecessary action based on chance variation and directs attention towards likely problems that require inspection and possible correction. The most important role of outlier testing is therefore to provide objective criteria for taking investigative or corrective action.

Outlier tests are also used in some circumstances to provide a degree of robustness. Although outright rejection of an extreme value on statistical grounds alone is not generally recommended (see Section 5.2.2), if there is good reason to believe that errors are likely, rejection of extreme outliers can, like robust statistics, prevent errors from unduly influencing results.

Practical Statistics for the Analytical Scientist: A Bench Guide, 2nd Edition
Stephen L R Ellison, Vicki J Barwick, Trevor J Duguid Farrant
© LGC Limited 2009
Published by the Royal Society of Chemistry, www.rsc.org

Outliers in Analytical Data 49

5.2.2 Action on Detecting Outliers

Before describing individual tests, it is useful to consider what action should be taken on the basis of outlier tests. A statistical outlier is only *unlikely* to arise by chance. A positive outcome from an outlier test is best considered as a signal to investigate the cause; usually, outliers should not be removed from the data set solely because of the result of a statistical test. However, experience suggests that human or other error is among the most common causes of extreme outliers. This experience has given rise to fairly widely used guidelines for acting on outlier tests on analytical data, based on the outlier testing and inspection procedure included in ISO 5725 Part 2 for processing interlaboratory data.[1] The main features are:

1. Test at the 95% and the 99% confidence level.
2. All outliers should be investigated and any errors corrected.
3. Outliers significant at the 99% level may be rejected unless there is a technical reason to retain them.
4. Outliers significant only at the 95% level (often termed *stragglers*) should be rejected only if there is an additional, technical reason to do so.
5. Successive testing and rejection is permissible, but not to the extent of rejecting a large proportion of the data.

This procedure leads to results which are not seriously biased by rejection of chance extreme values, but are relatively insensitive to outliers at the frequency commonly encountered in measurement work. Note, however, that this objective can be attained without outlier testing by using robust statistics where appropriate; this is the subject of Section 5.3.

Finally, it is important to remember that an outlier is only 'outlying' in relation to some prior expectation. The outlier tests in this chapter assume underlying normality. If the data were Poisson distributed, for example, many valid high values might be incorrectly rejected because they appear inconsistent with a *normal* distribution. It is also crucial to consider whether outlying values might represent genuine features of the population. For example, testing relatively inhomogeneous or granular materials will often show apparent outliers where different particles have very different analyte content. A well-studied example is the determination of aflatoxins in nuts, where a small portion – even one kernel in a laboratory sample – can contain hundreds of times more toxin than the majority. Rejecting the observation due to that kernel on statistical grounds would be entirely incorrect; the 'outlier' is genuinely representative of the population and it must be included in calculating mean values to obtain a correct decision. Similarly, an outlying observation in a process control environment is an important signal of a process problem; if all the outlying values were rejected, process control would be rendered ineffective. It follows that outlier testing needs careful consideration where the population characteristics are unknown or, worse, known to be non-normal.

5.2.3 The Dixon Tests

Dixon published a set of tests based on simple range calculations,[2] more recently updated by Rorabacher.[3] They have the advantage that all the necessary test statistics can easily be calculated by hand or with a very basic calculator. The simplest, often called Dixon's Q test, is particularly appropriate for small data sets (for example, up to about 10 observations).

5.2.3.1 The Dixon Q test

To carry out a Dixon Q test for both high and low outliers:

1. Rank the n results in ascending order to give the values x_1 to x_n.
2. Calculate the test statistic Q (r_{10} in Dixon's original nomenclature) from:

$$Q = r_{10} = \frac{x_2 - x_1}{x_n - x_1} \text{ (lowest observation)} \tag{5.1a}$$

$$Q = r_{10} = \frac{x_n - x_{n-1}}{x_n - x_1} \text{ (highest observation)} \tag{5.1b}$$

(Figure 5.1 shows the first of these schematically).

3. Compare the largest value of Q with the critical value in the table (see Appendix A, Table A.1) for the appropriate level of confidence and number of observations n.

If either of the calculated values for Q exceeds the critical value, the relevant observation is classed as an outlier at the chosen level of confidence.

Example

The weights of seven packets of marshmallows when recorded are shown in Table 5.1.
The outlier test was carried out using Dixon's Q test:

$$Q = \frac{x_2 - x_1}{x_7 - x_1} = \frac{150 - 147}{184 - 147} = 0.081 \text{ (lowest observation)}$$

$$Q = \frac{x_7 - x_6}{x_7 - x_1} = \frac{184 - 159}{184 - 147} = 0.676 \text{ (highest observation)}$$

The critical value for Q, or r_{10}, for seven data points ($n = 7$) is 0.568 at 95% confidence and 0.680 at 99% confidence (see Appendix A, Tables A.1a and A.1b). The largest calculated value of Q, 0.676, is therefore significant at the 95% but not at the 99% level of confidence. The packet weighing 184 g is therefore a 'straggler' – worth checking carefully but reasonably likely to arise by chance.

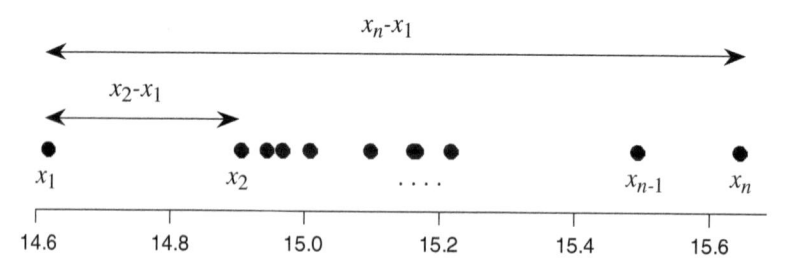

Figure 5.1 Schematic representation of Dixon's Q. Example data in the figure are random and units are arbitrary.

Table 5.1 Weights recorded for packets of marshmallow.

Weight (g)	159	153	184	153	156	150	147
Ordered data	147	150	153	153	156	159	184
Rank	1	2	3	4	5	6	7

Outliers in Analytical Data

Table 5.2 Test statistics for Dixon's outlier test.

Test statistic[a]	Application
$Q = r_{10} = \dfrac{x_2 - x_1}{x_n - x_1} \left(\text{OR } \dfrac{x_n - x_{n-1}}{x_n - x_1} \right)$	Test for single extreme value in a small data set. *Recommended* for n from 3 to 7.
$r_{11} = \dfrac{x_2 - x_1}{x_{n-1} - x_1} \left(\text{OR } \dfrac{x_n - x_{n-1}}{x_n - x_2} \right)$	Test for single extreme value in a small data set; unaffected by a single extreme value at the opposite end of the set. *Recommended* for n from 8 to 10.
$r_{12} = \dfrac{x_2 - x_1}{x_{n-2} - x_1} \left(\text{OR } \dfrac{x_n - x_{n-1}}{x_n - x_3} \right)$	Test for single extreme value; unaffected by up to two extreme values at the opposite end of the set.
$r_{20} = \dfrac{x_3 - x_1}{x_n - x_1} \left(\text{OR } \dfrac{x_n - x_{n-2}}{x_n - x_1} \right)$	Test for single extreme value; unaffected by one adjacent extreme value.
$r_{21} = \dfrac{x_3 - x_1}{x_{n-1} - x_1} \left(\text{OR } \dfrac{x_n - x_{n-2}}{x_n - x_2} \right)$	Test for single extreme value; unaffected by one adjacent value or an extreme value at the opposite end of the set. *Recommended* for n from 11 to 13.
$r_{22} = \dfrac{x_3 - x_1}{x_{n-2} - x_1} \left(\text{OR } \dfrac{x_n - x_{n-2}}{x_n - x_3} \right)$	Test for single extreme value; unaffected by one adjacent extreme value or up to two extreme values at the opposite end of the set. *Recommended* for n from 14 to 30.

[a]Formulae show the calculation for the lowest extreme value in the data set and, in parentheses, the corresponding tests for the highest observation in the data set.

5.2.3.2 Extended Tests

Dixon's tests include a variety of different test statistics, listed in Table 5.2 together with their recommended range of application. The recommended data set sizes are those given by Dixon. Using a particular test slightly outside the recommended range is not usually serious; for example, the Q (r_{10}) test has been recommended elsewhere for up to 10 observations. However, as the data set size increases, so does the risk of two extreme values 'masking' one another. For example, a pair of high values will not be marked as outliers by either the r_{10} or r_{11} tests. Following the recommendations helps to avoid this while retaining a reasonable probability of detecting aberrant values.

Some additional protection against suspected 'masking' can also be gained by choosing the test appropriately; for example, for intermediate sets (say, 5–10 values), r_{20} is unaffected by an adjacent observation and may be more appropriate than r_{11} if pairs of outliers are likely.

Application of several different tests in this series is, however, not generally advisable; one should choose one test appropriate to the size of the data set, taking account of any additional suspect values if necessary.

5.2.4 The Grubbs Tests

Three tests to detect outliers in a normal distribution were developed[4] and extended[5] by Grubbs. All use test statistics based on standard deviations. The first of these is a test for a single outlying value. The second tests for a pair of outlying values at opposite ends of the data set; the third checks for a pair of outlying values on the same side of the data set.

1. $$G'_{\text{low}} = \frac{\bar{x} - x_1}{s} \quad \text{or} \quad G'_{\text{high}} = \frac{x_n - \bar{x}}{s} \tag{5.2a}$$

2. $$G'' = \frac{x_n - x_1}{s} \tag{5.2b}$$

3. $$G'''_{low} = \frac{(n-3) \times s^2_{excluding\ 2\ lowest}}{(n-1) \times s^2} \quad \text{or} \quad G'''_{high} = \frac{(n-3) \times s^2_{excluding\ 2\ highest}}{(n-1) \times s^2} \qquad (5.2c)$$

Note that G''' becomes *smaller* as the suspected outliers become more extreme. Values of G''' *below* the critical value are therefore considered significant.

As with the Dixon test, the results must first be ranked in order $(x_1 < x_2 \ldots < x_n)$. The calculations in equations (5.2a–c) are carried out and the values obtained are compared with the critical values. The equations each use a different set of critical values, given in Appendix A, Table A.2.

For G' and G'', if the calculated value exceeds the critical value, the values tested are marked as outliers. For G''', in an exception to the usual interpretation, if the calculated value is *less than* the critical value, the test result is significant and the value is marked as an outlier. If necessary, extreme values are removed and the process repeated to identify further outliers.

The tests are often carried out in turn on the same data set if the single-outlier test is not significant, to ensure that the single-outlier test is not compromised by a second outlier (as would be detected by the other two tests). However, it is important to be aware that using all three Grubbs tests simultaneously will substantially increase the false-positive rate, typically by a factor of approximately 2 for data from a normal distribution.

Example
For illustration, the weights shown in Table 5.1 are tested using each of Grubbs' tests in turn:

1. G'

    The mean, $\bar{x}$, is 157.43 and the standard deviation s is 12.34. Given this,

    $$G'_{low} = \frac{157.43 - 147}{12.34} = 0.8452 \quad \text{and} \quad G'_{high} = \frac{184 - 157.43}{12.34} = 2.153$$

    Using Appendix A, Table A.2, the critical values for G' with $n = 7$ are 2.020 for 95% confidence and 2.139 for 99% confidence. Using Grubbs' test 1, therefore, the highest value of 184 is considered an outlier at 95% and at 99% confidence (compare this with the result for Dixon's test above; different outlier tests may give slightly different results for marginal values).

2. G''

    Using equation (5.2b):
    $$G'' = \frac{184 - 147}{12.34} = 2.998$$
    The critical values for $n = 7$ are 3.222 for 95% confidence and 3.338 for 99% confidence. The calculated value is less than either, so the two outermost values do not form a pair of outliers.

3. G'''

    The variance excluding the two lowest values is 171.5; excluding the two highest it is 11.7. For the whole data set, $s^2 = 152.29$. Using equation (5.2c):

    $$G'''_{low} = \frac{4 \times 171.5}{6 \times 152.29} = 0.751; \quad G'''_{high} = \frac{4 \times 11.7}{6 \times 152.29} = 0.051$$

The critical values for G''' are 0.0708 for 95% confidence and 0.0308 for 99% confidence. Remembering that these are *lower* critical values, we see that G'''_{high} is significant at the 95% level of confidence but not at the 99% level.

Note that once the single high outlier has been discovered, it is not unexpected to find that one or other of G'' or G''' also appears significant. In practice, with a single outlier already identified, one would not normally apply the tests for outlying pairs until the initial high outlier had been investigated or eliminated.

5.2.5 The Cochran Test

The Cochran test is designed to check for unusually high 'within-group' variation among several groups of data, such as laboratories in a collaborative trial. The test compares the highest individual variance with the sum of the variances for all the groups. The test statistic C is the ratio of the largest variance s^2_{max} to the sum of variances:

$$C = \frac{s^2_{max}}{\sum_{i=1}^{l} s_i^2} \tag{5.3}$$

where l is the number of groups. When each group contains only two replicate results, x_{i1} and x_{i2}, a slightly simpler calculation can be used:

$$C = \frac{d^2_{max}}{\sum_{i=1}^{l} d_i^2} \tag{5.4}$$

where $d_i = x_{i1} - x_{i2}$.

If C is larger than the relevant critical value (Appendix A, Table A.3), then the group can be classed as an outlier or a straggler as before. The critical value C_{crit} is based on the number of groups l and the number n in each group. If a few group sizes differ slightly, n is usually replaced by $\bar{n}$, the average number in each group. Groups containing only one observation, however, must be excluded. The critical values can also be calculated from critical values for the F distribution, using

$$C_{crit}(p_C, l, n) = \frac{1}{1 + (l-1)f} \tag{5.5}$$

where p_C is the desired level of confidence for the Cochran test (for example, 0.95), l is the number of groups, n is the number of observations in each group and f is the critical value of the F distribution for probability $p_F = (1 - p_C)/l$, with degrees of freedom $(l-1)(n-1)$ and $(n-1)$, respectively.

Warning: Most statistical software defines and calculates the F distribution such that a low probability p_F gives the lower critical value for F. For this convention, use $p_F = (1 - p_C)/l$ as above. Unfortunately, the FINV(p_F', v_1, v_2) functions common to MS Excel and OpenOffice provide the *upper* critical value for F for a *lower* value of p_F', so $p_F' = 1 - p_F$. If using the spreadsheet function to estimate Cochran critical values, use $p_F' = 1 - (1 - p)/l$ and not $(1 - p)/l$.

5.3 Robust Statistics

5.3.1 Introduction

Instead of rejecting outliers, robust statistics use methods which are less strongly affected by extreme values. A simple example of a robust estimate of a population mean is the median, which is essentially unaffected by the exact value of extreme observations. For example, the median of the data set (1,2,3,4,6) is identical with that of (1,2,3,4,60). The median, however, is substantially more variable than the mean when the data are not outlier-contaminated. A variety of estimators have accordingly been developed that retain a useful degree of resistance to outliers without unduly affecting performance on normally distributed data. A short summary of some estimators for means and standard deviations is given below. Robust methods also exist for analysis of variance, linear regression and other modelling and estimation approaches.

Most of the more sophisticated robust estimates require software. All the estimators described below are available in RSC Analytical Methods Committee software, freely available from the RSC website.[6] The Analytical Methods Committee have also published two detailed reports[7,8] and a useful Briefing Note[9] on robust statistics.

5.3.2 Robust Estimators for Population Means

5.3.2.1 Median and Trimmed Mean

The median (Chapter 4, Section 4.1.3.2) is widely available in software and very resistant to extreme values; up to half of the data may go to infinity without affecting the median value.

Another simple robust estimate is the so-called 'trimmed mean', which is the mean of the data set with some fraction (usually about 10–20%) of the most extreme values removed.

Both the median and the trimmed mean suffer from increases in variability for normally distributed data, the trimmed mean less so. The following two estimates, although requiring software methods, perform better for near-normal data with a modest incidence of outliers.

5.3.2.2 The A15 Estimate

Rather than removing the most extreme values, the A15 estimate works by giving less weight to values distant from the estimated mean. The calculation is iterative; A15 recalculates the mean value with numeric values x_i replaced with 'pseudovalues' z_i defined by

$$z_i = \begin{vmatrix} x_i & \text{if } \hat{X} - c\hat{s} < x_i < \hat{X} + c\hat{s} \\ \hat{X} \pm c\hat{s} & \text{otherwise} \end{vmatrix} \quad (5.6)$$

where c is a constant value, usually about 1.5, $\hat{X}$ the current estimate of the central value and $\hat{s}$ a robust estimate of the standard deviation (MAD_E is often used for $\hat{s}$; see Section 5.3.3.2). The calculation is repeated using the updated values of z until $\hat{X}$ stabilises. The procedure is therefore as follows:

1. Calculate a (robust) standard deviation $\hat{s}$.
2. Calculate an initial estimated central value $\hat{X}$.
3. Calculate 'pseudovalues' z_i from the data, using equation (5.6).
4. Update $\hat{X}$ by calculating the mean of the z_i values and repeat from step 2 until the value converges.

Steps 2–4 are illustrated in Figure 5.2. The initial estimates $\hat{X}_1$ and $\hat{s}$ are used to identify extreme values (shaded in Figure 5.2a). The extreme values are replaced with the values shown as shaded points in Figure 5.2b. The new data set is used to calculate the updated value $\hat{X}_2$. If the change is considered negligible, $\hat{X}_2$ is reported as the final estimate; if not, the cycle is repeated until successive values of $\hat{X}$ do not change significantly.

The replacement of the more extreme values with the modified 'pseudovalues' is equivalent to calculating a weighted mean in which the outer values have reduced weighting, but the inner values retain their usual weight.

Outliers in Analytical Data 55

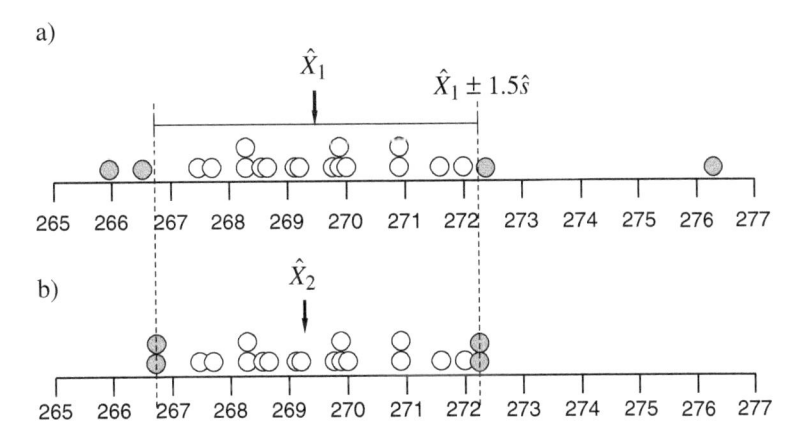

Figure 5.2 Calculating the A15 estimate. (a) Initial estimate $\hat{X}_1$ of mean value and identification of extreme values (shaded). (b) Replacement of extreme values and updated estimate $\hat{X}_2$ of mean value.

The A15 estimate is not very sensitive to the initial estimate of the mean (the median is usually used), but estimating the appropriate standard deviation is a problem. Usually, MAD_E is used unless it is zero, in which case the mean absolute deviation is used (this combination is sometimes called sMAD; see Section 5.3.3.3).

A15 is a very useful estimate of mean values – more stable than the median with 'good' data and fairly insensitive to outlying values.

5.3.2.3 Huber's Proposal 2

Also known as H15, this estimate operates in part in the same way as A15 (see above), recalculating the estimated mean value using equation (5.6) iteratively. However, for H15, the standard deviation estimate is also updated on each iteration; for H15, $\hat{s}$ is given by

$$\hat{s} = s(z)/\beta \qquad (5.7)$$

where $s(z)$ is the standard deviation of the (current) pseudovalues z_i and β is a correction factor based on the expected value of $s(z)$ for a standard normal distribution. H15 therefore provides simultaneous robust estimates of the mean and standard deviation.

Because the correction factor for the estimated standard deviation is based on the normal distribution, the H15 estimate of standard deviation can be treated as an estimate of the population standard deviation for normally distributed data in the absence of outliers.

5.3.3 Robust Estimates of Standard Deviation

5.3.3.1 Median Absolute Deviation

The median absolute deviation (MAD) is a fairly simple estimate that can be implemented in a spreadsheet. It is, as the name suggests, the median of absolute deviations from the data set median, calculated from

$$\mathrm{MAD} = \mathrm{median}\left(|x_i - \tilde{x}|_{i=1,2,\ldots,n}\right) \qquad (5.8)$$

with $\tilde{x}$ being the median.

The median absolute deviation is not directly comparable to the classical standard deviation; for a normal distribution, MAD $\approx \sigma/1.483$.

Example

Data (g)	5.6	5.4	5.5	5.4	5.6	5.3	5.2
Ranked data	5.2	5.3	5.4	5.4	5.5	5.6	5.6

Using equation (4.2) in Chapter 4, the median is equal to 5.4.

$|x_i - \tilde{x}|$	0.2	0.1	0.0	0.0	0.1	0.2	0.2
Ranked deviations	0.0	0.0	0.1	0.1	0.2	0.2	0.2

Applying equation (5.8) to the ranked deviations, the MAD is equal to 0.1.

5.3.3.2 MAD$_E$

The median absolute deviation is not an estimate of standard deviation, but if the underlying distribution is approximately normal, it can be modified to provide one. This is done by multiplying the MAD value by 1.483:

$$\text{MAD}_E = 1.483 \text{MAD} \tag{5.9}$$

For the example data in Section 5.3.3.1, MAD$_E$ = 0.15, essentially identical with the standard deviation of 0.15.

Notes:
1. In some statistical software, 'MAD' is already corrected to give the estimated standard deviation.
2. MAD and MAD$_E$ share the disadvantage that they both become zero if more than half of the data set are equal, perhaps because of excessive rounding or a large number of zero observations. This would, of course, be a problematic data set in any case.

5.3.3.3 Mean Absolute Deviation and sMAD

Because of the limitations of MAD$_E$, it is sometimes useful to use the *mean* absolute deviation instead of the median absolute deviation. For example, the mean absolute deviation for the data set in Section 5.3.3.1 is 0.114, rather than 0.1 for MAD. Although this is less robust than MAD$_E$, it does not become zero unless all the values in the data set are identical. sMAD is a compromise; if MAD$_E$ is non-zero, sMAD = MAD$_E$; if MAD$_E$ is zero, sMAD is the mean absolute deviation.

5.3.3.4 Huber's Proposal 2

Huber's proposal 2 (Section 5.3.2.3) generates a robust estimate of standard deviation as part of the procedure; this estimate is expected to be identical with the usual standard deviation for normally distributed data.

Outliers in Analytical Data 57

Examples
The data in Table 5.3 are used to illustrate the effect of the different estimators. Notice the marked outlier at 2.7 ng g^{-1} (Figure 5.3). The different robust estimates are listed in Table 5.4.

5.4 When to Use Robust Estimators

Robust estimators can be thought of as providing good estimates of the parameters for the 'good' data in an outlier-contaminated data set. They are appropriate when:

- The data are expected to be normally distributed. Here, robust statistics give answers very close to ordinary statistics.
- The data are expected to be normally distributed, but contaminated with occasional spurious values which are regarded as unrepresentative or erroneous and approximately symmetrically distributed around the population mean. Here, robust estimators are less affected by occasional extreme values and their use is recommended. Examples include setting up quality control (QC) charts from real historical data with occasional errors and interpreting interlaboratory study data with occasional problem observations.

Robust estimators are not recommended where:

- The data are expected to follow non-normal or skewed distributions, such as binomial and Poisson with low counts, chi-squared, *etc*. These generate extreme values with reasonable likelihood and robust estimates based on assumptions of underlying normality are not appropriate.

Table 5.3 Aflatoxin B$_1$ in a spice sample.

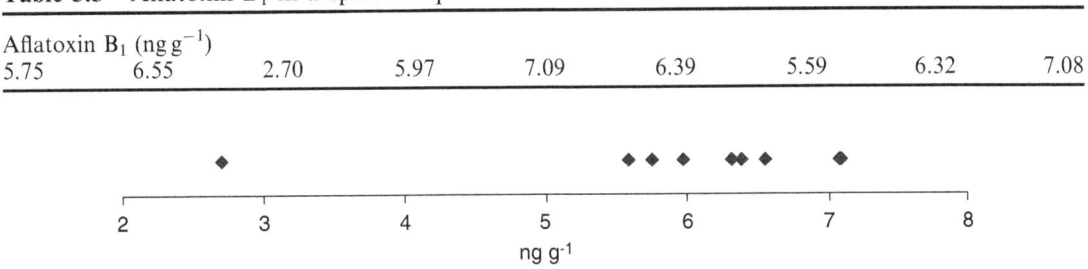

Aflatoxin B$_1$ (ng g^{-1})								
5.75	6.55	2.70	5.97	7.09	6.39	5.59	6.32	7.08

Figure 5.3 Dot plot of data in Table 5.3.

Table 5.4 Robust estimates from data in Table 5.3.

Estimate	Estimate value
Mean	5.94
Median	6.32
A15 mean	6.184
H15 mean	6.207
Standard deviation	1.32
MAD	0.570
MAD$_E$	0.845
sMAD	0.845
H15 standard deviation	0.752

- The data set shows evidence of multimodality or shows heavy skew.
- Summary statistics that represent the whole data distribution (including extreme values, outliers and errors) are required.

References

1. ISO 5725-2:1994, *Accuracy (Trueness and Precision) of Measurement Methods and Results – Part 2: Basic Method for the Determination of Repeatability and Reproducibility of a Standard Measurement Method*, International Organization for Standardization, Geneva, 1994.
2. W. J. Dixon, *Ann. Math. Stat.*, 1950, **21**, 488; 1951, **22**, 68.
3. D. B. Rorabacher, *Anal. Chem.*, 1991, **63**, 139.
4. F. E. Grubbs, *Ann. Math. Stat.*, 1950, **21**, 27.
5. F. E. Grubbs and G. Beck, *Technometrics*, 1972, **14**, 847.
6. RobStat Excel add-in and Minitab macros for robust statistics are available from the RSC AMC pages at http://www.rsc.org/amc/ (accessed 22 April 2008).
7. Analytical Methods Committee, *Analyst*, 1989, **114**, 1489.
8. Analytical Methods Committee, *Analyst*, 1989, **114**, 1693.
9. AMC Technical Brief No. 6, *Robust Statistics: a Method of Coping with Outliers*, Analytical Methods Committee, Royal Society of Chemistry, Cambridge, 2001; Available from http://www.rsc.org/amc/ (accessed 11 December 2008).

CHAPTER 6
Analysis of Variance

6.1 Introduction

Analysis of variance (ANOVA) is a statistical technique which allows variations within sets of data to be isolated and estimated. Consider a training exercise involving four analysts. Each analyst is asked to determine the concentration of lead in a water sample using the same standard procedure and the same instrumentation. Each analyst analyses five portions of the water sample. Their results are given in Table 6.1 and summarised in Figure 6.1.

Within the data set shown in Table 6.1 there are two sources of variation. First, we could calculate the standard deviation of the results produced by each analyst. This will represent the inherent random variability associated with the test method used to measure the concentration of lead in the water sample. In addition, we could determine the standard deviation of the mean values of the four sets of results. This will have a contribution from the random variation mentioned previously, plus any additional variation caused by the different analysts. Analysis of variance allows us to separate and estimate these sources of variation and then make statistical comparisons. In the example shown, we can use ANOVA to determine whether there is a significant difference among the means of the data sets produced by the analysts or whether the variation in the mean values can be accounted for by variation in the measurement system alone. In statistical terms, the hypotheses being tested are:

H_0: the population means of the groups of data are all equal;
H_1: the population means of the groups of data are not all equal.

ANOVA can be applied to any data set that can be grouped by particular *factors* (see Chapter 8, Section 8.1, for formal definitions of 'factors' and related terms). In the example above, one factor – the analyst – has changed from one set of data to the next. The particular analysts, Analyst 1 to Analyst 4, are the *levels* for the factor in this study. One-way (also known as single-factor) ANOVA is used when data can be grouped by a single factor. When there are two factors influencing the results, two-factor ANOVA is used.

The factors can represent either controlled (or 'fixed') effects, such as the times and temperatures involved in specifying gas chromatographic conditions or random effects, such as operator or 'run' effects. The basic calculations for ANOVA remain the same whether effects are controlled (fixed) or random, but we will see that for the more complex situations, the nature of the effects can sometimes influence the interpretation.

Table 6.1 Data from a staff training exercise on the determination of lead in water (results expressed in mg L^{-1}).

		Analyst	
1	*2*	*3*	*4*
64.6	65.6	64.0	68.6
65.5	67.5	62.6	68.0
69.2	65.9	64.8	60.0
65.6	63.6	63.5	70.1
63.5	63.8	68.7	60.8

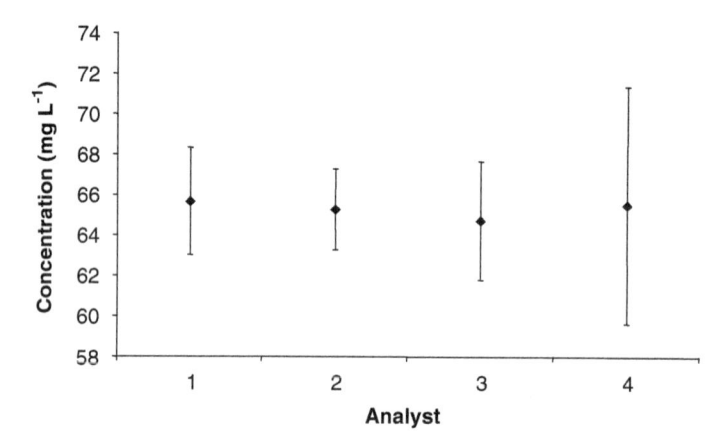

Figure 6.1 Summary of data presented in Table 6.1 (mean and 95% confidence interval shown).

6.2 Interpretation of ANOVA Tables

ANOVA involves the calculation and interpretation of a number of parameters. Detailed calculations for some important cases will be found in later sections and in the Appendix to this chapter. In practice, however, the calculations are best carried out using statistical software. This section introduces the typical output from ANOVA calculations, whether manual or using software, and shows how the results are interpreted.

6.2.1 Anatomy of an ANOVA Table

The general form of a results table from a one-way ANOVA, for a total of N observations in p groups, is shown in Table 6.2. Usually, for experimental data, each group includes the same number of observations. This tabular format is almost universally used for ANOVA output. It consists of a number of rows relating to different sources of variation and a number of columns containing calculated values related to each source of variance.

6.2.1.1 Rows in the ANOVA Table

Each line of the table relates to a different source of variation. The first line relates to variation between the means of the groups; the values are almost always either referred to as 'between-group' terms or are identified by the *grouping factor*. For example, data from different operators are

Analysis of Variance

Table 6.2 Results table from one-way analysis of variance.

Source of variation	Sum of squares	Degrees of freedom v	Mean square	F
Between-group	S_b	$p-1$	$M_b = S_b/(p-1)$	M_b/M_w
Within-group	S_w	$N-p$	$M_w = S_w/(N-p)$	
Total	$S_{tot} = S_b + S_w$	$N-1$		

'grouped by' operator; the relevant row in the table might then be labelled as an 'operator effect' or perhaps as a column or row number from a spreadsheet. The 'Total' line is not always given by software, but it is fairly consistently labelled 'Total' when present. Several different terms may be used to describe the within-group variation, 'within-group', 'residual', 'error' or 'measurement' being the most common.

6.2.1.2 Columns in the ANOVA Table

Sum of Squares

The sum of squares terms are calculated by adding a series of squared error terms. In the case of the within-group sum of squares term, we are interested in the differences between individual data points and the mean of the group to which they belong. For the total sum of squares term, it is the difference between the individual data points and the mean of all the data (the 'grand mean') that is of interest. The between-group sum of squares term is the difference between the total sum of squares and the within-group sum of squares.

Degrees of Freedom

For the one-way ANOVA in Table 6.2, the total number of data points is N and the number of groups of data is p. The total number of degrees of freedom is $N-1$, just as for a simple data set of size N. There are p different groups and therefore $p-1$ degrees of freedom for the between-group effect. The degrees of freedom associated with the within-group sum of squares term is the difference between these two values, $N-p$. Note that if each group of data contains the same number of replicates, n, then the degrees of freedom for the within-group sum of squares is equal to $p(n-1)$ and the total number of observations is $N = pn$.

Mean Squares

The mean squares are the key terms in classical ANOVA. They are variances, calculated by dividing the between- and within-group sum of squares by the appropriate number of degrees of freedom. In Table 6.2, M_b represents the between-group mean square term (sometimes denoted MS_B or M_1) and M_w represents the within-group mean square (sometimes denoted MS_W or M_0).

The mean squares are the values used in the subsequent test for significant differences between the group means. They also allow estimation of the *variance components*, that is, the separate variances for each different effect that contributes to the overall dispersion of the data.

To understand how, it is helpful to use an equation that describes how different effects contribute to each observation. For one-way ANOVA as normally used in analytical chemistry, this can be written

$$x_{ij} = \mu + \delta_i + e_{ij} \qquad (6.1)$$

This says that a data point x_{ij} (the jth observation in group i) is the sum of a (population) 'grand mean' μ, a deviation δ_i that gives group i its own true mean ($\mu + \delta_i$) and a random error e_{ij}.

The combination of an equation such as equation (6.1) and statements about the nature and variance of the different terms is called a *statistical model*.

Equation (6.1) tells us that the dispersion of all the observations is partly due to the dispersion of the deviations δ_i and partly due to the variance σ_w^2 associated with e. Using Equation (6.1) and knowing how the mean squares are calculated, it is possible to calculate how the mean squares depend on the within-group variance σ_w^2 and on the between-group variance σ_b^2 which describes the population from which the δ_i arise. For one-way ANOVA, the expected values of the mean squares are given in the following equations:

$$M_w: \sigma_w^2 \qquad (6.2a)$$

$$M_b: n\sigma_b^2 + \sigma_w^2 \qquad (6.2b)$$

where n is the number of observations in each group. These expressions are important because they allow separate estimation of the two different variances contributing to the overall dispersion. The variances are useful in determining some aspects of analytical method performance (see Chapter 9, Section 9.2.2.2); they also show how important each effect is and in turn how best to improve precision.

F (F Ratio)

The mean squares are compared using an *F*-test (see Chapter 4, Section 4.2.4) as indicated in Table 6.2. The hypotheses for the *F*-test in Table 6.2 are:

$$H_0: M_b = M_w$$
$$H_1: M_b > M_w$$

These hypotheses are equivalent to those concerning the group means mentioned in Section 6.1; if all the means are equal, the two mean squares should also be equal. The alternative hypothesis H_1 is $M_b > M_w$ because, looking at equation (6.2b) and remembering that population variances cannot be negative, the expected value of M_b cannot be less than the (true) within-group variance. The test is therefore a one-tailed test for whether M_b is greater than M_w and the correct test statistic is M_b/M_w. This is the value shown in the column F in Table 6.2. Where there are more sources of variance, the value of F calculated by software for each source of variance is usually the mean square for that source of variance divided by the residual mean square.

No value for F is given for the residual mean square, as there is no other effect with which it can usefully be compared.

F_{crit} and p-Value

Many ANOVA tables include one or two further columns, containing the critical value F_{crit} against which the calculated value of F is to be compared for a chosen significance level, and a *p*-value (see Chapter 4, Section 4.2.2.10) indicating the significance of the test. These are important for interpretation and are considered in the next subsection.

6.2.2 Interpretation of ANOVA Results

The mean squares M_b and M_w have degrees of freedom $v_b = p - 1$ and $v_w = N - p$, and these, together with the appropriate level of confidence, are used to determine the critical value for

Analysis of Variance

the *F*-test from the relevant one-tailed table as described in Chapter 4, Section 4.2.4. Interpretation is straightforward; if the calculated value of *F* exceeds the critical value for a particular source of variance, the null hypothesis is rejected and we conclude that the means of the individual groups of data are not equal. In other words, the variation among mean values cannot be explained plausibly by the measurement variation alone. Therefore, if the calculated value of *F* exceeds the critical value, the effect of the grouping factor – the between-group effect – is statistically significant at the chosen level of confidence.

If a *p*-value is given, a *p*-value *smaller* than the chosen significance level again indicates that the between-group effect is significant.

6.3 One-way ANOVA

6.3.1 Data for One-way ANOVA

One-way ANOVA is applicable when data can be subdivided into groups according to a single factor such as analyst, laboratory or a particular experimental condition. It can be used both to test for significant differences between the groups and to estimate separate within- and between-group variance components.

Consider a study where there are p different levels of a particular factor (that is, p groups of data) and n replicate measurements are obtained at each level. The total number of data points N is therefore equal to pn. A typical layout of the results is shown in Table 6.3.

Individual measurement results are represented by x_{ik} where $i = 1, 2, \ldots, p$ and $k = 1, 2, \ldots, n$. So x_{12}, for example, is the second measurement result ($k = 2$) obtained for the first group of data ($i = 1$).

6.3.2 Calculations for One-way ANOVA

6.3.2.1 Calculating the ANOVA Table

ANOVA calculations are best carried out using statistical software or a spreadsheet with statistical functions; manual calculation – including manually entered spreadsheet formulae – is very likely to introduce transcription or calculation errors. It is informative, however, to see how the ANOVA table is constructed.

The table requires three sums of squares. One, the total sum of squares S_{tot}, is a measure of the dispersion of the complete data set. This is obtained by summing the squared deviations of all the

Table 6.3 Results obtained in a study with p levels and n replicates per level.

Replicates	Levels				
	1	...	*i*	...	*p*
1	x_{11}		x_{i1}		x_{p1}
⋮					
k	x_{1k}		x_{ik}		x_{pk}
⋮					
n	x_{1n}		x_{in}		x_{pn}

data from the mean of all the data $\bar{\bar{x}}$ (the 'grand mean'):

$$S_{\text{tot}} = \sum_{i=1,p; k=1,n} (x_{ik} - \bar{\bar{x}})^2 \qquad (6.3\text{a})$$

The next, S_w, is a measure of the experimental precision, visible within each group and therefore termed the 'within-group sum of squares'. This can be obtained by summing all the squared deviations from the individual group means $\bar{x}_i$:

$$S_w = \sum_{i=1,p} \sum_{k=1,n} (x_{ik} - \bar{x}_i)^2 \qquad (6.3\text{b})$$

(the double summation implies 'sum within each group i and then add up all the group sums').

If the total sum of squares S_{tot} represents all of the dispersion and the within-group term represents the contribution of the repeatability within the groups, it sounds reasonable that their difference should tell us something about the contribution due to the differences between groups – and indeed that is exactly what happens. The 'between-group sum of squares' S_b is calculated as the difference of the other two:

$$S_b = S_{\text{tot}} - S_w \qquad (6.3\text{c})$$

These sums of squares are only indicators of the different contributions. By dividing S_b and S_w by their degrees of freedom, however, we obtain the respective mean squares as shown in Table 6.2. The mean squares can be shown to be independent estimates of particular variances and can therefore be used directly in F-tests.

In practice, the method above works fairly well in a spreadsheet. The frequent averaging and subtraction do, however, make for a rather inefficient hand calculation and a more efficient hand calculation involving fewer operations is provided in the Appendix to this chapter.

6.3.2.2 Calculating Variance Components for One-way ANOVA

It is often useful to estimate the relative sizes of the within- and between-group variances σ_w^2 and σ_b^2. For example, these terms are used in calculating repeatability and reproducibility (Chapter 9, Sections 9.2.1 and 9.2.2). Equations (6.2a) and (6.2b) show how the expected mean squares are related to the (true) within- and between-group variances. Rearranging, we can calculate the estimated within- and between-group variances (s_w^2 and s_b^2, respectively) using the equations in Table 6.5. Note that we have used s^2 to denote the estimated variances; the Greek symbols, as before, refer to the unknown population parameters that we wish to estimate. The corresponding standard deviations s_w and s_b are simply the square roots of the variances.

In practice, the difference $M_b - M_w$, required for the calculation of the between-group variance s_b^2, is sometimes negative. The usual action is then to set s_b^2 to zero.

Example

One-way ANOVA is applied to the data from the training exercise shown in Table 6.1 to determine if there is a significant difference between the data produced by the different analysts. The detailed calculations are shown (using the manual method) in the Appendix to this chapter; the completed ANOVA table is shown in Table 6.4.

The interpretation of one-way ANOVA does not depend on whether the grouping factor represents a controlled (fixed) effect or a random effect. To test for a significant difference between

Analysis of Variance

Table 6.4 ANOVA table for data presented in Table 6.1.

Source of variation	Sum of squares	Degrees of freedom v	Mean square	F
Between-group	$S_b = 2.61$	$v_b = p - 1 = 3$	$M_b = S_b/(p-1) = 0.87$	$M_b/M_w = 0.10$
Within-group	$S_w = 140.48$	$v_w = N - p = 16$	$M_w = S_w/(N-p) = 8.78$	
Total	$S_{tot} = 143.09$	$v_{tot} = N - 1 = 19$		

Table 6.5 Variance component estimates for one-way ANOVA.

Variance component	Estimate
Within-group variance σ_w^2	$s_w^2 = M_w$
Between-group variance σ_b^2	$s_b^2 = \dfrac{M_b - M_w}{n}$

groups, we compare the calculated value of F in Table 6.4 with the critical value. The critical value for F for a one-tailed test at significance level $\alpha = 0.05$ and numerator and denominator degrees of freedom 3 and 16, respectively, is 3.239 (see Appendix A, Table A.5a). The calculated value of F is less than the critical value, so we conclude that there is no significant difference between the data produced by the different analysts. In other words, the variation in the data obtained can be accounted for by the random variability of the test method used to determine the concentration of lead in the water sample.

If we wish, we can also estimate the within- and between-group variance components as shown in Table 6.5. These are estimates of population variances; in this case, the within-run precision of the method as described by the repeatability variance and the variance of the deviations δ_i (now corresponding to analyst bias) in the population of analysts as a whole. We would typically estimate these variances as part of method validation or method performance studies (see Chapter 9, Section 9.2.2.2).

The estimated within-group variance in our example is equal to M_w, 8.78, corresponding to a standard deviation of $\sqrt{8.78} = 2.96$ mg L^{-1}. The between-group variance is estimated from $(M_b - M_w)/n$, with $n = 5$. Since M_b is less than M_w, the estimated between-group variance is negative; in this situation, it is usual to report negative variance estimates as indicating a zero variance, so our estimated between-group variance in this case is taken as zero. This is entirely consistent with the finding that the between-group effect is not statistically significant.

6.4 Two-factor ANOVA

6.4.1 Applications of Two-factor ANOVA

As its name suggests, two-factor ANOVA, also often called *two-way ANOVA*, is used when there are two factors that can influence the result of a measurement. In the training exercise described previously, all the analysts made their measurements of the concentration of lead using the same instrument. If there were actually three instruments available in the laboratory, we could ask each analyst to repeat their measurements on all three instruments. The two factors for the ANOVA would therefore be analyst and instrument. We can use two-factor ANOVA to investigate whether either (or both) of the factors have a significant influence on the results obtained. It is also possible to determine whether there is any interaction between the factors, that is, whether one factor changes the effect of another.

There are two types of experimental design which lend themselves to analysis by two-factor ANOVA – factorial (or cross-classified) designs and hierarchical (or nested) designs. Both designs are described in detail in Chapter 8 (see Sections 8.4.3 and 8.4.4). The use and interpretation of ANOVA for each are described in general terms below. Since two-factor ANOVA is virtually always carried out using statistical software or a spreadsheet with statistical functions, detailed calculations are deferred to the Appendix to this chapter.

6.5 Two-factor ANOVA With Cross-classification

Note that throughout Section 6.5, 'two-factor ANOVA' refers to two-factor ANOVA with cross-classification unless stated otherwise.

6.5.1 Two-factor ANOVA for Cross-classification Without Replication

6.5.1.1 Data for Two-factor ANOVA Without Replication

A cross-classified design involves making at least one measurement for every combination of the different levels of the factors being studied. If there is only one observation for each combination of factor levels, the design is cross-classified without replication. The typical layout is shown in Table 6.6.

In the table, Factor 1 (columns) has levels $i = 1, 2, ..., p$ and Factor 2 (rows) has levels $j = 1, 2, ..., q$. The individual results obtained for each combination of factors are represented by x_{ij}.

6.5.1.2 Results Table for Two-factor ANOVA Without Replication

The layout of the ANOVA table is shown in Table 6.7.

As with one-way ANOVA, the mean square terms can be used to estimate the residual variance (σ_{res}^2) and, if the particular factor levels are assumed to be drawn from a larger population with population variances σ_{col}^2 and σ_{row}^2 for Factor 1 and Factor 2, the respective column and row

Table 6.6 Cross-classified design.

Factor 2 levels	Factor 1 levels				
	1	...	i	...	p
1	x_{11}		x_{i1}		x_{p1}
⋮ j	x_{1j}		x_{ij}		x_{pj}
⋮ q	x_{1q}		x_{iq}		x_{pq}

Table 6.7 ANOVA table for two-factor ANOVA without replication (cross-classification).

Source of variation	Sum of squares	Degrees of freedom v	Mean square	F
Between columns (Factor 1)	S_2	$p - 1$	$M_2 = S_2/(p-1)$	M_2/M_0
Between rows (Factor 2)	S_1	$q - 1$	$M_1 = S_1/(q-1)$	M_1/M_0
Residual	S_0	$(p-1)(q-1)$	$M_0 = S_0/(p-1)(q-1)$	
Total	$S_2 + S_1 + S_0$	$N - 1$		

Analysis of Variance

variance components σ_{col}^2 and σ_{row}^2. The expected values of the mean squares are as follows:

$$M_0: \sigma_{res}^2 \qquad (6.4a)$$

$$M_1: p\sigma_{row}^2 + \sigma_{res}^2 \qquad (6.4b)$$

$$M_2: q\sigma_{col}^2 + \sigma_{res}^2 \qquad (6.4c)$$

6.5.1.3 F-Tests in Two-factor ANOVA Without Replication

Note that each mean square above contains contributions only from the residual variance and from its own population variance. For each factor, therefore, an F-test can be carried out to determine whether the mean square (M_2 or M_1) is significantly greater than the residual mean square (M_0). If the F-test gives a significant result, then we conclude that the factor has a significant effect on the measurement results.

Example

The data in Table 6.8 are taken from a study of a new method to determine the amount of a low-calorie sweetener in different foodstuffs. As part of method development, several laboratories determined the recovery of the sweetener from different sample types. The mean recoveries reported by four of the laboratories on four samples are shown in Table 6.8. The results are also shown in Figure 6.2, plotted by (a) laboratory and (b) sample.

It seems from the consistent ordering of laboratory results for each sample that the different laboratories achieve different recoveries. However, we should first check whether any of these observations could reasonably be chance variation. To do this, we use two-factor ANOVA.

The ANOVA table is shown in Table 6.9. Detailed calculations are given in the Appendix to this chapter.

The critical value for the F-test for the effect of different laboratories uses a significance level $\alpha = 0.05$, corresponding to 95% confidence, with degrees of freedom $v_1 = 3$ and $v_2 = 9$. From Table A.5a in Appendix A we find that the critical value is 3.863. For the test of the effect of different samples, the critical value is the same, as we again have 3 and 9 degrees of freedom for numerator and denominator, respectively. For the laboratories, the calculated F ratio substantially exceeds the critical value; the samples' F ratio does not. We therefore conclude that recoveries for the different laboratories are significantly different at the 95% level of confidence, but that there is insufficient evidence to conclude that the recovery depends on sample type.

As for one-way ANOVA, the interpretation of two-factor ANOVA without replication is unaffected by whether the factors are controlled (fixed) or random.

Table 6.8 Reported mean recoveries (%) for a sweetener in four food samples.

Laboratory	Sample			
	1	2	3	4
1	99.5	83.0	96.5	96.8
2	105.0	105.5	104.0	108.0
3	95.4	81.9	87.4	86.3
4	93.7	80.8	84.5	70.3

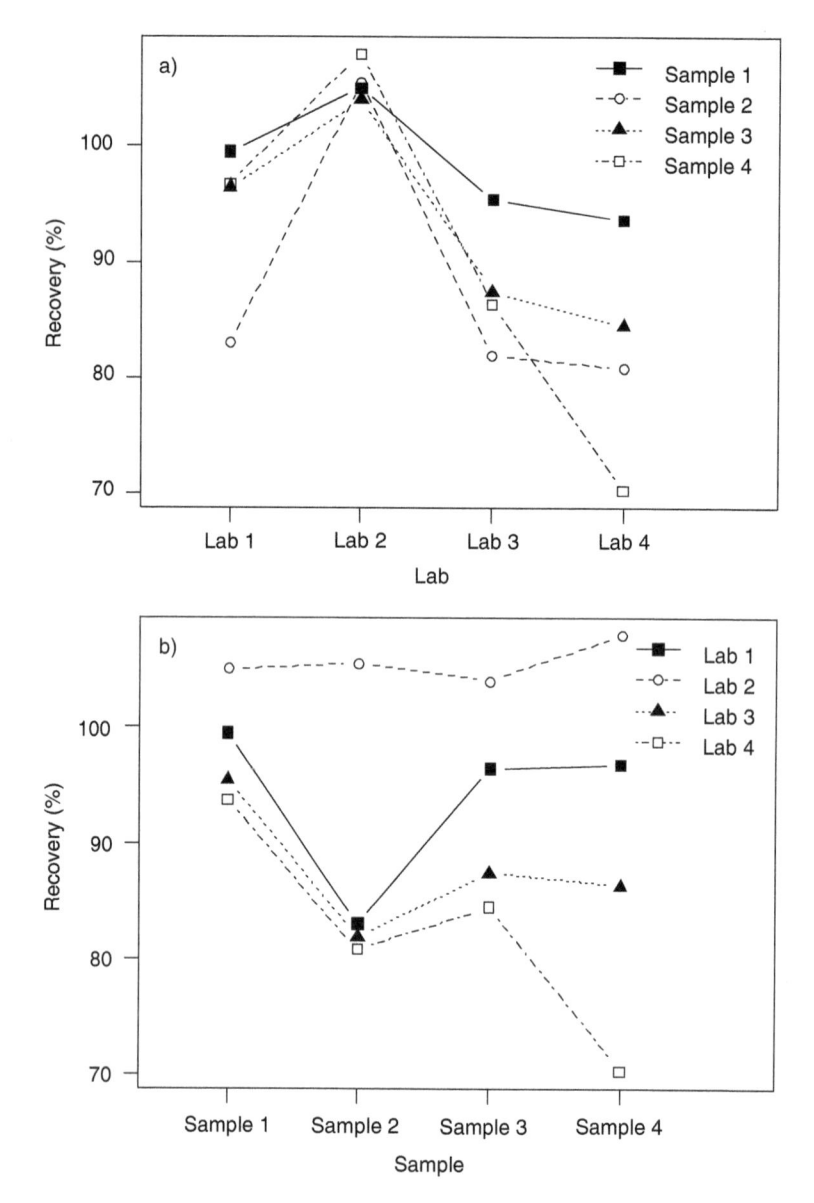

Figure 6.2 Summary of data presented in Table 6.8. (a) grouped by laboratory with lines and symbols representing different samples; (b) grouped by sample with lines and symbols representing different laboratories.

Table 6.9 ANOVA table for the data presented in Table 6.8.

Source of variation	Sum of squares	Degrees of freedom v	Mean square	F
Between columns (samples)	247.4	3	82.5	2.46
Between rows (laboratories)	1201.7	3	400.6	11.92
Residual	302.6	9	33.6	
Total	1751.8	15		

Analysis of Variance

6.5.2 Two-factor ANOVA for Cross-classification With Replication

6.5.2.1 *Data for Two-factor ANOVA With Replication*

The basic study described in the previous section can be extended by obtaining replicate results for each combination of factor levels. If Factor 1 has levels $i = 1, 2, \ldots, p$, Factor 2 has levels $j = 1, 2, \ldots, q$ and $k = 1, 2, \ldots, n$ replicate results are obtained for each factor combination, individual measurement results are represented by x_{ijk}, as shown in Table 6.10. In a table such as this, each combination of the two factors is often called a 'cell', so the difference between Tables 6.6 and 6.10 is that each cell in the former contains only one observation, whereas cells in Table 6.10 each have n replicates.

6.5.2.2 *The Concept of Interaction*

If the effects of two factors are independent, their influence on the measurement results remains the same regardless of the value (level) of the other factor. However, if the effect on the measurement results due to one factor depends on the level of the other factor, we say that there is an *interaction* between the factors. This is illustrated in Figure 6.3.

The practical result of an interaction is a different effect on each individual cell mean. In two-factor ANOVA with only one observation per cell, there is no way of separating this effect from

Table 6.10 Cross-classified design with replication.

Factor 2 levels	Factor 1 levels				
	1	...	i	...	p
1	$x_{111}, \ldots, x_{11n}$		$x_{i11}, \ldots, x_{i1n}$		$x_{p11}, \ldots, x_{p1n}$
⋮					
j	$x_{1j1}, \ldots, x_{1jn}$		$x_{ij1}, \ldots, x_{ijn}$		$x_{pj1}, \ldots, x_{pjn}$
⋮					
q	$x_{1q1}, \ldots, x_{1qn}$		$x_{iq1}, \ldots, x_{iqn}$		$x_{pq1}, \ldots, x_{pqn}$

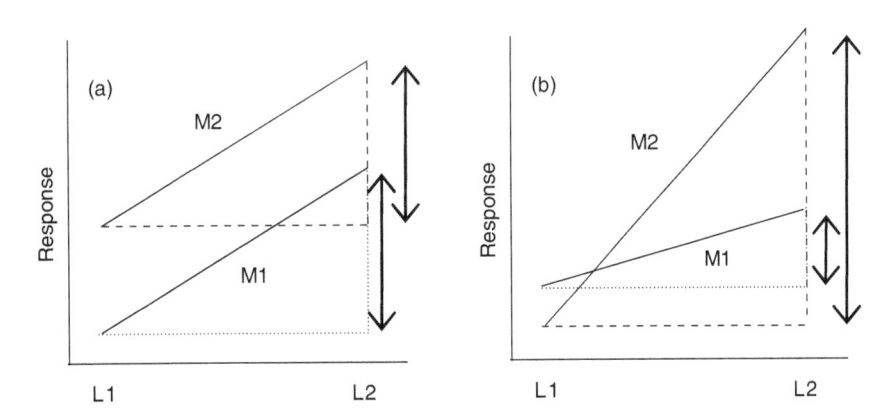

Figure 6.3 The figure shows the effect of two factors, L and M, on the response from a test method. Each factor has two levels (L1, L2 and M1, M2, respectively). In (a), there is no interaction between the factors. The change in response on going from L1 to L2 is the same for both M1 and M2. In (b), there is an interaction between factors L and M. The change in response when factor L changes from L1 to L2 depends on whether factor M is at level M1 or M2. In (b), when factor M is at level M2 the change in response on changing from L1 to L2 is much greater than when factor M is at level M1.

Table 6.11 ANOVA table for two-factor ANOVA with replication (cross-classification).

Source of variation	Sum of squares	Degrees of freedom v	Mean square	F
Between columns	S_3	$p-1$	$M_3 = S_3/(p-1)$	M_3/M_0
Between rows	S_2	$q-1$	$M_2 = S_2/(q-1)$	M_2/M_0
Interaction	S_1	$(p-1)(q-1)$	$M_1 = S_1/(p-1)(q-1)$	M_1/M_0
Residual	S_0	$N-pq$	$M_0 = S_0/(N-pq)$	
Total	$S_3 + S_2 + S_1 + S_0$	$N-1$		

random variation. With replicate measurement results in each cell, however, we can obtain an estimate of the within-group or repeatability term in addition to an estimate of the cell mean. This allows us to investigate whether there is any interaction between the effects of the two individual factors (often called *main effects*).

6.5.2.3 Results Table for Two-factor ANOVA With Replication

The layout of a typical results table for two-factor ANOVA with replication is shown in Table 6.11. It differs from the previous two-factor table by the addition of an extra row; instead of just the residual term (sometimes again called the within-group term, as in one-way ANOVA), there is also a row for the interaction, with its own sum of squares, degrees of freedom, mean square and F ratio.

6.5.2.4 F-Tests in Two-factor ANOVA With Replication

The interpretation of two-factor ANOVA with replication is affected by whether the effects under study are controlled (fixed effects) or random. The default interpretation of the two-factor ANOVA table with interactions assumes fixed effects and we describe this first; partly because it is the default used by most software and partly because it is the more common situation in analytical science. The random-effects case is discussed further in Section 6.5.2.6.

In the default analysis, an F ratio is calculated for the row, column and interaction effects by dividing the respective mean squares by the residual mean square, as in column 'F' of Table 6.11. These are compared with the critical values derived from the one-tailed table for F. If the critical value is exceeded (or the associated p-value is smaller than the chosen significance level), the effect is deemed significant.

The order in which the effects are checked is, however, important, because our conclusions about the individual factors depend on the presence or absence of interaction between them. Interpretation of two-factor ANOVA with replication should always start with the interaction term, as indicated in Figure 6.4. From the interaction term, there are two possibilities:

1. Interaction not significant
    If the interaction is not significant, any row and column effects are approximately consistent – the effect of each factor is not strongly dependent on the other. If the two factors are non-interacting, it is sensible and useful to ask whether either or both of the row and column effects are significant in isolation. This is done, for the default interpretation provided by most software, by comparing the row and column mean squares with the residual mean square using the calculated F or p-values. A value of F which exceeds the critical value is taken as evidence that the associated effect is, at the chosen level of confidence, significantly greater than can be accounted for by the residual (or within-group) variance.

Analysis of Variance

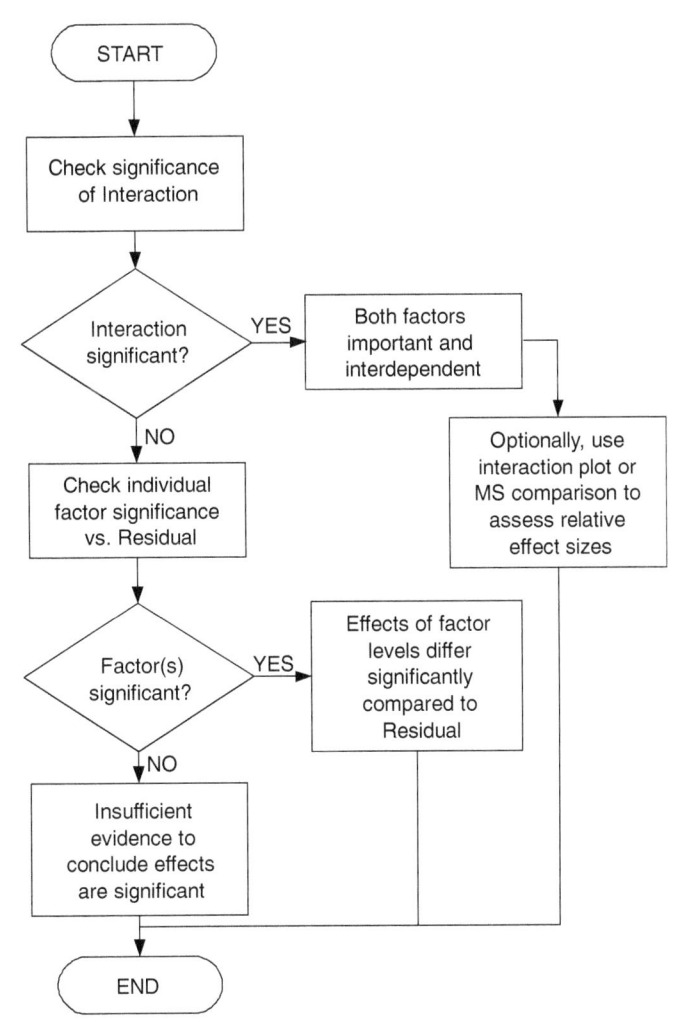

Figure 6.4 Interpretation of two-factor ANOVA with replication.

2. Interaction significant
    If the interaction is significant, it should be concluded that both the 'main effects' (between-row and between-column) are important, since the interaction term results from their combined influence. This remains true whether the individual factors show a significant *F*-test result or not. A significant interaction also indicates that the effects of the two factors are interdependent.

With a significant interaction term, it is often unwise to generalise about the individual factors; if their effects are interdependent, we cannot predict the effect of one factor unless we also know the level of the other. It therefore makes little sense to examine the significance of the individual factors. It may, however, be useful to ask about the relative size of the interaction effect and the row or column effects. For example, in a solvent extraction study across different sample types, we would certainly be interested to know whether one solvent generally extracts more than others or if there is a fairly consistent ordering of solvent effectiveness, even if the difference is not perfectly consistent across sample types. In such a situation, we expect a moderately significant interaction as well as a significant mean square for the factor 'solvent' and it would help to understand which is dominant.

An indication of the relative size of the row and/or column effects and the interaction can be gained by:

- Inspecting the mean squares for the individual factors and comparing them with the mean square for the interaction term. Large differences in mean square indicate that one or other is likely to dominate.
- Examining an 'interaction plot', discussed in detail in Section 6.5.2.5.

This interpretation methodology is summarised in Figure 6.4.

6.5.2.5 Creating and Interpreting Interaction Plots

A basic interaction plot is simply a plot of the means of each 'cell' in the data table, plotted by row or column. Usually, levels of the other factor are identified by drawing a line between cell means with the same factor level. For example, Figure 6.6 shows the two possible interaction plots derived from Table 6.12 (their interpretation is discussed in the example below). For spreadsheet users, the plot is essentially a 'line plot' of a table of cell means, with grouping by either rows or columns as desired. In some statistical software, interaction plots also include a confidence interval on each cell mean.

Three extreme cases can be distinguished in interaction plots. A single dominant main effect will appear as near-parallel, essentially horizontal lines when plotted against the minor factor (Figure 6.5a). When both factors are important but do not interact strongly, the lines remain approximately parallel but are no longer simply horizontal (Figure 6.5b). A dominant interaction shows markedly non-parallel lines, often crossing (Figure 6.5c). Combinations of effects generate intermediate plots, such as Figure 6.5d, which shows approximately equal individual factor effects combined with a strong interaction term. The lines are not parallel, but consistently higher or lower responses may be visible for one or more levels, as in the figure.

Note that when all effects are insignificant, the plot will often resemble Figure 6.5c, because of the random variation due to the residual variance. If confidence intervals are included on the plotted cell means as an indication of their precision – as in some statistical software – it may be clearer from the plot whether the variation can be explained by the residual term alone. Otherwise, it is important to consider the plots and the ANOVA table together.

6.5.2.6 'Random Effects' and Two-factor ANOVA With Replication

The comparisons on the individual factor effects in Section 6.5.2.4 are valid when we can assume that those effects are so-called 'fixed effects', such as the quantitative factors typically used to control experimental conditions (extraction times, column temperatures, *etc.*) or at least that the conclusions are to be limited to the specific levels chosen in the experiment. Sometimes, however, sources of variation in analytical science are better considered as *random effects*, that is, the effects of different analysts, runs, *etc.*, can be assumed to be randomly drawn from a wider population and we wish to draw inferences about the population as a whole. This difference has no effect on one-way ANOVA or on two-factor ANOVA without replication. However, it does affect the interpretation of two-factor and higher ANOVA with calculated interaction terms. This is because, unlike the fixed-effect situation, when the factors represent random effects any underlying interaction also contributes to the expected mean squares for the main effects M_2 and M_3, causing

Analysis of Variance

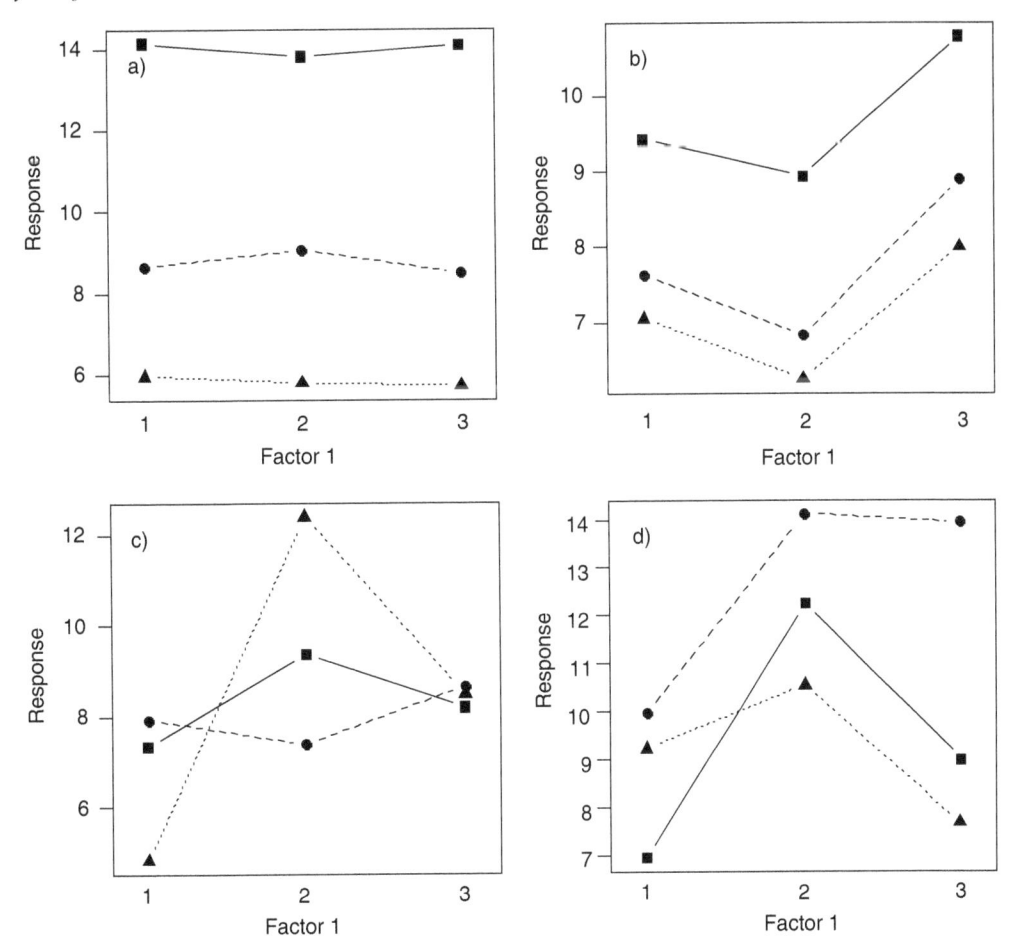

Figure 6.5 Examples of interaction plots. The figure shows cell means for a 3 × 3 design, plotted against Factor 1 and identified by line type and point shape according to the level of Factor 2. The different plots show examples of (a) a single dominant factor (Factor 2 in this case); (b) both individual factors important but no significant interaction; (c) a dominant interaction term; (d) a combination of appreciable main effects and a strong interaction.

Table 6.12 Results from a study of the effect of heating temperature and time on the determination of the fibre content of animal feedstuffs (results expressed as mg g^{-1}).

Time (min)	Temperature (°C)		
	500	*550*	*600*
30	27.1	27.2	27.0
	27.2	27.0	27.1
	27.1	27.1	27.1
60	27.3	27.1	27.1
	27.1	27.1	27.1
	27.2	27.0	27.0
90	27.0	27.2	27.0
	27.1	27.2	27.1
	27.1	27.3	26.9

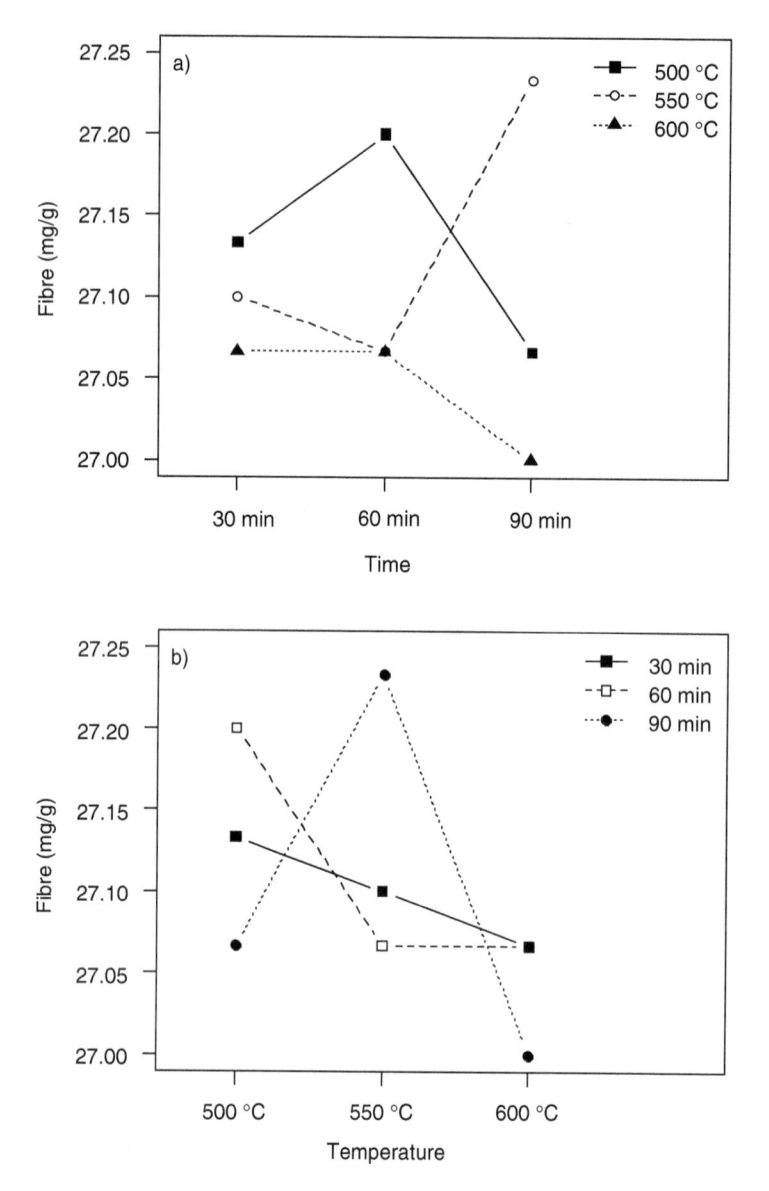

Figure 6.6 Interaction plots of the data shown in Table 6.12 (each point represents the mean of n=3 replicate results).

Table 6.13 ANOVA table for data presented in Table 6.12.

Source of variation	Sum of squares	Degrees of freedom v	Mean square	F	F_{crit}	p-Value
Between columns (temperature)	0.0474	2	0.02371	4.26	3.555	0.030
Between rows (time)	0.0007	2	0.00037	0.07	3.555	0.936
Interaction	0.0815	4	0.02037	3.66	2.928	0.024
Residual	0.1000	18	0.00556			
Total	0.2296	26				

Analysis of Variance

them to be larger than M_0, *even if the relevant main effect is itself negligible*. In fact, the expected main-effects mean squares include the expected interaction mean square in addition to a contribution from the individual factor. For this reason, the recommended method of interpreting two-factor ANOVA when the effects are believed to be random is always to compare M_2 and M_3 with the interaction mean square M_1, rather than with M_0. This can be done by calculating $F_2 = M_2/M_1$ and $F_3 = M_3/M_1$ and, using the appropriate degrees of freedom for each mean square, comparing these with the critical values for F for a one-tailed test. (This F-test is not appropriate for the fixed-effects case, as the main-effects mean square need not be larger than the interaction term in the fixed-effects case; this is why Figure 6.4 and Section 6.5.2.4 do not use an F-test to compare main-effect and interaction mean squares.)

In practice, however, not all ANOVA software provides the relevant comparisons automatically. Perhaps more importantly, the analyst may well be unsure about whether to treat effects as random or fixed. Fortunately, the random-effects assumption only affects inferences about the main effects and then only if there is a real underlying interaction. Since the interpretation described in Figure 6.4 only considers formal tests on main effects when the interaction is not significant, this is rarely a serious problem. At worst, an undetected, and therefore usually modest, underlying interaction may be mistaken for a similarly modest main effect, triggering further investigation.

For most practical purposes, therefore, the interpretation methodology given in Section 6.5.2.4 is sufficient for applications in analytical method development and validation.

Example

A test method for determining the content of fibre in animal feedstuffs involves heating the sample at a particular temperature for a defined period. A study is planned to determine whether changing the temperature or time causes a significant change in the results obtained. Portions of a carefully homogenised sample of a feedstuff were analysed at three different temperatures and times. Three determinations were carried out for each combination of temperature and time. The results are shown in Table 6.12. In this example, Factor 1 is temperature with $p = 3$, Factor 2 is time with $q = 3$ and $n = 3$.

The mean results are shown in Figure 6.6, plotted by (a) time and (b) temperature.

The ANOVA table is given in Table 6.13. The calculations are given in the Appendix to this chapter.

Following the interpretation procedure described in Section 6.5.2.4, we first compare the interaction mean square with the residual mean square:

$$F = \frac{0.02037}{0.00556} = 3.664$$

The critical value for F (one-tailed test, $\alpha = 0.05$, $v_1 = 4$, $v_2 = 18$) is 2.928. The interaction term is therefore significant at the 95% confidence level. We conclude that there are significant differences between results at different time–temperature combinations and that the effects of the two factors are interdependent.

As the interaction term is significant, we now accept that both the time and temperature are important. However, we may want to know whether the individual time or temperature effects are large compared with the interaction term or whether the interaction itself is the dominant effect. Following the guidance in Section 6.5.2.4, we compare the mean squares for the two main effects, time and temperature, with that for the interaction term. Neither is much greater than the interaction mean square (and one is much smaller), suggesting that neither individual effect dominates; the variation caused by their combined effect is apparently the dominant effect on the result. This is confirmed by inspection of the interaction plots in Figure 6.6. In both figures, the mean values vary appreciably and the lines are not only far from parallel (as expected with a significant interaction), but they overlap considerably; no particular time or temperature consistently produces high or low results.

We therefore conclude that time and temperature are both important, but that their effects are interdependent and neither is individually dominant.

Finally, note that because the experimental factors are controlled (that is, fixed effects), the additional considerations of Section 6.5.2.6 do not apply.

6.6 Two-factor ANOVA for Nested Designs (Hierarchical Classification)

6.6.1 Data for Two-factor ANOVA for Nested Designs

Consider a training exercise where four analysts are based in different laboratories and in each laboratory the analyst has access to three instruments. The arrangement is shown in Figure 6.7. Although the analysts would use different instruments in their own laboratories, they would not travel to the other laboratories to use all the other instruments available; instruments are 'nested' within analysts. This type of study would be a nested (or hierarchical) design, as described in Chapter 8, Section 8.4.3.

Replicate measurements can also be made on each instrument. If each analyst carries out n replicate determinations using each of the three instruments in their laboratory, the results can be represented by x_{ijk} ($i = 1, 2, \ldots, p$ analysts; $j = 1, 2, \ldots, q$ instruments per analyst, $k = 1, 2, \ldots, n$ replicates). There are then two nested factors with additional replication.

6.6.2 Results Table for Two-factor ANOVA for Nested Designs

The layout of a typical ANOVA table for a two-factor nested design is shown in Table 6.14. The most obvious difference between the ANOVA calculations for nested and cross-classified designs is

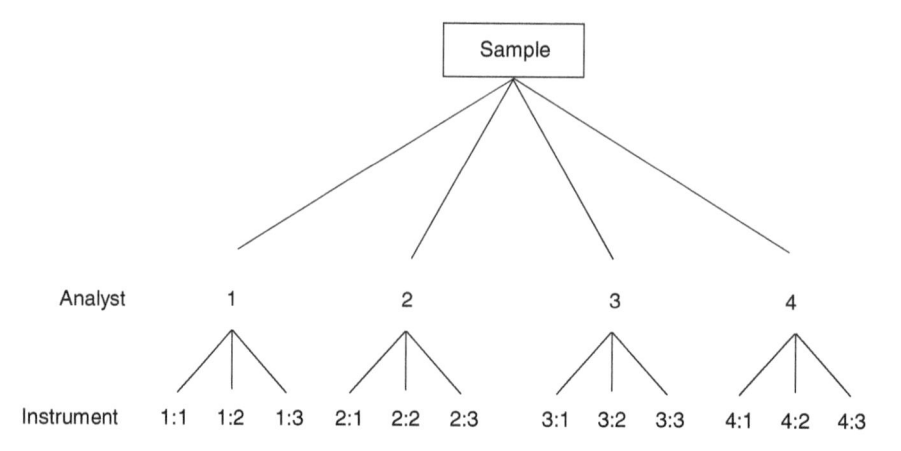

Figure 6.7 A nested (hierarchical) experimental design.

Analysis of Variance

Table 6.14 ANOVA table for two-factor ANOVA with replication (hierarchical classification).

Source of variation	Sum of squares	Degrees of freedom v	Mean square
Between analysts	S_2	$p-1$	$M_2 = S_2/(p-1)$
Between instruments	S_1	$p(q-1)$	$M_1 = S_1/p(q-1)$
Residual	S_0	$N-pq$	$M_0 = S_0/(N-pq)$
Total	$S_2 + S_1 + S_0$	$N-1$	

that because of the nesting, there is no possibility of calculating an interaction term from a two-factor nested design. The ANOVA table therefore includes no interaction term.

6.6.3 Variance Components

Hierarchical or nested designs are most often used to investigate random effects, so we work on that basis. The most useful information about such effects is usually the variance contributed by each factor. These contributions, usually termed 'variance components', can be calculated from the mean squares in the ANOVA table.

The mean square terms are related to the residual variance (σ^2_{res}) and the variances due to changing the two factors ($\sigma^2_{analyst}$ and $\sigma^2_{instrument}$). Their expected values are as follows:

$$M_0 : \sigma^2_{res} \tag{6.5a}$$

$$M_1 : n\sigma^2_{instrument} + \sigma^2_{res} \tag{6.5b}$$

$$M_2 : nq\sigma^2_{analyst} + n\sigma^2_{instrument} + \sigma^2_{res} \tag{6.5c}$$

We can therefore obtain estimates of the different variance components (designated s^2_{res}, $s^2_{instrument}$ and $s^2_{analyst}$ to distinguish the estimates from the population variances) as follows:

$$s^2_{res} = M_0 \tag{6.6a}$$

$$s^2_{instrument} = \frac{M_1 - M_0}{n} \tag{6.6b}$$

$$s^2_{analyst} = \frac{M_2 - M_1}{nq} \tag{6.6c}$$

Conventionally, where one of the differences in equations (6.6b) and (6.6c) is negative, the associated variance component is set to zero.

6.6.4 *F*-Tests for Two-factor ANOVA on Nested Designs

Usually, it is sufficient to estimate the different variance components from a hierarchical design and compare them to determine which are important and which are not. It is, however, sometimes useful to have an objective test for the significance of a particular effect. The mean squares above show that each mean square includes contributions from all the sources of variance 'below' it in the hierarchical design. It follows that if we wish to ask about the effect of a particular factor, we must test each mean square against the mean square immediately below it in the ANOVA table. For each factor, therefore, an F-test is carried out to determine whether the mean square (M_2 or M_1) is significantly greater than the mean square immediately below it in the table (M_1 or M_0, respectively). The F-test should use the one-tailed value for F. If the F-test gives a significant result, then we conclude that the factor has a significant effect on the measurement results.

Note that ANOVA software usually calculates F ratios with respect to the residual mean square by default.

Example

As part of a training exercise, four analysts working at different sites within a company are asked to make two replicate measurements of a quality control sample on each of three different instruments available in their laboratory. The results are given in Table 6.15.

The ANOVA table is given in Table 6.16. The calculations are given in the Appendix to this chapter.

The variance components are calculated as follows:

$$s^2_{res} = M_0 = 5.11$$

$$s^2_{instrument} = \frac{M_1 - M_0}{n} = \frac{10.86 - 5.11}{2} = 2.88$$

$$s^2_{analyst} = \frac{M_2 - M_1}{nq} = \frac{202.74 - 10.86}{2 \times 3} = 31.98$$

Table 6.15 Results from a training exercise (results expressed in mg L^{-1}).

Analyst 1			Analyst 2			Analyst 3			Analyst 4		
I1	I2	I3	I1	I2	I3	I1	I2	I3	I1	I2	I3
47.4	42.2	45.6	50.2	52.1	57.2	43.7	40.3	47.1	51.9	52.3	50.5
44.7	42.2	44.8	54.1	56.8	58.2	46.6	41.3	45.2	57.4	56.8	54.1

Table 6.16 ANOVA table for data presented in Table 6.15.

Source of variation	Sum of squares[a]	Degrees of freedom v	Mean square[a]	F[b]	F$_{crit}$[b]
Between analysts	608.23	3	202.74	18.67	4.066
Between instruments	86.90	8	10.86	2.13	2.849
Residual	61.35	12	5.11		
Total	756.48	23			

[a]Figures are shown to two decimal places.
[b]The F ratios are calculated against the mean square on the following row as described in the text and the critical values use the associated degrees of freedom. This differs from the default usually found in ANOVA software, which tests all effects against the residual mean square.

Analysis of Variance

Clearly, the instrument effect is comparatively small, whereas the between-analyst variance is very large.

These are the best estimates of the variances, but to check whether these effects could reasonably have arisen by chance, we follow up with suitable *F*-tests:

Comparing M_1 with M_0 to test the significance of the effect of using different instruments:

$$F = \frac{10.862}{5.113} = 2.124$$

The critical value for $F(\alpha = 0.05, v_1 = 8, v_2 = 12)$ is 2.849. The effect of the analysts using different instruments to make the measurements is therefore not statistically significant at the 95% confidence level.

Comparing M_2 with M_1 to check the effect of changing analyst:

$$F = \frac{202.743}{10.86} = 18.67$$

The critical value for $F(\alpha = 0.05, v_1 = 3, v_2 = 8)$ is 4.066. The effect of having different analysts carry out the measurements is therefore significant at the 95% confidence level. Given the very large variance component, this should not be surprising.

6.7 Checking Assumptions for ANOVA

6.7.1 Checking Normality

Although the calculations of the mean squares and of the variance components, such as the between-group variance σ_b^2 in equation (6.2b), are unaffected by non-normality, *p*-values depend on the assumption that the underlying random errors are approximately normally distributed.

The simplest method of checking normality is to review the residuals for the data set. For ANOVA, the residuals are the differences between the observations and their expected value, usually the group mean. For example, for the data set in Table 6.1, the mean for the first analyst is 65.68 and the residuals are (−1.08, −0.18, 3.52, −0.08, −2.18). For two-factor ANOVA with replication, the residuals are the deviations from cell means. For two-factor ANOVA without replication, the residuals are more complicated; the residual r_{ij} for row *j* and column *i* is calculated from

$$r_{ij} = x_{ij} - \bar{x}_{\bullet j} - \bar{x}_{i \bullet} + \bar{x}_{\bullet \bullet} \tag{6.7}$$

where $\bar{x}_{\bullet j}$ is the corresponding row mean, $\bar{x}_{\bullet i}$ is the column mean and $\bar{x}_{\bullet \bullet}$ is the mean of all the data.

Once the residuals have been calculated, the residuals should be checked for normality using dot plots, normal probability plots (Chapter 2, Section 2.10) or any of the methods mentioned in Section 3.7 in Chapter 3.

Example

For the data set in Table 6.1, the residuals are calculated by subtracting the group means to obtain the list of residuals in Table 6.17. A normal probability plot is shown in Figure 6.8. The data fall approximately along a straight line, giving no reason to suspect serious departure from normality, although the two low values might merit examination as possible outliers.

Table 6.17 Residuals for a one-way ANOVA using the data in Table 6.1.

| \multicolumn{4}{c}{Analyst} |
1	2	3	4
−1.08, −0.18, 3.52, −0.08, −2.18	0.32, 2.22, 0.62, −1.68, −1.48	−0.72, −2.12, 0.08, −1.22, 3.98	3.10, 2.50, −5.50, 4.60, −4.70

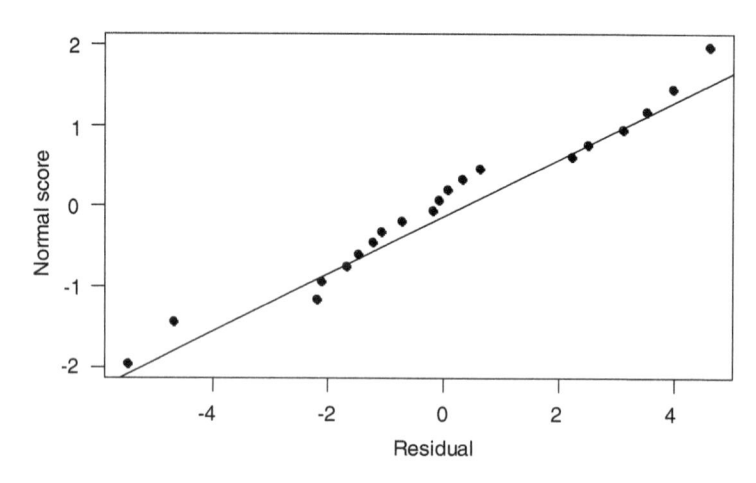

Figure 6.8 Checking normality for a one-way ANOVA. The figure shows the normal probability plot for the residuals shown in Table 6.17.

6.7.2 Checking Homogeneity of Variance – Levene's Test

ANOVA assumes that the residual variance is the same for all the groups examined, that is, that the variance is homogeneous. Substantial differences among the variances can cause the analysis to miss important between-group effects. Review of the data using box plots or similar graphical tools (Chapter 2) will usually give a good indication of likely failures in this assumption, but it is sometimes useful to check whether apparently different variances could reasonably arise by chance. A convenient test which can be implemented using spreadsheets and one-way ANOVA is Levene's test.

Levene's test uses the absolute deviations from group means or medians. Larger variances cause the deviations from group means (or medians) to increase and this in turn causes the mean of the absolute deviations to increase. A test for significantly different mean absolute deviation among the groups will indicate significantly different variances. The differences among the group means can be tested using one-way ANOVA.

The procedure is therefore as follows:

1. Calculate the medians $\tilde{x}_i$ for each group of data (calculating the means is also valid, but less common).
2. Calculate the absolute deviations $d_{ik} = |x_{ik} - \tilde{x}_i|$.
3. Carry out a one-way ANOVA on the absolute deviations in their corresponding groups.
4. A calculated value for F in excess of the critical value indicates a significant difference between the groups and hence significantly different variances.

Analysis of Variance

Table 6.18 Absolute deviations for a one-way ANOVA example.

	Analyst			
	1	2	3	4
Group median	65.5	65.6	64.0	68.0
Absolute deviations	0.9, 0.0, 3.7, 0.1, 2.0	0.0, 1.9, 0.3, 2.0, 1.8	0.0, 1.4, 0.8, 0.5, 4.7	0.6, 0.0, 8.0, 2.1, 7.2

Table 6.19 ANOVA table for Levene's test.

	Sum of squares	Degrees of freedom v	Mean square	F	p-Value	F_{crit}
Analyst	19.01	3	6.337	1.210	0.338	3.239
Residuals	83.79	16	5.237			

Note that Levene's test is clearly approximate, as the absolute deviations are not expected to be normally distributed. The test nonetheless performs sufficiently well for most practical purposes.

Example
Figure 6.1 shows some modest differences in dispersion between the groups of data in Table 6.1. Levene's test is used to check whether the different dispersions are statistically significant or could reasonably occur by chance. Using the data from Table 6.1, the medians and calculated deviations are shown in Table 6.18.

The one-way ANOVA table for these data is given as Table 6.19. The p-value is over 0.05 and F is well under the critical value; there is therefore no evidence of significant differences in variance between the analysts.

6.8 Missing Data in ANOVA

Missing data do not represent a serious problem in one-way ANOVA; most software copes well with variable numbers of observations in each group and the conclusions are largely unaffected. For nested designs or for two-factor and higher order ANOVA without interactions, imbalance adversely affects interpretation. Some simpler software will (perhaps sensibly) not accept ANOVA data with missing values at all. More sophisticated software will calculate tables, but the results will depend on the order in which effects are specified unless special techniques are used. Good statistical software will provide appropriate methods for nested and main-effects ANOVA with missing data; for example, ANOVA based on so-called 'Type II' sums of squares will handle these cases adequately. For two-factor and higher order ANOVA with interactions, however, missing data can be a very serious problem indeed. With all cells populated, but with different numbers of observations in each, so-called 'Type III' sums of squares is the most widely recommended approach, although many statisticians would increasingly advocate an approach based on 'linear modelling' and comparison of successive models. Even with either of these approaches, interpretation may be problematic and it would be prudent to consult an experienced statistician. With complete cells missing, interpretation becomes increasingly unsafe and it becomes essential and not merely prudent to consult a statistician for advice on the options available.

Appendix: Manual Calculations for ANOVA

A Note on Numerical Accuracy

The intermediate sums of squares used in the most efficient manual calculations are often very large and the calculations usually involve differences between these large numbers. Where the range of the data is small compared with the largest value, there can be a very serious risk of losing the most important digits because of rounding, whether manually or in software. To avoid this, it is often useful to 'code' the data, that is, to subtract a convenient amount from all the raw observations. Using spreadsheets or similar general-purpose software, mean centring by subtracting the mean from the data is convenient and usually slightly more effective. Either strategy can greatly reduce the size of intermediate sums with no effect on the resulting ANOVA tables. As a rough guide for manual ANOVA calculations, coding or mean-centring is useful where the range of values is less than about 20% of the maximum value, and strongly recommended where the range is below 5% of the maximum value. If neither coding nor mean centring is used in manual calculations, it is crucially important to retain all available digits at least until the final sums of squares for the ANOVA table are calculated.

The worked examples below illustrate coding where appropriate.

One-way ANOVA

One of the most efficient ways of calculating the sums of squares for one-way ANOVA by hand is given here. A number of intermediate summations are required:

$$S_1 = \frac{\sum_{i=1}^{p}\left(\sum_{k=1}^{n} x_{ik}\right)^2}{n} \tag{6.8a}$$

$$S_2 = \sum_{i=1}^{p}\sum_{k=1}^{n} x_{ik}^2 \tag{6.8b}$$

$$S_3 = \frac{\left(\sum_{i=1}^{p}\sum_{k=1}^{n} x_{ik}\right)^2}{N} \tag{6.8c}$$

where N is the total number of observations.

If the individual groups of data contain moderately different numbers of replicates, then equation (6.8a) can, for a reasonably good approximation during hand calculation, be replaced with

$$S_1' = \sum_{i=1}^{p} \frac{\left(\sum_{k=1}^{n_i} x_{ik}\right)^2}{n_i} \tag{6.9}$$

where n_i is the number of replicates obtained in group i. These values are used to calculate the sum of squares terms as shown in Table 6.20.

Once the sums of squares have been calculated, the degrees of freedom and mean squares are calculated as in Table 6.2.

Analysis of Variance

Table 6.20 Calculation of sum of squares terms in one-way ANOVA.

Source of variation	Sum of squares
Between-group	$S_b = S_1 - S_3$
Within-group	$S_w = S_2 - S_1$
Total	$S_{tot} = S_b + S_w = S_2 - S_3$

Table 6.21 Data from Table 6.1 after coding.

\multicolumn{4}{c}{Analyst}
1	2	3	4
4.6	5.6	4.0	8.6
5.5	7.5	2.6	8.0
9.2	5.9	4.8	0.0
5.6	3.6	3.5	10.1
3.5	3.8	8.7	0.8

Table 6.22 One-way ANOVA calculations: Summation of data presented in Table 6.21.

\multicolumn{4}{c}{Analyst}	
1	2	3	4	
$\sum_{k=1}^{n} x_{1k} = 28.4$	$\sum_{k=1}^{n} x_{2k} = 26.4$	$\sum_{k=1}^{n} x_{3k} = 23.6$	$\sum_{k=1}^{n} x_{4k} = 27.5$	$\sum_{i=1}^{p}\sum_{k=1}^{n} x_{ik} = 105.9$

Example

The following example uses the data in Table 6.1. The data are first coded by subtracting 60, to give the adjusted data in Table 6.21.

The required group and total sums are shown in Table 6.22. In this example, $p = 4$ and $n = 5$. The calculations required to evaluate the sum of squares terms are as follows:

$$S_1 = \frac{\sum_{i=1}^{p}\left(\sum_{k=1}^{n} x_{ik}\right)^2}{n} = \frac{28.4^2 + 26.4^2 + 23.6^2 + 27.5^2}{5} = 563.35$$

$$S_2 = \sum_{i=1}^{p}\sum_{k=1}^{n} x_{ik}^2 = 4.6^2 + 5.5^2 + 9.2^2 + 5.6^2 + 3.5^2 + 5.6^2 + 7.5^2 + 5.9^2 + 3.6^2 + 3.8^2$$
$$+ 4.0^2 + 2.6^2 + 4.8^2 + 3.5^2 + 8.7^2 + 8.6^2 + 8.0^2 + 0.0^2 + 10.1^2 + 0.8^2$$
$$= 703.83$$

$$S_3 = \frac{\left(\sum_{i=1}^{p}\sum_{k=1}^{n} x_{ik}\right)^2}{N} = \frac{105.9^2}{20} = 560.74$$

Note that without coding, the largest intermediate value would be 85411.83 instead of 703.83.
The ANOVA table for this example is calculated as in Table 6.23.

Table 6.23 ANOVA table for data presented in Table 6.1.

Source of variation	Sum of squares[a]	Degrees of freedom v	Mean square	F
Between-group	$S_b = S_1 - S_3$ $= 563.35 - 560.74 = \mathbf{2.61}$	$v_b = p - 1 = 3$	$M_b = S_b/(p-1)$ $= 2.61/3 = \mathbf{0.87}$	M_b/M_w $= 0.87/8.78$ $= \mathbf{0.10}$
Within-group	$S_w = S_2 - S_1$ $= 703.83 - 563.35 = \mathbf{140.48}$	$v_w = N - p = 16$	$M_w = S_w/(N-p)$ $= 140.48/16 = \mathbf{8.78}$	
Total	$S_{tot} = S_b + S_w$ $= S_2 - S_3 = 703.83 - 560.74$ $= \mathbf{143.09}$	$v_{tot} = N - 1 = 19$		

[a]Intermediate values shown after coding.

Table 6.24 Cross-classified design.

Factor 2 levels	Factor 1 levels				
	1	...	i	...	p
1	x_{11}		x_{i1}		x_{p1}
⋮					
j	x_{1j}		x_{ij}		x_{pj}
⋮					
q	x_{1q}		x_{iq}		x_{pq}

Two-way ANOVA: Cross-classification Without Replication

Table 6.24 shows the layout and nomenclature for two-factor ANOVA without replication. Factor 1 has levels $i = 1, 2, \ldots, p$ and Factor 2 has levels $j = 1, 2, \ldots, q$. The individual results obtained for each combination of factors are represented by x_{ij}.

Calculation of Sum of Squares and Mean Square Terms

The summations required to carry out the analysis of variance are shown below:

(i)
$$\frac{\sum_{i=1}^{p}\left(\sum_{j=1}^{q} x_{ij}\right)^2}{q} \tag{6.10a}$$

(ii)
$$\frac{\sum_{j=1}^{q}\left(\sum_{i=1}^{p} x_{ij}\right)^2}{p} \tag{6.10b}$$

(iii)
$$\sum_{i=1}^{p}\sum_{j=1}^{q} x_{ij}^2 \tag{6.10c}$$

(iv)
$$\frac{\left(\sum_{i=1}^{p}\sum_{j=1}^{q} x_{ij}\right)^2}{N} \tag{6.10d}$$

where N is the total number of observations.

The ANOVA table is shown in Table 6.25.

Analysis of Variance

Table 6.25 ANOVA table for two-factor ANOVA without replication (cross-classification).

Source of variation	Sum of squares	Degrees of freedom v	Mean square	F
Between columns (Factor 1)	$S_2 = $ (i) $-$ (iv)	$p-1$	$M_2 = S_2/(p-1)$	M_2/M_0
Between rows (Factor 2)	$S_1 = $ (ii) $-$ (iv)	$q-1$	$M_1 = S_1/(q-1)$	M_1/M_0
Residuals	$S_0 = [$(iii)$ + (iv)] - [(i) + (ii)]$	$(p-1)(q-1)$	$M_0 = S_0/(p-1)(q-1)$	
Total	$S_2 + S_1 + S_0 = $ (iii) $-$ (iv)	$N-1$		

Table 6.26 Two-factor ANOVA, cross-classification without replication: summation of data presented in Table 6.8.

Laboratory	Sample 1	Sample 2	Sample 3	Sample 4	
1	99.5	83.0	96.5	96.8	$\sum_{i=1}^{p} x_{i1} = 375.8$
2	105.0	105.5	104.0	108.0	$\sum_{i=1}^{p} x_{i2} = 422.5$
3	95.4	81.9	87.4	86.3	$\sum_{i=1}^{p} x_{i3} = 351.0$
4	93.7	80.8	84.5	70.3	$\sum_{i=1}^{p} x_{i4} = 329.3$
	$\sum_{j=1}^{q} x_{1j} = 393.6$	$\sum_{j=1}^{q} x_{2j} = 351.2$	$\sum_{j=1}^{q} x_{3j} = 372.4$	$\sum_{j=1}^{q} x_{4j} = 361.4$	$\sum_{i=1}^{p}\sum_{j=1}^{q} x_{ij} = 1478.6$

Example

The example uses the sweetener recovery data described in Section 6.5.1. For this example, the data cover a range of about 40, compared with a maximum of 108. Coding therefore has little effect and the calculations are performed on the raw data (Table 6.26).

The calculations required to evaluate the sum of squares terms are as follows:

(i)
$$\frac{\sum_{i=1}^{p}\left(\sum_{j=1}^{q} x_{ij}\right)^2}{q} = \frac{393.6^2 + 351.2^2 + 372.4^2 + 361.4^2}{4} = 136888.53$$

where q is the number of *rows*;

(ii)
$$\frac{\sum_{j=1}^{q}\left(\sum_{i=1}^{p} x_{ij}\right)^2}{p} = \frac{375.8^2 + 422.5^2 + 351.0^2 + 329.3^2}{4} = 137842.85$$

where p is the number of *columns*;

(iii)
$$\sum_{i=1}^{p}\sum_{j=1}^{q} x_{ij}^2 = 99.5^2 + 83.0^2 + 96.5^2 + 96.8^2 + 105.0^2 + 105.5^2 + 104.0^2 + 108.0^2$$
$$+ 95.4^2 + 81.9^2 + 87.4^2 + 86.3^2 + 93.7^2 + 80.8^2 + 84.5^2 + 70.3^2$$
$$= 138392.88$$

(iv)
$$\frac{\left(\sum_{i=1}^{p}\sum_{j=1}^{q}x_{ij}\right)^2}{N} = \frac{1478.6^2}{16} = 136641.12$$

The ANOVA table for this example is shown in Table 6.27.

Two-factor ANOVA: Cross-classification With Replication

Calculation of Sum of Squares and Mean Square Terms

The summations required to carry out a two-factor ANOVA for cross-classification with replication are as follows. Table 6.28 shows the layout and nomenclature.

(i)
$$\frac{\sum_{i=1}^{p}\left(\sum_{j=1}^{q}\sum_{k=1}^{n}x_{ijk}\right)^2}{qn} \tag{6.11a}$$

(ii)
$$\frac{\sum_{j=1}^{q}\left(\sum_{i=1}^{p}\sum_{k=1}^{n}x_{ijk}\right)^2}{pn} \tag{6.11b}$$

Table 6.27 ANOVA table for the data presented in Table 6.8.

Source of variation	Sum of squares	Degrees of freedom ν	Mean square	F
Between columns (samples)	$S_2 =$ (i) − (iv) $= 136888.53 - 136641.12$ $= 247.4$	$p - 1 = 3$	$M_2 = S_2/(p-1)$ $= 247.4/3$ $= 82.5$	M_2/M_0 $= 82.5/33.6$ $= 2.46$
Between rows (laboratories)	$S_1 =$ (ii) − (iv) $= 137842.85 - 136641.12$ $= 1201.7$	$q - 1 = 3$	$M_1 = S_1/(q-1)$ $= 1201.7/3$ $= 400.6$	M_1/M_0 $= 400.6/33.6$ $= 11.92$
Residuals	$S_0 = [$(iii) + (iv)$] - [$(i) + (ii)$]$ $= (138392.88 + 136641.12) -$ $(136888.53 + 137842.85) = 302.6$	$(p-1)(q-1) = 9$	$M_0 = S_0/(p-1)(q-1)$ $= 302.6/9 = 33.6$	
Total	$S_2 + S_1 + S_0$ $=$ (iii) − (iv) $= 138392.88 - 136641.12$ $= 1751.8$	$N - 1 = 15$		

Table 6.28 Two-factor ANOVA, cross-classification, with replication.

Factor 2 levels	Factor 1 levels[a]				
	1	...	i	...	p
1	$x_{111}, ..., x_{11n}$		$x_{i11}, ..., x_{i1n}$		$x_{p11}, ..., x_{p1n}$
$\vdots$					
j	$x_{1j1}, ..., x_{1jn}$		$x_{ij1}, ..., x_{ijn}$		$x_{pj1}, ..., x_{pjn}$
$\vdots$					
q	$x_{1q1}, ..., x_{1qn}$		$x_{iq1}, ..., x_{iqn}$		$x_{pq1}, ..., x_{pqn}$

[a] The replicate number is denoted $k = 1, ..., n$, so a single observations is denoted x_{ijk}.

Analysis of Variance

Table 6.29 ANOVA table for two-factor ANOVA with replication (cross-classification).

Source of variation	Sum of squares	Degrees of freedom v	Mean square
Between columns (Factor 1)	$S_3 = (i) - (v)$	$p - 1$	$M_3 = S_3/(p-1)$
Between rows (Factor 2)	$S_2 = (ii) - (v)$	$q - 1$	$M_2 = S_2/(q-1)$
Interaction	$S_1 = [(iii) + (v)] - [(i) + (ii)]$	$(p-1)(q-1)$	$M_1 = S_1/(p-1)(q-1)$
Residual	$S_0 = (iv) - (iii)$	$N - pq$	$M_0 = S_0/(N-pq)$
Total	$S_3 + S_2 + S_1 + S_0 = (iv) - (v)$	$N - 1$	

(iii)
$$\frac{\sum_{i=1}^{p}\sum_{j=1}^{q}\left(\sum_{k=1}^{n} x_{ijk}\right)^2}{n} \tag{6.11c}$$

(iv)
$$\sum_{i=1}^{p}\sum_{j=1}^{q}\sum_{k=1}^{n} x_{ijk}^2 \tag{6.11d}$$

(v)
$$\frac{\left(\sum_{i=1}^{p}\sum_{j=1}^{q}\sum_{k=1}^{n} x_{ijk}\right)^2}{N} \tag{6.11e}$$

where N is the total number of observations.

The ANOVA table is shown in Table 6.29.

Example
The example uses the data on the effect of heating temperature and time on the determination of fibre described in Section 6.5.2. Because the range of the data is very small compared with the maximum value (less than 2%), the data have been coded by subtracting 27 to give the revised data in Tables 6.30 and 6.31. Note that one adjusted value is negative.

The calculations required to evaluate the sum of squares terms are as follows:

(i) $$\frac{(1.2^2 + 1.2^2 + 0.4^2)}{(3 \times 3)} = \frac{3.04}{9} = 0.33778$$

(ii) $$\frac{(0.9^2 + 1.0^2 + 0.9^2)}{(3 \times 3)} = \frac{2.62}{9} = 0.29111$$

(iii) $$\frac{\begin{pmatrix} 0.4^2 + 0.3^2 + 0.2^2 + 0.6^2 + 0.2^2 \\ + 0.2^2 + 0.2^2 + 0.7^2 + 0.0^2 \end{pmatrix}}{3} = \frac{1.26}{3} = 0.42$$

(iv) $$\begin{pmatrix} 0.1^2 + 0.2^2 + 0.0^2 + 0.2^2 + 0.0^2 + 0.1^2 + 0.1^2 + 0.1^2 + 0.1^2 \\ + 0.3^2 + 0.1^2 + 0.1^2 + 0.1^2 + 0.1^2 + 0.1^2 + 0.2^2 + 0.0^2 + 0.0^2 \\ + 0.0^2 + 0.2^2 + 0.0^2 + 0.1^2 + 0.2^2 + 0.1^2 + 0.1^2 + 0.3^2 + (-0.1)^2 \end{pmatrix} = 0.52$$

Table 6.30 Fibre data from Table 6.12 after coding.

Time (min)	Temperature (°C)		
	500	550	600
30	0.1 0.2 0.1	0.2 0.0 0.1	0.0 0.1 0.1
60	0.3 0.1 0.2	0.1 0.1 0.0	0.1 0.1 0.0
90	0.0 0.1 0.1	0.2 0.2 0.3	0.0 0.1 −0.1

Table 6.31 Two-factor ANOVA, cross-classification with replication: summation of data presented in Table 6.30.

Time (min) ($q=3$)	Temperature (°C) ($p=3$)			Total
	500 ($i=1$)	550 ($i=2$)	600 ($i=3$)	
30 ($j=1$)	$\sum_{k=1}^{n} x_{11k} = 0.4$	$\sum_{k=1}^{n} x_{21k} = 0.3$	$\sum_{k=1}^{n} x_{31k} = 0.2$	$\sum_{i=1}^{p}\sum_{k=1}^{n} x_{i1k} = 0.9$
60 ($j=2$)	$\sum_{k=1}^{n} x_{12k} = 0.6$	$\sum_{k=1}^{n} x_{22k} = 0.2$	$\sum_{k=1}^{n} x_{32k} = 0.2$	$\sum_{i=1}^{p}\sum_{k=1}^{n} x_{i2k} = 1.0$
90 ($j=3$)	$\sum_{k=1}^{n} x_{13k} = 0.2$	$\sum_{k=1}^{n} x_{23k} = 0.7$	$\sum_{k=1}^{n} x_{33k} = 0.0$	$\sum_{i=1}^{p}\sum_{k=1}^{n} x_{i3k} = 0.9$
Total	$\sum_{j=1}^{q}\sum_{k=1}^{n} x_{1jk} = 1.2$	$\sum_{j=1}^{q}\sum_{k=1}^{n} x_{2jk} = 1.2$	$\sum_{j=1}^{q}\sum_{k=1}^{n} x_{3jk} = 0.4$	$\sum_{i=1}^{p}\sum_{j=1}^{q}\sum_{k=1}^{n} x_{ijk} = 2.8$

(v)
$$\frac{2.8^2}{(3 \times 3 \times 3)} = \frac{7.84}{27} = 0.29037$$

The ANOVA table is given in Table 6.32. Note that without coding, the largest intermediate value would be 19834.72, requiring 9- or 10-digit numerical precision for reasonable accuracy in the between-rows sum of squares.

Two-factor ANOVA for Nested Designs (Hierarchical Classification)

Calculation of Sum of Squares and Mean Square Terms

The summations required to carry out a two-factor ANOVA for a nested design are as follows. Table 6.33 shows the layout and nomenclature.

(i)
$$\frac{\sum_{i=1}^{p} \left(\sum_{j=1}^{q} \sum_{k=1}^{n} x_{ijk} \right)^2}{qn} \tag{6.12a}$$

Analysis of Variance

Table 6.32 ANOVA table for data presented in Table 6.12.

Source of variation	Sum of squares[a]	Degrees of freedom v	Mean square[b]
Between columns (temperature)	$S_3 =$ (i) − (v) = 0.33778 − 0.29037 = **0.04741**	$p − 1 = 2$	$M_3 = S_3/(p−1)$ = 0.04741/2 = **0.02371**
Between rows (time)	$S_2 =$ (ii) − (v) = 0.29111 − 0.29037 = **0.00074**	$q − 1 = 2$	$M_2 = S_2/(q−1)$ = 0.00074/2 = **0.00037**
Interaction	$S_1 =$ [(iii) + (v)] − [(i) + (ii)] = (0.42 + 0.29037) − (0.33778 + 0.29111) = **0.08148**	$(p−1)(q−1) = 4$	$M_1 = S_1/(p−1)(q−1)$ = 0.08148/4 = **0.02037**
Residual	$S_0 =$ (iv) − (iii) = 0.52 − 0.42 = **0.1000**	$N − pq = 18$	$M_0 = S_0/(N−pq)$ = 0.1/18 = **0.00556**
Total	$S_3 + S_2 + S_1 + S_0 =$ (iv) − (v) = 0.52 − 0.29037 = **0.22963**	$N − 1 = 26$	

[a]Intermediate values shown after coding.
[b]To 5 decimal places.

Table 6.33 Two-factor nested ANOVA (hierarchical classification).

		\multicolumn{9}{c}{Factor 1 levels}
		1	...		i			p		
Factor 2 levels		1,1, ...	1,j, ...	1,q	i,1, ...	i,j, ...	i,q	p,1, ...	p,j, ...	p,q
Replicate numbers	1	x_{111}	x_{1j1}	x_{1q1}	x_{i11}	x_{ij1}	x_{iq1}	x_{p11}	x_{pj1}	x_{pq1}
	$\vdots$									
	k	x_{11k}	x_{1jk}	x_{1qk}	x_{i1k}	x_{ijk}	x_{iqk}	x_{p1k}	x_{pjk}	x_{pqk}
	$\vdots$									
	n	x_{11n}	x_{1jn}	x_{1qn}	x_{i1n}	x_{ijn}	x_{iqn}	x_{p1n}	x_{pjn}	x_{pqn}

(ii) $$\frac{\sum_{i=1}^{p}\sum_{j=1}^{q}\left(\sum_{k=1}^{n}x_{ijk}\right)^2}{n} \qquad (6.12b)$$

(iii) $$\sum_{i=1}^{p}\sum_{j=1}^{q}\sum_{k=1}^{n}x_{ijk}^2 \qquad (6.12c)$$

(iv) $$\frac{\left(\sum_{i=1}^{p}\sum_{j=1}^{q}\sum_{k=1}^{n}x_{ijk}\right)^2}{N} \qquad (6.12d)$$

where N is the total number of observations.

The ANOVA table is shown in Table 6.34.

Table 6.34 ANOVA table for two-factor nested ANOVA (hierarchical classification).

Source of variation	Sum of squares	Degrees of freedom v	Mean square
Factor 1	$S_2 = $ (i) − (iv)	$p - 1$	$M_2 = S_2/(p-1)$
Factor 2	$S_1 = $ (ii) − (i)	$p(q - 1)$	$M_1 = S_1/p(q-1)$
Residual	$S_0 = $ (iii) − (ii)	$N - pq$	$M_0 = S_0/(N-pq)$
Total	$S_2 + S_1 + S_0 = $ (iii) − (iv)	$N - 1$	

Example

The example uses the data in the training exercise described in Section 6.6. Factor 1 in Table 6.33 corresponds to analysts and Factor 2 to instruments, nested within analyst. Since the data have a range of about 18 and a maximum value of about 58, coding is not expected to be useful and the calculation is performed on the raw data (Table 6.35).

The calculations required to evaluate the sum of squares terms are as follows:

(i)
$$\frac{(266.9^2 + 328.6^2 + 264.2^2 + 323.0^2)}{(3 \times 2)} = \frac{353344.2}{6} = 58890.70167$$

(ii)
$$(92.1^2 + 84.4^2 + 90.4^2 + 104.3^2 + 108.9^2 + 115.4^2 + 90.3^2 + 81.6^2 + 92.3^2 + 109.3^2 + 109.1^2 + 104.6^2)/2 = 117955.19/2 = 58977.595$$

(iii)
$$47.4^2 + 44.7^2 + 42.2^2 + 42.2^2 + 45.6^2 + 44.8^2 + 50.2^2 + 54.1^2 + 52.1^2$$
$$+ 56.8^2 + 57.2^2 + 58.2^2 + 43.7^2 + 46.6^2 + 40.3^2 + 41.3^2 + 47.1^2$$
$$+ 45.2^2 + 51.9^2 + 57.4^2 + 52.3^2 + 56.8^2 + 50.5^2 + 54.1^2 = 59038.95$$

(iv)
$$\frac{(1182.7)^2}{24} = \frac{1398779.29}{24} = 58282.47042$$

The final calculations for the ANOVA table are given in Table 6.36.

Note that in this example, coding by subtracting 40 would have reduced the largest intermediate values from about 59 000 to about 2800. This would not provide a great improvement in numerical accuracy.

Table 6.35 Two-factor ANOVA, nested design: Summation of data in Table 6.15.

	Analyst 1			Analyst 2			Analyst 3			Analyst 4		
Instrument	1:1	1:2	1:3	2:1	2:2	2:3	3:1	3:2	3:3	4:1	4:2	4:3
$\sum_{k=1}^{n} x_{ijk}$	92.1	84.4	90.4	104.3	108.9	115.4	90.3	81.6	92.3	109.3	109.1	104.6
$\sum_{j=1}^{q}\sum_{k=1}^{n} x_{ijk}$	266.9			328.6			264.2			323.0		
$\sum_{i=1}^{p}\sum_{j=1}^{q}\sum_{k=1}^{n} x_{ijk}$	1182.7											

Analysis of Variance

Table 6.36 ANOVA table for data presented in Table 6.15.

Source of variation	Sum of squares	Degrees of freedom v	Mean square
Between analysts	$S_2 =$ (i) $-$ (iv) $= 58890.702 - 58282.470$ $= $ **608.23**	$p - 1 = 3$	$M_2 = S_2/(p-1)$ $= 608.23/3 = $ **202.743**
Between instruments	$S_1 =$ (ii) $-$ (i) $= 58977.595 - 58890.702$ $= $ **86.893**	$p(q-1) = 8$	$M_1 = S_1/p(q-1)$ $= 86.893/8 = $ **10.862**
Residual	$S_0 =$ (iii) $-$ (ii) $= 59038.950 - 58977.595$ $= $ **61.355**	$N - pq = 12$	$M_0 = S_0/(N-pq)$ $= 61.355/12 = $ **5.113**
Total	$S_2 + S_1 + S_0 =$ (iii) $-$ (iv) $= 59038.950 - 58282.470 = $ **756.480**	$N - 1 = 23$	

CHAPTER 7
Regression

7.1 Linear Regression

7.1.1 Introduction to Linear Regression

Linear regression is used to establish or confirm a relationship between two variables. In analytical chemistry, linear regression is commonly used in the construction of calibration functions required for techniques such as gas chromatography and atomic absorption spectrometry, where a linear relationship is expected between the analytical response and the concentration of the analyte.

Any two data sets of equal size can be plotted against each other on a scatter plot to see if a relationship exists between corresponding pairs of results. In analytical chemistry, the data sets are often instrument response and analyte concentration. If there is reason to believe that the value of one variable depends on the value of the other, the former is known as the dependent variable. The dependent variable is conventionally plotted on the y-axis of the scatter plot. In the instance of a calibration curve for a determination, the instrument response will be the dependent variable as its value will depend on the concentration of the analyte present. The (known) analyte concentration in the calibration solutions is therefore described as the independent variable. In chemical analysis, a relationship between these two variables is essential. The object of regression is to establish the relationship in terms of an equation and to study other aspects of the calibration.

The general equation which describes a fitted straight line can be written as

$$y = a + bx \qquad (7.1)$$

where b is the gradient of the line and a is its intercept with the y-axis. The method of least-squares linear regression is used to establish the values of a and b. The 'best fit' line obtained from least-squares linear regression is the line which minimises the sum of the squared differences between the observed and fitted values for y. The signed difference between an observed value (y) and a fitted value ($\hat{y}$) is known as a residual. The most common form of regression is of y on x. This assumes that the x values are known exactly and the only error occurs in the measurement of y.

7.1.2 Assumptions in Linear Regression

A simple least-squares linear regression of y on x relies on a number of assumptions. Substantial violation of any of the assumptions will usually require special treatment. The key assumptions are

Regression

the following:

- The errors in the x values should be negligible.
- For calculating confidence intervals and drawing inferences, the error associated with the y values must be normally distributed. If there is doubt about the normality, it may be sufficient to replace single y values with averages of three or more values of y for each value of x, as mean values tend to be normally distributed even where individual results are not.
- The variance of the error in the y values should be constant across the range of interest, that is, the standard deviation should be constant. Simple least-squares regression gives equal weight to all points; this will not be appropriate if some points are much less precise than others.
- Both the x and y data must be continuous valued and not restricted to integers, truncated or categorised (for example, sample numbers, days of the week).

Visual inspection of the raw data and residuals is among the most effective ways of checking many of these assumptions for linear regression in analytical measurement. Inspection of the raw data is described in Section 7.1.3. The calculation and inspection of residuals are discussed in more detail in Section 7.1.5.

7.1.3 Visual Examination of Regression Data

Before carrying out the linear regression calculations, it is good practice to examine a scatter plot of the data for possible outliers and points of influence. An outlier is a result which is inconsistent with the rest of the data set (see Chapter 5). A point of influence is a data point which has a disproportionate effect on the position of the regression line. A point of influence may be an outlier, but may also be caused by poor experimental design (see Section 7.1.10). Points of influence can have one of two effects on a regression line – high leverage or bias. Figure 7.1a shows high leverage caused by the uneven distribution of the data points. Ideally, the x values should be approximately evenly spaced across the range of interest to avoid this problem. Figure 7.1b illustrates the effect of leverage caused by an outlier. High leverage will affect both the gradient and the intercept of the line.

An outlier in the middle of the data range will cause the intercept to be biased as the line will be shifted either upwards or downwards, as shown in Figure 7.2.

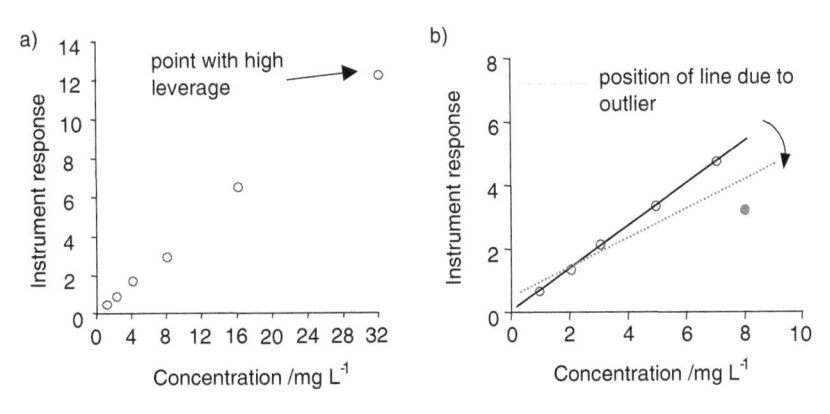

Figure 7.1 Points of influence – leverage. (a) High leverage due to uneven data distribution; (b) effect of an outlier with high leverage.

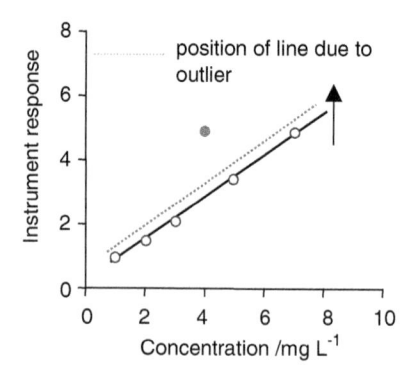

Figure 7.2 Points of influence – bias. Bias caused by an outlying point.

7.1.4 Calculating the Gradient and Intercept

Least-squares linear regression is usually carried out using statistical software or the software associated with a particular analytical instrument. For manual calculation, the relevant equations for obtaining estimates of the gradient and intercept, together with their associated standard deviations (also often called the standard errors), are given in this section.

The gradient b and intercept a of the best fit straight line are calculated from the following equations:

$$b = \frac{\sum_{i=1}^{n}[(x_i - \bar{x})(y_i - \bar{y})]}{\sum_{i=1}^{n}(x_i - \bar{x})^2} \tag{7.2}$$

$$a = \bar{y} - b\bar{x} \tag{7.3}$$

where $\bar{x}$ and $\bar{y}$ are the means of the x and y values, respectively.

Once the gradient and intercept have been calculated, it becomes possible to calculate both the residuals (which should be inspected as described in Section 7.1.5) and the residual standard deviation, $s_{y/x}$, which is required for the calculation of a number of regression statistics. The residual standard deviation is a measure of the deviation of the data points from the regression line and is calculated using the equation

$$s_{y/x} = \sqrt{\frac{\sum_{i=1}^{n}(y_i - \hat{y}_i)^2}{n-2}} \tag{7.4}$$

where y_i is the observed value at a given x value (for example, an observed instrument response for a particular analyte concentration), $\hat{y}_i$ is the value of y calculated for a given x value using the equation of the line (that is, the predicted y value) and n is the number of pairs of data used in the regression. Each difference $(y_i - \hat{y}_i)$ is one residual. Note that the divisor in the residual standard deviation [equation (7.4)] is $n-2$. Here, $n-2$, the number of observations minus the number of parameters estimated (a and b), is the number of degrees of freedom for the residual standard deviation.

Regression

The standard deviation for the estimate of the gradient, s_b, is calculated using the equation

$$s_b = \frac{s_{y/x}}{\sqrt{\sum_{i=1}^{n}(x_i - \bar{x})^2}} \qquad (7.5)$$

and the standard deviation for the estimate of the intercept, s_a, is calculated using the equation

$$s_a = s_{y/x} \sqrt{\frac{\sum_{i=1}^{n} x_i^2}{n \sum_{i=1}^{n}(x_i - \bar{x})^2}} \qquad (7.6)$$

The confidence intervals for the estimates of the gradient and the intercept are calculated from equations (7.7) and (7.8), respectively:

$$b \pm t \times s_b \qquad (7.7)$$

$$a \pm t \times s_a \qquad (7.8)$$

where t is the two-tailed Student's t value (from Table A.4 in Appendix A) at the required significance level (typically $\alpha = 0.05$) and with degrees of freedom $v = n - 2$.

Example
The response of an instrument is determined for calibration solutions with six different concentrations. The data are shown in Table 7.1 and plotted in Figure 7.3.
 The mean of the x values is $\bar{x} = 5$ and the mean of the y values is $\bar{y} = 51.67$.
 The values required to calculate the gradient and intercept for the least-squares straight line are given in Table 7.2.
 The gradient is calculated using equation (7.2):

$$b = \frac{\sum_{i=1}^{n}[(x_i - \bar{x})(y_i - \bar{y})]}{\sum_{i=1}^{n}(x_i - \bar{x})^2} = \frac{708.0}{70} = 10.114$$

The intercept is calculated using equation (7.3):

$$a = \bar{y} - b\bar{x} = 51.67 - (10.114 \times 5) = 1.100$$

Table 7.1 Regression data from an instrument calibration experiment.

Concentration (x)	0	2	4	6	8	10
Response (y)	0	24	41	60	82	103

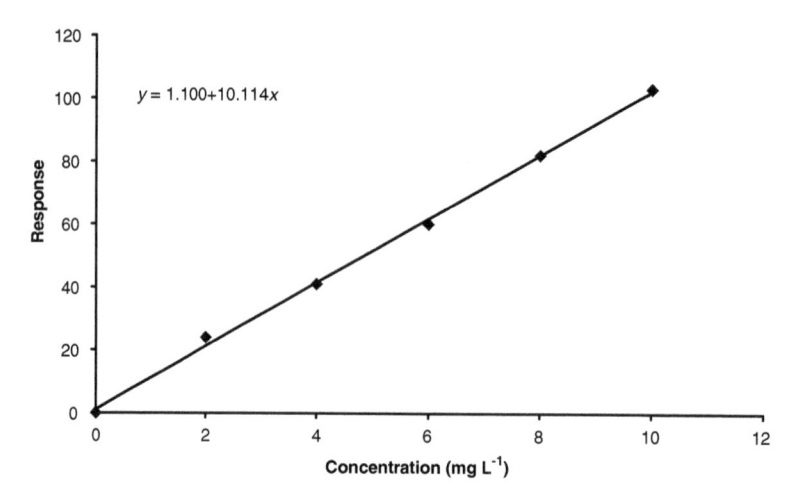

Figure 7.3 Scatter plot of data presented in Table 7.1.

Table 7.2 Calculations required for the estimation of the gradient, intercept and correlation coefficient for the data shown in Table 7.1.

$(x_i - \bar{x})$	$(x_i - \bar{x})^2$	$(y_i - \bar{y})$	$(y_i - \bar{y})^2$	$(x_i - \bar{x})(y_i - \bar{y})$
−5	25	−51.67	2669.4	258.3
−3	9	−27.67	765.4	83.0
−1	1	−10.67	113.8	10.7
1	1	8.33	69.4	8.3
3	9	30.33	920.1	91.0
5	25	51.33	2635.1	256.7
$\sum_{i=1}^{n}$	70		7173.3	708.0

The equation of the least-squares straight line for the data in Table 7.1 is therefore

$$y = 1.100 + 10.114x$$

To calculate the standard deviation of the gradient and the intercept, we first need to calculate the residual standard deviation. The residuals are shown in Table 7.3.

The residual standard deviation is calculated using equation (7.4):

$$s_{y/x} = \sqrt{\frac{\sum_{i=1}^{n}(y_i - \hat{y}_i)^2}{n-2}} = \sqrt{\frac{12.42}{4}} = 1.762$$

The standard deviation for the estimate of the gradient is calculated using equation (7.5):

$$s_b = \frac{s_{y/x}}{\sqrt{\sum_{i=1}^{n}(x_i - \bar{x})^2}} = \frac{1.762}{\sqrt{70}} = 0.211$$

Regression

The standard deviation for the estimate of the intercept is calculated using equation (7.6):

$$s_a = s_{y/x} \sqrt{\frac{\sum_{i=1}^{n} x_i^2}{n \sum_{i=1}^{n}(x_i - \bar{x})^2}} = 1.762 \sqrt{\frac{220}{6 \times 70}} = 1.275$$

7.1.5 Inspecting the Residuals

Plotting the residuals can help to identify problems with poor or incorrect curve fitting. If there is a good fit between the data and the regression model, the residuals should be distributed randomly about zero, as shown in Figure 7.4. This chart is a plot of the residual values given in Table 7.3.

Deviations from a random distribution can indicate problems with the data. Some examples are shown in Figure 7.5. Figure 7.5a shows an ideal residual plot. The residuals are scattered approximately randomly around zero and there is no trend in the spread of residuals with concentration. Figure 7.5b shows the pattern of residuals that is obtained if the standard deviation of the y-values increases with analyte concentration. Figure 7.5c illustrates a typical residual plot that is obtained when a straight line is fitted through data that follow a non-linear trend. Figure 7.5d shows a possible pattern of residuals when the regression line has been incorrectly forced through zero. Figure 7.5e shows evidence of close grouping at each value of x, a form of correlation among

Table 7.3 Residuals calculated for the data in Table 7.1.

y_i	$\hat{y}_i$ ($\hat{y}_i = a + bx_i$)	$(y_i - \hat{y}_i)$	$(y_i - \hat{y}_i)^2$
0	1.10	−1.10	1.21
24	21.33	2.67	7.14
41	41.56	−0.56	0.31
60	61.78	−1.78	3.18
82	82.01	−0.01	0.00
103	102.24	0.76	0.58
			$\sum_{i=1}^{n} = 12.42$

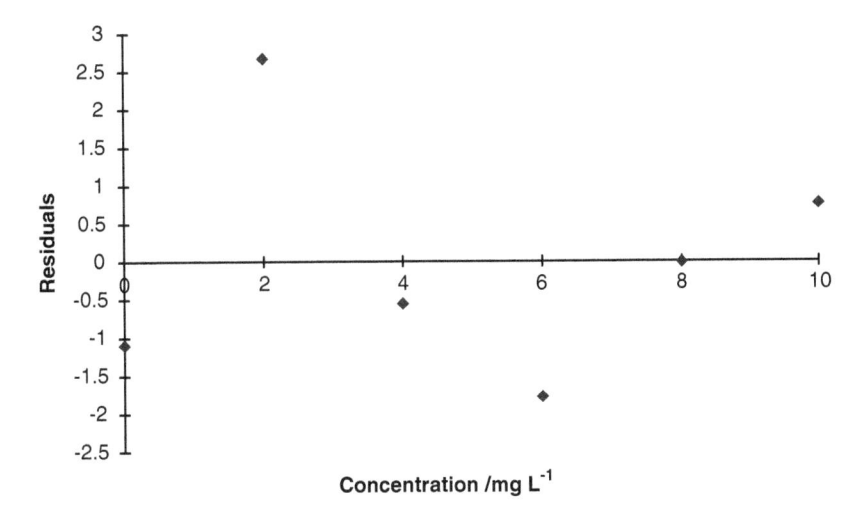

Figure 7.4 Residuals plot for the data presented in Table 7.1.

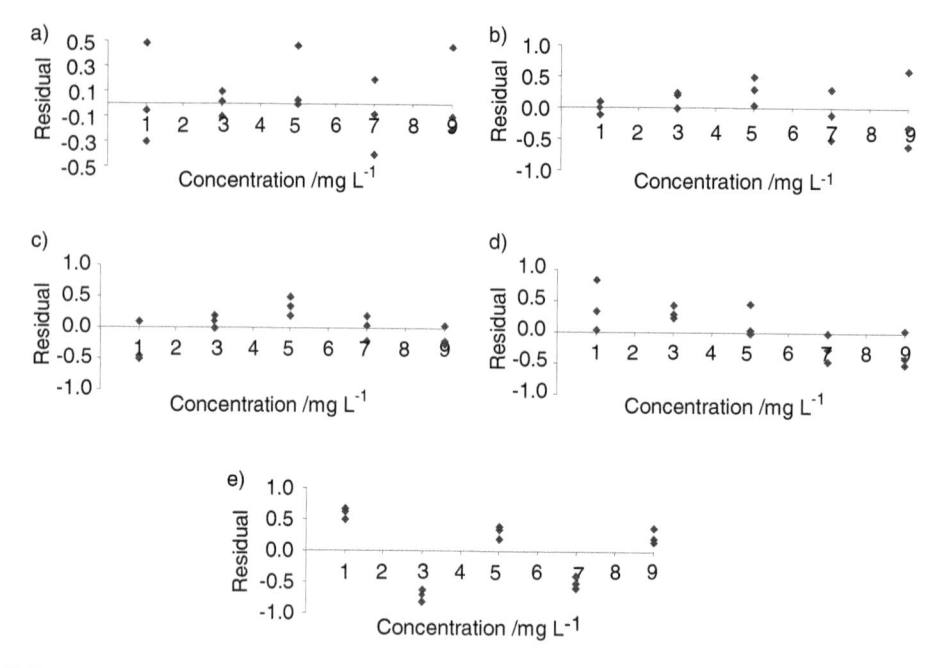

Figure 7.5 Examples of residuals plots. (a) Ideal – random distribution of residuals about zero; (b) standard deviation increases with concentration; (c) curved response; (d) intercept incorrectly set to zero; (e) correlation among residuals.

residuals. The significance of this effect can be checked by applying ANOVA to the residuals (see Section 7.1.9). Perhaps the most common cause of the pattern shown in Figure 7.5e is the practice of replicating analyses of each standard solution (see Section 7.1.10.2); replicate analyses can group very closely but other errors affecting each group, such as volumetric errors (or in the case of certified reference materials, matrix effects), may cause larger deviations from the line. Where such an effect is found, prediction standard errors and other parameter uncertainties are often misleadingly good; more accurate estimates of the uncertainties are obtained by regression using the group means instead of the individual data points. Figure 7.5b–e should all cause concern as the pattern of the residuals is clearly not random; Figure 7.5e should dictate the use of group means to obtain correct uncertainty estimates, although the use of group means will prevent further tests of goodness of fit. The issue of replication in calibration experiments is discussed in more detail in Section 7.1.10.2.

Finally, if there is a suspicion of non-normality in the residuals, a normal probability plot of the residuals may be a useful additional check; see Chapter 2, Section 2.10.

7.1.6 The Correlation Coefficient

7.1.6.1 Calculating the Correlation Coefficient

The correlation coefficient can be obtained from software or calculated using the equation

$$r = \frac{\sum_{i=1}^{n}[(x_i - \bar{x})(y_i - \bar{y})]}{\sqrt{\left[\sum_{i=1}^{n}(x_i - \bar{x})^2\right]\left[\sum_{i=1}^{n}(y_i - \bar{y})^2\right]}} \quad (7.9)$$

Regression

Example

The correlation coefficient for the data given in Table 7.1 is calculated using equation (7.9). The relevant intermediate calculations are given in Table 7.2.

$$r = \frac{\sum_{i=1}^{n}[(x_i - \bar{x})(y_i - \bar{y})]}{\sqrt{\left[\sum_{i=1}^{n}(x_i - \bar{x})^2\right]\left[\sum_{i=1}^{n}(y_i - \bar{y})^2\right]}} = \frac{708}{\sqrt{70 \times 7173}} = 0.9992$$

7.1.6.2 Interpretation of the Correlation Coefficient

The correlation coefficient, r, measures the degree of linear association between the y and x variables. The value of r will be in the range ± 1. The closer $|r|$ is to 1, the stronger is the correlation between the variables. The correlation coefficient cannot generally be taken as a measure of linearity or even of a significant relationship except in particular circumstances,[1] so it is important to interpret the r value in conjunction with a plot of the data; r can be usefully interpreted as indicative of good linearity only when data are reasonably evenly distributed along the x-axis, with no serious anomalies. For predictions made from a calibration curve to have small uncertainties, r needs to be very close to 1; for example, for approximately 10% relative uncertainties in prediction at the centre of the calibration range, r needs to be about 0.99; for 1%, r needs to be better than 0.9999 (see reference 1 for a quantitative analysis). Note that a non-linear trend is often easy to spot even in a calibration giving $r \approx 0.999$.

A low r value does not necessarily mean that there is no relationship; a useful *non*-linear relationship between the y and x values would not necessarily lead to high linear correlation coefficients.

7.1.7 Uncertainty in Predicted Values of x

Once the linear regression has been carried out and the equation of the best fit straight line established, the equation can be used to obtain predicted values for x from experimentally determined values of y. A predicted value ($\hat{x}$) is obtained using the equation

$$\hat{x} = \frac{\bar{y}_0 - a}{b} \qquad (7.10)$$

where $\bar{y}_0$ is the mean of N repeated measurements of y.

There will be an uncertainty associated with $\hat{x}$ which can be calculated from the equation

$$s_{\hat{x}} = \frac{s_{y/x}}{b}\sqrt{\frac{1}{N} + \frac{1}{n} + \frac{(\bar{y}_0 - \bar{y})^2}{b^2 \sum_{i=1}^{n}(x_i - \bar{x})^2}} \qquad (7.11)$$

where $s_{\hat{x}}$ is sometimes referred to as the standard error of prediction for $\hat{x}$. Note that the uncertainty associated with predicted values is at a minimum at the centroid of the best fit line; that is, at the point $(\bar{x}, \bar{y})$. Although uncertainties increase at the extremes of the data range, the difference is often unimportant in usable calibration functions as shown in Figure 7.6. Designing a calibration study to minimise the uncertainty in predicted values of x is discussed in detail in Section 7.1.10.1.

The confidence interval for $\hat{x}$ is given by the equation

$$\hat{x} \pm t \times s_{\hat{x}} \tag{7.12}$$

Example

The equation of the best fit straight line for the data in Table 7.1 was calculated as $y = 1.100 + 10.114x$. The response for a test sample was 80. The predicted concentration for this sample is therefore

$$\hat{x} = \frac{80 - 1.100}{10.114} = 7.80 \, \text{mg L}^{-1}$$

The standard error of prediction for this predicted value of x is calculated using the equation

$$s_{\hat{x}} = \frac{s_{y/x}}{b} \sqrt{\frac{1}{N} + \frac{1}{n} + \frac{(\bar{y}_0 - \bar{y})^2}{b^2 \sum_{i=1}^{n}(x_i - \bar{x})^2}} = \frac{1.762}{10.114} \sqrt{\frac{1}{1} + \frac{1}{6} + \frac{(80 - 51.67)^2}{10.114^2 \times 70}} = 0.197$$

The 95% confidence interval for the predicted value is

$$\hat{x} \pm t \times s_{\hat{x}} = 7.80 \pm 2.776 \times 0.197 = 7.8 \pm 0.55 \, \text{mg L}^{-1}$$

where 2.776 is the two-tailed Student's t value for significance level $\alpha = 0.05$ and degrees of freedom $v = n - 2$.

Figure 7.6 shows the 95% confidence interval for a range of predicted x values.

7.1.8 Interpreting Regression Statistics from Software

7.1.8.1 Statistics Describing Gradient and Intercept Estimates

When software is used to carry out linear least-squares regression, some or all of the statistics in Table 7.4 will often be obtained.

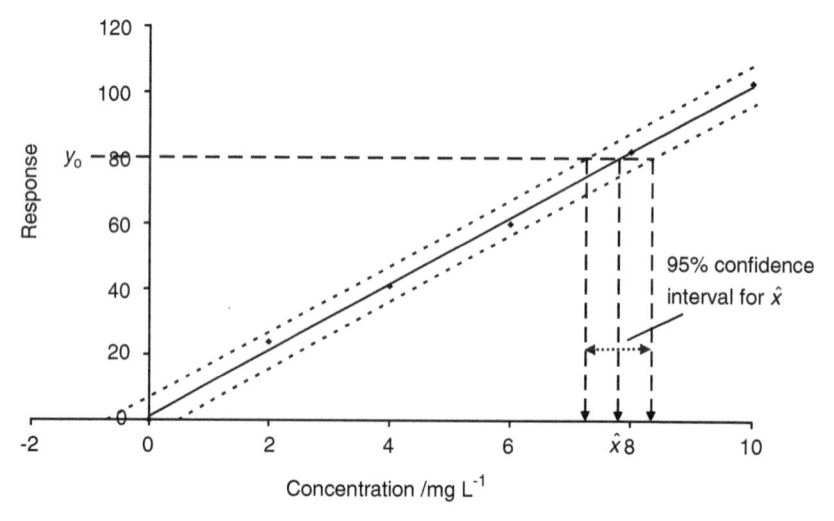

Figure 7.6 95% confidence interval for predicted values of x (calibration data from Table 7.1).

Regression

Table 7.4 Regression coefficients for data presented in Table 7.1.

	Coefficients	Standard deviation	t-Statistic	p-Value	Confidence interval Lower 95%	Upper 95%
Intercept	1.100	1.275	0.86	0.44	−2.44	4.64
Gradient	10.114	0.211	48.03	1.12×10^{-6}	9.53	10.70

Calculation of the regression coefficients and their associated standard deviations and confidence intervals was discussed in Section 7.1.4. The t statistic and p-value relate to a Student's t-test comparing whether the gradient and intercept are significantly different from zero. The relevant hypotheses are, for the gradient:

$$H_0: b = 0$$
$$H_1: b \neq 0$$

and for the intercept:

$$H_0: a = 0$$
$$H_1: a \neq 0$$

A test for significant slope is more useful in exploratory analysis than in calibration. In a calibration experiment, the gradient should be significantly different from zero as it is essential that the y and x values are highly correlated. The p-value should therefore be very small (far less than the usual 0.05) and the t value should be very much greater than the critical value if the calibration is to be useful. For exploratory analysis, on the other hand, it is usually important to test whether there is any relationship at all; for this purpose, it is usually sufficient to compare the calculated value of t for the gradient with the critical value or to compare the p-value with 0.05 in the usual way.

For calibration, once it has been established that a linear fit of the data is satisfactory, it is often useful to test whether the intercept, a, is significantly different from zero. If the intercept is not significantly different from zero, the t value will be less than the critical value and the p-value will be greater than the specified significance level. Whether the regression line can reasonably be assumed to pass through zero can also be judged by inspecting the confidence interval for the intercept. If this spans zero, then the intercept is not statistically different from zero. In the example shown, the p-value is greater than 0.05 and the 95% confidence interval for the intercept spans zero, so we can conclude that the intercept is not significantly different from zero.

7.1.8.2 ANOVA Tables for Linear Regression

Regression software may also return an ANOVA table similar to that shown in Table 7.5.

The total variation of measured values of y about the mean value $\bar{y}$ can be divided into two components:

1. The differences between the actual values and the fitted line. These lead to the residual standard deviation discussed in Section 7.1.4 and calculated using equation (7.4). In Table 7.5, this source of variation is represented by the 'Residual' row.
2. The differences between the fitted (that is, predicted) values of y and the mean value $\bar{y}$. This source of variation is usually associated with the regression itself and represented by the 'Regression' row in Table 7.5.

In short, the total variation is composed of the dispersion in y values represented by the fitted line and the remaining dispersion shown by the residuals. Where a significant correlation between the y

Table 7.5 ANOVA table for the linear regression of data presented in Table 7.1.

	Sum of squares[a]	Degrees of freedom v	Mean square	F	p-Value
Regression	7160.5	1	7160.5	2309.8	1.12×10^{-6}
Residual	12.4	4	3.1		
Total	7173.3	5			

[a]Sums of squares calculated manually using equations in Table 7.6. Values may differ slightly from values obtained from software due to rounding.

Table 7.6 ANOVA calculations for least-squares linear regression.

Source of variation	Sum of squares	Degrees of freedom v	Mean square	F
Regression	$S_1 = b^2 \times \sum_{i=1}^{n}(x_i - \bar{x})^2$	1	$M_1 = S_1/1$	M_1/M_0
Residual	$S_0 = \sum_{i=1}^{n}(y_i - \hat{y}_i)^2$	$n - 2$	$M_0 = S_0/(n-2)$	
Total	$\sum_{i=1}^{n}(y_i - \bar{y})^2$	$n - 1$		

and the x values exists, the majority of the variation in y values can be accounted for by the equation of the line that has been fitted to the data.

The two sources of variation can be separated and compared using ANOVA (see Chapter 6). The relevant calculations are given in Table 7.6.

The mean square terms are obtained by dividing the sum of squares terms by their associated degrees of freedom. Note that M_0 is the square of the residual standard deviation.

The critical value for F is the one-tailed F-value for the required significance level and the degrees of freedom associated with the regression (1 for simple linear regression) and the residuals $(n-2)$. For the example shown in Table 7.5 the critical value (obtained from Table A.5a in Appendix A) is 7.709 for $\alpha = 0.05$, $v_1 = 1$ and $v_2 = 4$.

Where a relation between the y and x values is expected, the residual sum of squares should be small compared with the regression sum of squares, because if the regression line fits the data well, the residual standard deviation will be small compared with the range of the data. Therefore, the regression mean square, M_1, is expected to be substantially greater than the residual mean square, M_0. If this is the case then the calculated value of F will be large compared with the critical value for F and the p-value will be very small.

As with the significance test for the gradient (which should show the same significance level), the significance test in this ANOVA table is a test for a significant relationship, not a test for a useful calibration or for linearity. A very strongly significant value of F should be expected in a calibration experiment.

7.1.9 Testing for Non-linearity

It is important to check that a calibration is linear. As mentioned previously, regression statistics such as the correlation coefficient do not test linearity well; more specific tests for *non*-linearity are more effective. The issue of designing a suitable study to allow checks for non-linearity is discussed in Section 7.1.10.4. The following checks are useful.

7.1.9.1 *Inspection of the Residuals*

Non-linearity can usually be identified by examining a scatter plot of the data and, more especially, a plot of the residuals. Section 7.1.5 gives further detail.

Regression

7.1.9.2 F-Test of Residual Standard Deviation Against Repeatability Standard Deviation

The residual standard deviation can be compared with an independent estimate of the repeatability of the *y*-values at a single *x*-value using an *F*-test as shown in equation (7.13). The estimate of s_r can be obtained, for example, from replicate analysis of a calibration standard.

$$F = \frac{s_{y/x}^2}{s_r^2} \qquad (7.13)$$

The null hypothesis for the test is H_0: $s_{y/x} = s_r$ and the alternative hypothesis is H_1: $s_{y/x} > s_r$. The test is therefore a one-tailed test. The estimate of $s_{y/x}$ will have $n - 2$ degrees of freedom, where n is the number of pairs of data in the regression data set. At the 95% confidence level, the appropriate critical value for *F* is obtained from tables for $\alpha = 0.05$, $v_1 =$ degrees of freedom for $s_{y/x}$ and $v_2 =$ degrees of freedom for s_r. If the calculated value for *F* exceeds the critical value, the null hypothesis is rejected. This indicates that the residuals are more widely dispersed than can be accounted for by random error. This could be evidence of non-linearity but a significant result could also occur if, for example, one or two observations were biased by other factors. Studying the scatter plot and plot of the residuals will help decide between the two.

7.1.9.3 ANOVA Applied to Residuals

If observations are replicated at each value of *x*, applying one-way ANOVA (see Chapter 6, Section 6.3) to the residuals using the *x* values as the grouping factor can warn of non-linearity. A significant between group mean square indicates that the group means deviate from the line more than would be expected from the repeatability alone (represented by the within-group mean square). This may point to significant non-linearity. Inspection of the residuals is still advisable, however, as a variety of effects can cause a significant between group effect in the residuals; see the discussion of Figure 7.5e in Section 7.1.5.

7.1.9.4 Testing for Significant Higher Order Terms

Another approach to evaluating non-linear data is to fit a more complex (higher order) equation to the data and determine whether the new equation is a better representation of the data. This is discussed in Section 7.2.3.

7.1.10 Designing Linear Calibration Experiments

This section discusses some of the main experimental design considerations for calibration based on linear regression. The principal issues are:

- the number of different concentrations included in the calibration;
- the number of replicates for each concentration;
- the best distribution of *x* values, that is, the range of concentrations of the calibration solutions and the intervals between them.

It is important to be clear as to the objectives of any experimental design. In linear regression for calibration, the principal objective must be to generate predicted values of *x* (for example, the concentration of an analyte in a test solution) with minimum uncertainty in the range of interest. However, other factors are important. For example, it is often desirable to provide a visible check on linearity in addition to providing for interpolation. It is also important to consider whether analytical variation is constant across the range or changes with concentration (see Section 7.1.2); frequently, but not universally, the dispersion of observations increases with concentration. Finally, the purpose may not be calibration at all, but linearity verification, which is considered separately in Section 7.1.10.4).

7.1.10.1 Minimising Prediction Uncertainty

The standard error of prediction for a predicted value of x (for example, concentration) is calculated from equation (7.11). Minimising the standard error can be achieved by:

1. maximising N, the number of observations on the test sample;
2. maximising n, the total number of independent observations during calibration;
3. minimising $(\bar{y}_0 - \bar{y})$;
4. maximising $\sum(x_i - \bar{x})^2$.

Optimising the number of replicate observations, N
This is largely a simple decision based on the required uncertainty. For $N = 1$, the term $1/N$ will often dominate the standard error of prediction near the centroid of the calibration. Clearly, if this is sufficient for the task, no further increase is necessary. If not, however, increasing N is potentially useful, as Figure 7.7 shows. The largest standard error of prediction is for $N = 1$. As N increases from

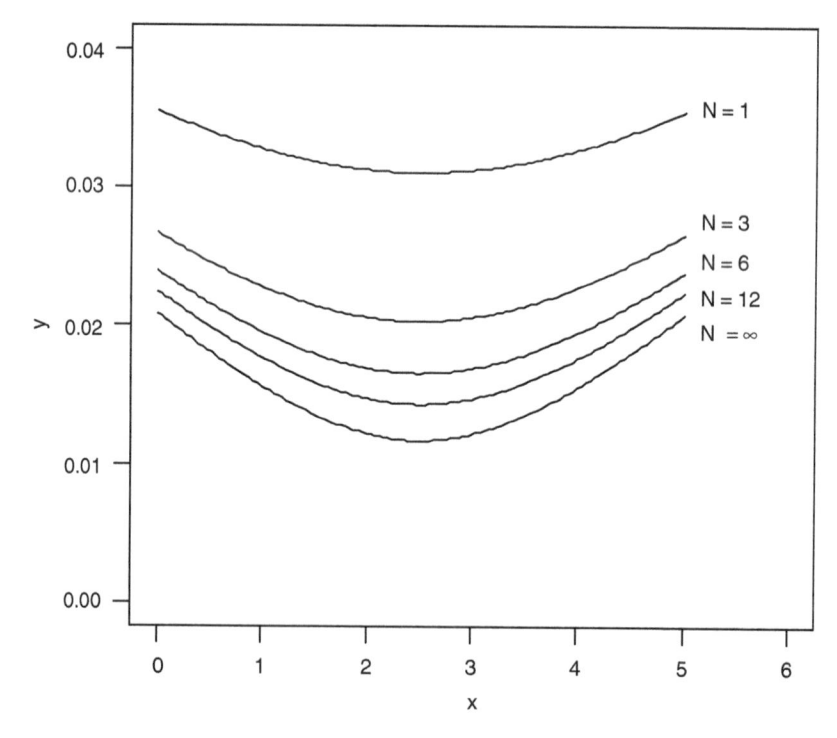

Figure 7.7 Effect of increasing number of observations on a test sample in linear calibration. The figure shows the standard error of prediction as a function of x for a linear regression on six equally spaced observations ($x = 0-5$) with residual standard deviation 0.115 and a best fit of $y = 0.061 + 3.989x$. Successive curves labelled $N = 1$ to $N = \infty$ show the same curve for different numbers of observations on each test sample.

Regression

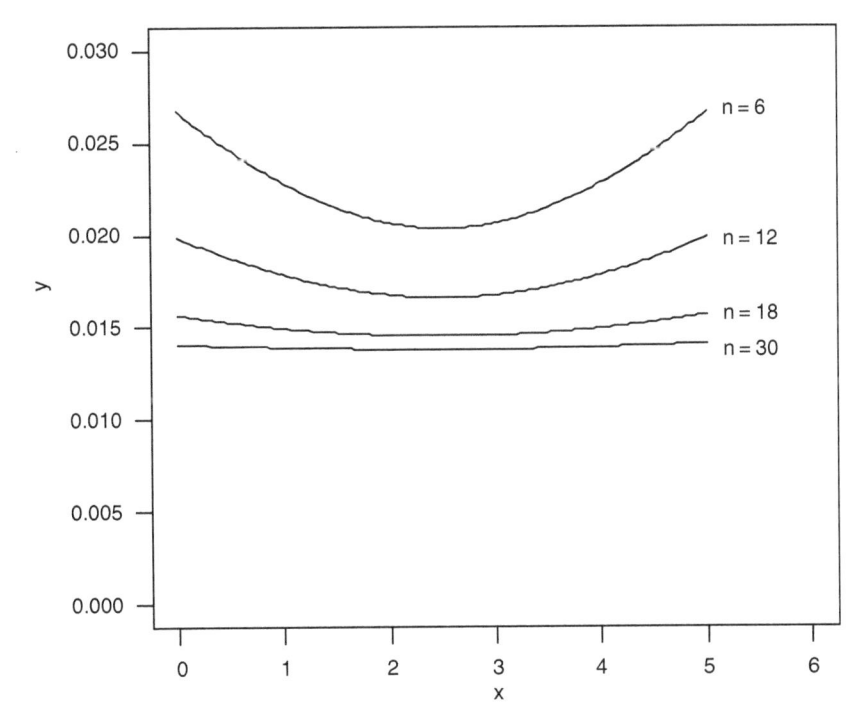

Figure 7.8 Effect of increasing number of calibration observations. The figure shows the standard error of prediction for the same regression as in Figure 7.7, with increasing number n of (x, y) observations during calibration. N was held constant at 3.

1 to 3, the standard error of prediction drops by a worthwhile 30%, but further increases are less and less effective. It follows that choosing values of N between 1 and about 6 gives potentially useful improvements in prediction uncertainty, but diminishing returns accompany further increases in N.

Increasing the total number of independent observations during calibration, n
Increasing the total number of *independent* observations during calibration is also beneficial to an extent. This is shown in Figure 7.8. However, with six observations already present, increasing the number of observations during calibration has relatively modest effects unless N is also increased substantially. Based on Figure 7.8, using more than 10 independent observations is unlikely to provide worthwhile benefits compared with 5–6 observations. The lower limit for number of observations is set largely by the available degrees of freedom, which is $n - 2$. With $n = 3$ or 4, the residual standard deviation has only one or two degrees of freedom and the resulting 95% confidence intervals must be based on $t = 12.71$ or 4.30, respectively. Retaining at least three degrees of freedom (*i.e.* a minimum of $n = 5$) is therefore desirable when the calibration uncertainty is practically significant.

Minimising $(\bar{y}_0 - \bar{y})$
Figure 7.7 shows very clearly that the standard error of prediction is at a minimum at the centre of the regression. This follows from the fact that the value of $(\bar{y}_0 - \bar{y})$ is at a minimum here; the standard error of prediction is therefore always at a minimum at the centroid of the calibration data (weighted centroid, if using weighted regression). It follows that planning the calibration so that the centroid is closest to the concentrations of most interest will deliver the smallest uncertainties.

Maximising $\sum(x_i - \bar{x})^2$
The sum $\sum(x_i - \bar{x})^2$ is affected by the distribution of the observations in the calibration experiment. For unweighted regression, maximising this sum generates the smallest uncertainties for a given

number of observations. It can be shown that this is achieved in two circumstances: first, when the calibration range is maximised, and second, when exactly half of the n observations are at the upper extreme of the calibration range and half at the lower extreme, that is, for a two-point calibration with appropriate replication. Even an increase to three distinct x values, with one-third of the observations at the minimum, centre and maximum values, respectively, increases the prediction uncertainty significantly. However, before unreservedly adjusting all calibration curves to two-point calibration with maximum range, it is worth considering other factors. The most important is the need for some check on the linearity of the calibration and on the correctness of the various dilutions used in making up calibration solutions. Clearly, a two-point calibration relies very heavily on the assumption of linearity, but includes no such check. Using very large calibration ranges can only compound this problem. It is accordingly essential to plan for at least one, and preferably at least two, QC observations within the calibration range if using two-point calibration. For many analysts, it is practically simpler to include additional materials in the calibration. It follows that while the best uncertainty can be delivered by a two-point calibration strategy, this must be balanced against a need for more stringent QC and validation. It may also be necessary to consider the complexity of the calibration materials; for example, in standard additions, the effort of preparing multiple blends may be considerable, making a two-point strategy more attractive subject to appropriate QC. Further, for standard additions, the extrapolations involved cause the technique to perform very poorly indeed for small calibration ranges; it is accordingly recommended to add three to five times the original analyte concentration if carrying out a standard additions experiment[2].

Another common pattern used in calibration is a successive dilution, resulting in logarithmically decreasing concentrations (for example, 0, 1, 2, 4 and 8 mg L^{-1}). This is simple and has the advantage of providing a high upper calibrated level, which may be useful in analysing routine samples that occasionally show high values. However, the layout has several disadvantages. First, errors in dilution are multiplied at each step, increasing the volumetric uncertainties and, perhaps worse, increasing the risk of an undetected gross dilution error (especially if the analyst commits the cardinal sin of using one of the calibration solutions as a QC sample!). Second, the highest concentration point has high leverage (Section 7.1.3); errors at the high concentration will cause potentially large variation in results. Third, departures from linearity are easier to detect with fairly evenly spaced points. In general, therefore, equally spaced calibration points are to be preferred.

7.1.10.2 Replication in Calibration

Replication is useful if replicates are genuinely independent. It improves precision by increasing n and provides additional checks on the calibration solution preparations and on the precision at different concentrations. Independent replication is accordingly a viable method of increasing n when the best possible performance is desired. Given that increasing the number of independent concentrations has little practical effect (and actually increases the uncertainty slightly compared with a few widely spaced points), independent replication can be recommended as a method of improving uncertainties. However, replication suffers from an important drawback. Most analysts have seen the effect of simply injecting a calibration solution twice; the plotted residuals appear in close pairs and are clearly not independent. This is essentially useless for improving precision. Worse, it artificially increases the number of degrees of freedom for simple linear regression, giving a misleadingly small prediction interval. Ideally, therefore, replicated observations should be entirely independent; using different calibration solutions if at all possible. Otherwise, it is better to first examine replicate injections to check for outlying differences and then to calculate the calibration based on the mean value of y for each distinct concentration.

There is one side effect of replication that may be useful. If means of replicates are taken, the distribution of errors in the means tends to the normal distribution as the number of replicates

Regression

increases, regardless of the parent distribution. The distribution of the mean of as few as three replicates is very close to the normal distribution even with fairly extreme departures from normality. Averaging three or more replicates can therefore provide more accurate statistical inference in critical cases where non normality is suspected.

7.1.10.3 Summary of Linear Calibration Design Recommendations

If the calibration uncertainty is small compared with other effects in routine analysis, the calibration design is not critical. However, if calibration uncertainties are important, the following guidelines can markedly improve the resulting confidence interval for predicted values:

1. Use at least five independent observations.
2. Space the concentrations approximately evenly.
3. Ensure that the mean concentration for the calibration materials is close to the levels of interest.
4. Include independent replicates if necessary to improve precision.
5. Increase the number of replicate observations on test samples to at least three.
6. For standard additions methods in particular, consider two-point calibration with a wide calibration range, but ensure good QC and validation.

7.1.10.4 Planning Linearity Checks

Linearity checks use various statistical tests for departures from linearity. Among these tests are:

- ANOVA on the residuals to check for significant between-concentration effects (see Section 7.1.9.3).
- Tests for significance of higher order terms in the regression (Section 7.2.3).
- Checks on normality of residual errors.

These tests all benefit substantially from higher levels of replication and additional concentrations compared with the requirements for calibration. ANOVA in particular can only be carried out if observations are made at least in duplicate. The power of the ANOVA is also poor for few groups, but increases with the number of independent groups and replicates. For example, to obtain 80% power for detection of a between-group variance when the true between- and within-group variances are equal requires at least three replicates and eight groups.

For linearity checks, therefore, the following recommendations, added to those for linear calibration, can increase confidence very considerably:

- Increase the number of independent concentrations to at least seven.
- Use at least two, and preferably three or more, replicates at each concentration.

7.1.11 Two Common Mistakes

Some mistakes are so common in routine application of linear regression that it is worth describing them so that they can be avoided:

1. *Incorrectly forcing the regression through zero.* Some software allows a regression to be forced through zero (for example, by specifying removal of the intercept or ticking a 'Constant is

zero' option). This is valid *only* with good evidence to support its use, for example, if it has been previously shown that the intercept is not significant, using methods such as those in Section 7.1.8. Otherwise, interpolated values at the ends of the calibration range will be incorrect – often very seriously so near zero.

2. *Including the point (0,0) in the regression when it has not been measured.* Sometimes it is argued that the point ($x=0$, $y=0$) should be included in the regression, usually on the grounds that $y = 0$ is the expected response at $x = 0$. This is entirely fallacious; it is simply inventing data. To include invented data is always bad practice in any case; it also has adverse effects on the statistical interpretation. Adding an invented point at (0,0) will cause the fitted line to move closer to (0,0), making the line fit the data more poorly near zero and also making it more likely that a real non-zero intercept will go undetected (because the calculated intercept will be smaller). The *only* circumstance in which a point (0,0) can validly be added to a regression data set is when a standard at zero concentration has been included and the observed response is either zero or is too small to detect and can reasonably be interpreted as zero.

7.2 Polynomial Regression

7.2.1 Polynomial Curves and Non-linearity

Not all data can be represented by a linear relationship. Figure 7.9 shows data that obviously follow a non-linear relationship.

The linear model, $y = a + bx$, described in Section 7.1.1 is also known as a first-order polynomial. A second-order polynomial or quadratic function, contains an extra term, in x^2:

$$y = a + bx + cx^2 \qquad (7.14)$$

A quadratic model has $n-3$ degrees of freedom as the model contains three parameters, a, b and c. The additional term provides for simple curvature in the fitted line.

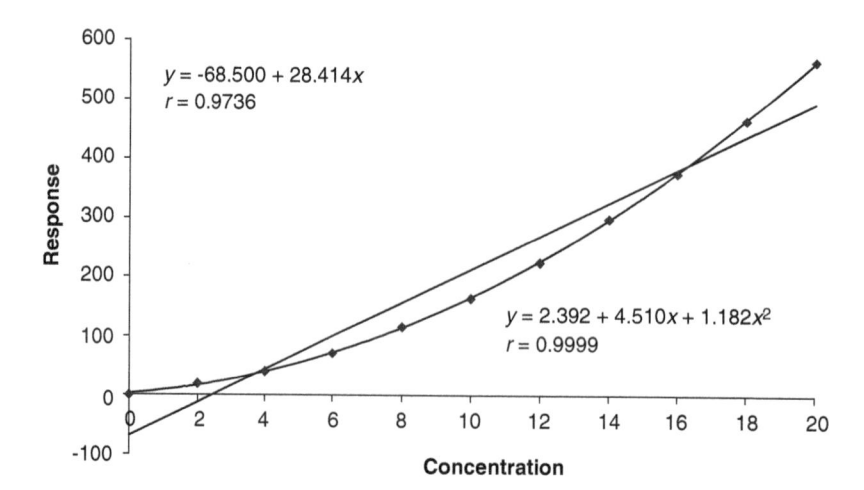

Figure 7.9 Scatter plot showing both first-order (linear) and second-order (quadratic) polynomial trend lines.

7.2.2 Fitting a Quadratic (Second-order Polynomial)

Fitting a second-order polynomial is most readily carried out using statistical software. The Appendix to this chapter describes the necessary calculations to find the regression coefficients a, b and c manually.

Statistical software will generally produce a table of regression coefficients and their standard errors. It may also produce an ANOVA table (the manual calculations of which are also given in the Appendix to this chapter). These can both be used to check the significance of the different terms (see Section 7.2.3). Note that the correlation coefficient, r, should not be used to judge whether a more complex equation is a better fit of the data than a simple linear model. When an additional term is added to a regression equation, the correlation coefficient r will always improve, even if the predictive value is poorer.

7.2.3 Using Polynomial Regression for Checking Linearity

Section 7.1.9 indicated that polynomial regression can be used to test for significant non-linearity in a proposed linear calibration. The principle is straightforward: calculate a quadratic (or higher) polynomial fit to the same data and inspect the regression coefficients and standard errors. If the quadratic coefficient [c in equation (7.14)] is significantly different from zero, there is evidence of systematic curvature and it should be concluded that there is a statistically significant departure from linearity. Alternatively, inspecting the ANOVA table (as in the Appendix to this chapter) provides an equivalent check.

Appendix: Calculations for Polynomial Regression

Calculating the Coefficients

A second-order polynomial (quadratic function) is described by the equation

$$y = a + bx + cx^2 \tag{7.14}$$

The coefficients a, b and c are calculated from the following equations:

$$bS_{xx} + cS_{xx^2} = S_{xy} \tag{7.15}$$

$$bS_{xx^2} + cS_{x^2x^2} = S_{x^2y} \tag{7.16}$$

$$a = \bar{y} - b\bar{x} - \frac{c\sum_{i=1}^{n} x_i^2}{n} \tag{7.17}$$

where the S terms are defined as follows:

$$S_{xy} = \sum_{i=1}^{n} x_i y_i - \left(\frac{\sum_{i=1}^{n} x_i \times \sum_{i=1}^{n} y_i}{n} \right) \tag{7.18a}$$

$$S_{xx} = \sum_{i=1}^{n} x_i^2 - \left[\frac{\left(\sum_{i=1}^{n} x_i\right)^2}{n} \right] \tag{7.18b}$$

$$S_{yy} = \sum_{i=1}^{n} y_i^2 - \left[\frac{\left(\sum_{i=1}^{n} y_i\right)^2}{n}\right] \tag{7.18c}$$

$$S_{x^2y} = \sum_{i=1}^{n} x_i^2 y_i - \left(\frac{\sum_{i=1}^{n} x_i^2 \times \sum_{i=1}^{n} y_i}{n}\right) \tag{7.18d}$$

$$S_{xx^2} = \sum_{i=1}^{n} x_i^3 - \left(\frac{\sum_{i=1}^{n} x_i \times \sum_{i=1}^{n} x_i^2}{n}\right) \tag{7.18e}$$

$$S_{x^2x^2} = \sum_{i=1}^{n} x_i^4 - \left[\frac{\left(\sum_{i=1}^{n} x_i^2\right)^2}{n}\right] \tag{7.18f}$$

Example

The terms required to evaluate the regression coefficients for the data in Table 7.7 are calculated using equations (7.18a–f):

$$S_{xy} = \sum_{i=1}^{n} x_i y_i - \left(\frac{\sum_{i=1}^{n} x_i \times \sum_{i=1}^{n} y_i}{n}\right) = 35802 - \left(\frac{110 \times 2342}{11}\right) = 12382$$

Table 7.7 Data for scatter plot shown in Figure 7.9.

	x_i	y_i	x_i^2	x_i^3	x_i^4	y_i^2	$x_i y_i$	$x_i^2 y_i$
	0	0	0	0	0	0	0	0
	2	19	4	8	16	361	38	76
	4	40	16	64	256	1600	160	640
	6	71	36	216	1296	5041	426	2556
	8	116	64	512	4096	13456	928	7424
	10	164	100	1000	10000	26896	1640	16400
	12	225	144	1728	20736	50625	2700	32400
	14	299	196	2744	38416	89401	4186	58604
	16	376	256	4096	65536	141376	6016	96256
	18	466	324	5832	104976	217156	8388	150984
	20	566	400	8000	160000	320356	11320	226400
Sums	110	2342	1540	24200	405328	866268	35802	591740

Regression

$$S_{xx} = \sum_{i=1}^{n} x_i^2 - \left[\frac{\left(\sum_{i=1}^{n} x_i\right)^2}{n}\right] = 1540 - \left(\frac{110^2}{11}\right) = 440$$

$$S_{yy} = \sum_{i=1}^{n} y_i^2 - \left[\frac{\left(\sum_{i=1}^{n} y_i\right)^2}{n}\right] = 866268 - \left(\frac{2342^2}{11}\right) = 367634.9$$

$$S_{x^2y} = \sum_{i=1}^{n} x_i^2 y_i - \left(\frac{\sum_{i=1}^{n} x_i^2 \times \sum_{i=1}^{n} y_i}{n}\right) = 591740 - \left(\frac{1540 \times 2342}{11}\right) = 263860$$

$$S_{xx^2} = \sum_{i=1}^{n} x_i^3 - \left(\frac{\sum_{i=1}^{n} x_i \times \sum_{i=1}^{n} x_i^2}{n}\right) = 24200 - \left(\frac{110 \times 1540}{11}\right) = 8800$$

$$S_{x^2x^2} = \sum_{i=1}^{n} x_i^4 - \left[\frac{\left(\sum_{i=1}^{n} x_i^2\right)^2}{n}\right] = 405328 - \left(\frac{1540^2}{11}\right) = 189728$$

The regression coefficients are calculated using equations (7.15)–(7.17):

$$bS_{xx} + cS_{xx^2} = S_{xy}, \quad \text{therefore} \quad 440b + 8800c = 12382$$

$$bS_{xx^2} + cS_{x^2x^2} = S_{x^2y}, \quad \text{therefore} \quad 8800b + 189728c = 263860$$

Solving these simultaneous equations gives $b = 4.5104$ and $c = 1.1815$. Therefore:

$$a = \bar{y} - b\bar{x} - \frac{c\sum_{i=1}^{n} x_i^2}{n} = 212.9091 - (4.5104 \times 10)$$
$$- \left(1.1815 \times \frac{1540}{11}\right) = 2.392$$

The quadratic equation is therefore

$$y = 2.392 + 4.510x + 1.182x^2$$

Table 7.8 Analysis of variance calculations for second-order polynomial.

Source of variation	Sum of squares	Degrees of freedom v
Due to regression	$S_1 = bS_{xy} + cS_{x^2y}$	2
Residuals	$S_0 = S_{yy} - (bS_{xy} + cS_{x^2y})$	$n-3$
Total	S_{yy}	$n-1$

Table 7.9 Testing the significance of the x^2 term in a second-order polynomial.

Source of variation	Sum of squares	Degrees of freedom v	Mean square	F
Due to linear regression	$S_x = b_0 S_{xy}$	1		
Due to addition of the x^2 term	$S_{x^2} = bS_{xy} + cS_{x^2y} - b_0 S_{xy}$	1	$M_{x^2} = S_{x^2}/1$	M_{x^2}/M_0
[Subtotal] Due to quadratic regression	$S_1 = bS_{xy} + cS_{x^2y}$	2		
Residuals	$S_0 = S_{yy} - (bS_{xy} + cS_{x^2y})$	$n-3$	$M_0 = S_0/(n-3)$	
Total	S_{yy}	$n-1$		

Table 7.10 ANOVA table for quadratic regression on data shown in Table 7.7.

Source of variation	Sum of squares	Degrees of freedom v	Mean square	F
Due to linear regression	$b_0 S_{xy} = 28.141 \times 12382$ $= \mathbf{348441.9}$	1		
Due to addition of the x^2 term	$bS_{xy} + cS_{x^2y} - b_0 S_{xy}$ $= (4.5104 \times 12382)$ $+ (1.1815 \times 263860)$ $- (28.141 \times 12382)$ $= \mathbf{19156.5}$	1	$19156.5/1$ $= \mathbf{19156.5}$	$19156.5/4.56715$ $= \mathbf{4194.41}$
[Subtotal] Due to quadratic regression	$bS_{xy} + cS_{x^2y}$ $= (4.5104 \times 12382) + (1.1815 \times 263860)$ $= \mathbf{367598.4}$			
Residuals	$S_{yy} - (bS_{xy} + cS_{x^2y})$ $= 367634.9 - (4.5104 \times 12382)$ $+ (1.1815 \times 263860)$ $= \mathbf{36.5}$	$n-3=8$	$36.5/8 = \mathbf{4.56715}$	
Total	$S_{yy} = \mathbf{367634.9}$	$n-1=10$		

ANOVA Tables for Polynomial Regression

As for linear regression, an ANOVA table can be constructed as shown in Table 7.8.

The significance of the x^2 term can be evaluated as shown in Table 7.9, where b_0 is the gradient obtained from the least-squares linear regression. If the calculated value of F exceeds the appropriate one-tailed critical value for F ($v_1 = 1$, $v_2 = n-3$), we can conclude that the x^2 term is significant.

Table 7.11 Completed ANOVA table for second-order polynomial of data shown in Table 7.7.

Source of variation	Sum of squares	Degrees of freedom v	Mean square	F
Due to linear regression	348441.9	1		
Due to addition of the x^2 term	19156.5	1	19156.5	4194.4
[Subtotal] Due to quadratic regression	367598.4			
Residuals	36.5	8	4.6	
Total	367634.9	10		

Example
The calculations for the ANOVA table showing the significance of the x^2 term for the data shown in Table 7.7 are shown in Table 7.10; the completed table is shown as Table 7.11.

The critical value for F (one-tailed, $\alpha = 0.05$, $v_1 = 1$, $v_2 = 8$) is 5.318 (see Appendix A, Table A.5a). The calculated value of F for the comparison of the x^2 mean square term with the residual mean square exceeds the critical value. We can therefore conclude that the x^2 term is significant and that the second-order polynomial is a better representation of the data compared with a linear model.

References

1. S. L. R. Ellison, *Accred. Qual. Assur.*, 2006, **11**, 146.
2. M. Thompson and S. L. R. Ellison, *Accred. Qual. Assur.*, 2005, **10**, 82.

CHAPTER 8
Designing Effective Experiments

'To consult a statistician after an experiment is finished is often merely to ask him to conduct a post mortem examination. He can perhaps say what the experiment died of.'

R. A. Fisher
Presidential Address to the
First Indian Statistical Congress, 1938

8.1 Some New Terminology

Before discussing experimental design, it is helpful to introduce some new terms which will be relevant throughout the discussion. The following terms are particularly important.

Factor: In experimental design, any quantity or condition which may affect the result of the experiment. For example, in a study of operator effects, 'operator' would be considered a factor.

Factor level: A particular value for a factor. For example, a particular operator in an operator effect study is one level of the factor 'operator'; a particular temperature in a temperature study is one level of the factor 'temperature'.

Treatment: A particular combination of factor levels. For example, a digestion temperature of 90 °C in a temperature study is a particular treatment; a digestion time of 15 min in 1 M HCl in a time–acid concentration study is a particular treatment.

Response: The measured output of a process or experiment. For example, analyte recovery, measured concentration or chromatography retention time.

Effect: The change in response caused by changes in level of a factor or factors.

Type I error: Incorrect rejection of the null hypothesis (see also Chapter 4, Section 4.2.2). Often this could also be called a false positive, as it results in a significant finding when there is no true effect. The (desired) probability of a Type I error is often denoted α.

Type II error: Incorrect acceptance of the null hypothesis; an insignificant test result when the null hypothesis is false. For example, a finding of no significant bias when in fact there is a small bias. This may often be considered a false negative. The (desired) probability of a Type II error is often denoted β.

Power (of a test): The probability of correctly rejecting the null hypothesis when it is false. Equal to 1 minus the Type II error probability, that is, $1 - \beta$.

8.2 Planning for Statistical Analysis

8.2.1 Measuring the Right Effect

Most measurements represent straightforward application of a measuring device or method to a test item. However, many experiments are intended to test for the presence or absence of some specific 'treatment' effect, such as the effect of changing a measurement method or adjusting a manufacturing method. For example, one might wish to assess whether a reduction in sample preconditioning time had an effect on measurement results. In these cases, it is important that the experiment measures the intended effect and not some external 'nuisance' effect. For example, measurement systems often show significant changes from day to day or operator to operator. To continue the preconditioning example, if test items for 'short preconditioning' were obtained by one operator and for 'long preconditioning' by a different operator, operator effects might be misinterpreted as a significant conditioning effect. This is called *confounding*; the conditioning effect is said to be confounded with the operator effect. Ensuring that nuisance parameters do not interfere with the result of an experiment is one of the aims of good experimental design and is discussed in Section 8.6.

8.2.2 Single- *Versus* Multi-factor Experiments

A second, but often equally important, aim of experimental design is to minimise the cost of an experiment. For example, a naïve study to investigate six possible effects might investigate each individually, using, say, three replicate measurements at each of two levels for each effect, a total of 36 measurements. If the measurement has a standard deviation σ, each comparison between levels will have a precision of $\sqrt{\sigma^2/3 + \sigma^2/3} = \sigma\sqrt{2/3}$, or about 0.82σ. Yet in Chapter 9, Section 9.7.1, we will see that a single experiment can estimate the effects of up to seven factors simultaneously and because each difference is calculated from two sets of four runs each, the precision for each comparison would be expected to be $\sigma/\sqrt{2}$, or about 0.7σ. Hence careful experimental designs which vary all parameters simultaneously can obtain the same information, with better precision, in this case with less than one-quarter of the number of experiments.

8.3 General Principles

Experimental design is a substantial topic and a range of reference texts and software are available. Some of the basic principles of good design are, however, summarised below:

1. *Arrange experiments for cancellation or comparison*
   The most precise and accurate measurements seek to cancel out sources of bias. If planning an experiment to test for significance of an effect, experiments should compare the effect of different levels of the factor directly wherever possible, rather than measuring the different effects independently on different occasions; difference and ratio experiments are usually more sensitive to changes than separate 'absolute' measurements. Similarly, in planning calibrations, simultaneous measurement of test item and calibrant reduces calibration differences; examples include the use of internal standards in chemical measurement and the use of comparator instruments in gauge block calibration.
2. *Control if you can; randomise if you can't*
   A good experimenter will identify the main sources of bias and control them. For example, if temperature is an issue, temperature should be controlled as far as possible. If direct control is impossible, the statistical analysis should include the 'nuisance' parameter. 'Blocking' – systematic allocation of test items to different strata – can help reduce bias and is described in

more detail in Section 8.6.3. Where an effect is known but cannot be controlled, and also to guard against unknown systematic effects, randomisation should always be used. For example, measurements should always be made in random order within runs as far as possible (although the order should be recorded to allow trends to be identified) and test items of the same type should be assigned randomly to treatments in planned experiments.

3. *Plan for replication or obtain independent uncertainty estimates*

    Without knowledge of the precision available and more generally of the uncertainty, the experiment cannot be interpreted. Statistical tests generally rely on comparison of an effect with some estimate of the uncertainty of the effect, usually based on observed precision. Therefore, experiments should always include some replication to allow precision to be estimated or should provide for additional information on the uncertainty.

4. *Design for statistical analysis*

    An experiment should always be planned with a specific method of statistical analysis in mind. Otherwise, despite the considerable range of tools available, there is too high a risk that no statistical analysis will be applicable. One particular issue in this context is that of 'balance'. Many experiments test several parameters simultaneously. If more data are obtained on some combinations than others, it may be impossible to separate the different effects. This applies particularly to two-factor or higher order analysis of variance, in which interaction terms are not generally interpretable with unbalanced designs. Imbalance can be tolerated in some types of analysis, but not in all. For that reason, it is best to plan for fully balanced experiments, in which each level of a factor appears an equal number of times.

8.4 Basic Experimental Designs for Analytical Science

Experimental design has attracted considerable attention and a wide range of experimental design strategies are available. The designs that can be analysed using the methods in this book are described here; some useful, but more complex, designs are listed in Section 8.7 for further information.

8.4.1 Simple Replication

The simplest experiment is a simple replicated series of observations (Figure 8.1a). The experiment is simply a series of observations on a single test material. It is most useful for estimating precision and, if a reference value is available, bias. Section 8.5 considers the number of observations necessary in such designs; Chapter 9, Sections 9.2 and 9.3, describe precision and bias determination for validation in more detail. Statistical analysis of simple replicated observations will usually use the basic summary statistics described in Chapter 4, possibly supplemented by significance tests (particularly *t*-tests as in Chapter 4, Section 4.2.3).

8.4.2 Linear Calibration Designs

Commonly used for instrument calibration and linearity tests (Chapter 7 and Chapter 9, Section 9.6), these designs include observations at a range of levels of some quantitative factor, usually analyte concentration (Figure 8.1b). The design may include replication at each level. Analysis generally involves linear regression (Chapter 7, Section 7.1). By their nature, these designs require a quantitative factor; regression through arbitrary levels of a qualitative or categorical factor is meaningless. Because of their relevance to linear regression, detailed discussion of linear calibration designs can be found in Chapter 7 and will not be repeated here.

Designing Effective Experiments 117

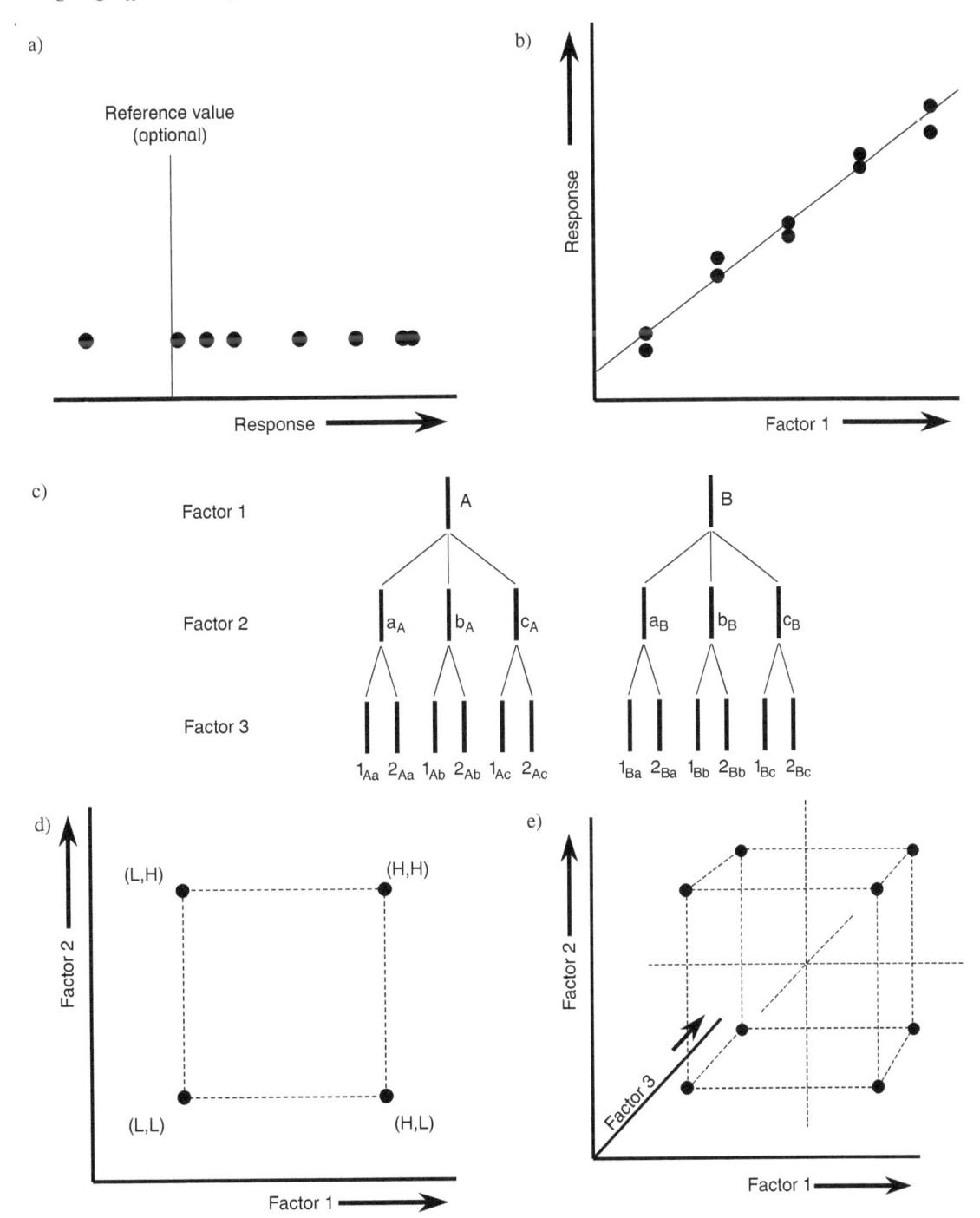

Figure 8.1 Basic experimental designs. (a) Simple replication with (optional) reference value; (b) linear calibration design; (c) a three-factor nested design; (d) a two-factor, two-level full factorial (2^2) design; (e) a three-factor, two-level full factorial (2^3) design.

8.4.3 Nested Designs

A nested design (sometimes called a hierarchical design) is an 'experimental design in which each level (potential setting, value or assignment of a factor) of a given factor appears in only a single level of any other factor' (ISO 3534-3, 2.6[1]). For example, in a homogeneity study, multiple samples from a particular container are nested within that container and replicate analyses are nested within subsamples. A given replicate analysis can only come from one possible subsample and container. The factors may be qualitative (categorical) or quantitative, although the former is more common. A *balanced* nested design is one in which each group at the same level of the hierarchy contains the same number of observations (confusingly, the term 'level' is often used to describe a particular depth of nesting in hierarchical designs; in that terminology, each *hierarchical* level corresponds to a different factor and the groups at that hierarchical level correspond to what would normally be called the different levels of the same factor). Figure 8.1c illustrates a possible balanced nested design for three factors. Note that the factor labelling has been subscripted to make the nested nature of the design clear. In practice, within-group labelling is often ambiguous and care needs to be taken to distinguish nested designs from factorial or 'crossed' designs.

Although perhaps as accurately considered as simple grouped data, the simplest example of a nested design is a set of replicate observations at each level of a single experimental factor, for example, replicate observations for each of several sets of extraction conditions or replicate observations on each of several days. Single-factor nested designs such as this can be analysed by one-way ANOVA (Chapter 6, Section 6.3). The analysis of two-factor nested designs is described in Chapter 6, Section 6.6. The same principles apply to essentially any depth of nesting.

Nested designs are most often used for estimating precision under various conditions of measurement. For example, repeatability and reproducibility standard deviations are commonly derived from a single-factor nested design (Chapter 9, Section 9.2.2.2), and intermediate precision may also be derived if multiple analytical runs are included in each laboratory carrying out a study.

8.4.4 Factorial Designs

Factorial designs are designs in which some or all levels for one factor occur in conjunction with two or more levels of another factor. Any two factors for which all combinations of levels appear are sometimes described as *crossed*. If all possible combinations of levels actually appear in the experiment, the design is called a *full factorial* design. The factor levels (not the responses) are shown schematically for two- and three-factor two-level designs in Figure 8.1d and e. Full factorials with the same number of levels for every factor have their own abbreviated nomenclature; a design with f factors each with p levels is called a p^f factorial. For example, Figure 8.1d shows a 2^2 factorial and Figure 8.1e a 2^3 factorial layout. Factorial designs are not, however, limited to designs with the same number of levels for each factor; a *mixed level design* with one three-level factor and one four-level factor is still a full factorial design if all combinations of factor levels appear.

Factorial designs are most commonly used for testing for the significance of changes in experimental parameters or other factors. Factors may be qualitative (for example, analytical sample type) or quantitative (for example, time, temperature) but are always restricted to discrete levels in the experiment. Usually, only a small number of levels are used; two levels for each factor are usually sufficient, although for two-factor experiments involving multiple sample types, more levels are common. The results are usually assessed using two-factor or (for more than two factors) higher order variants of ANOVA.

The number of observations in full factorial designs grows very fast with the number of factors. Even for two-level factors, whereas a two-factor design includes only $2^2 = 4$ treatments, a five-factor

Designing Effective Experiments 119

full factorial design needs $2^5=32$ treatments and additional replication is required to test properly for all possible effects. For moderate to large numbers of factors (four or more), most experimenters seek fractional factorials or other more economical designs (see Section 8.7 for a brief description).

8.5 Number of Samples

One of the most common questions in experimental design is, 'How many samples or observations must I use?'. Sadly, and perhaps surprisingly, there is no general answer to this question. However, given some additional information and some assumptions, it is possible to estimate a minimum number of observations for some types of analytical investigations. The following sections describe some of the most common cases.

8.5.1 Number of Samples for a Desired Standard Deviation of the Mean

Sometimes it is important to increase the number of observations so that the precision for *mean* results is improved to meet some external requirement. For example, in quality control applications, it is common to choose a measurement method with an uncertainty of not more than one-third of the process standard deviation.

Given a standard deviation for single observations of s, equation (4.5) in Chapter 4 gives the standard deviation $s(\bar{x})$ of the mean of n observations as $s/\sqrt{n}$. From this, the minimum number of observations n_{min} is given by

$$n_{min} \geq \left[\frac{s}{s(\bar{x})}\right]^2 \quad (8.1)$$

Equation (8.1) is also valid if both s and $s(\bar{x})$ are expressed as relative standard deviations.

Example

A drug manufacturer intends to produce tablets with the active ingredient controlled to within ±5% of the intended content (at 95% confidence). To monitor this production process, they request an assay method with an uncertainty of 0.83% or better, expressed as a relative standard deviation (this is based on taking ±5% as a 95% interval, dividing by 2 to obtain the approximate process standard deviation as 2.5% and dividing by 3 to obtain a 3:1 ratio of process to assay standard deviation). An analyst develops a method which provides within-run precision for single observations on different samples from the bulk of 1.7%, as a relative standard deviation. Assuming that the uncertainty is dominated by the within-run precision, how many samples should be analysed to obtain a relative uncertainty less than 0.83%?

The desired relative standard deviation of the mean is 0.83%; the observed relative standard deviation is 1.7%. Using equation (8.1),

$$n_{min} = \left(\frac{1.7}{0.83}\right)^2 = 4.195$$

With a minimum number of about 4.2, the analyst may decide that five samples should be taken to guarantee better precision than required, or may agree with the process engineers that a likely precision of 0.85% expected for four samples is sufficient for the purpose.

8.5.2 Number of Samples for a Given Confidence Interval Width

Sometimes it is useful to aim for a particular confidence interval width. This is more complex than selecting a number of observations for a required standard deviation of the mean. A confidence interval for a mean value is calculated using equation (4.19), Chapter 4 as:

$$\bar{x} \pm t\left(\frac{s}{\sqrt{n}}\right)$$

Unfortunately, this depends on the particular value of Student's t, which in turn depends on the number of samples n. The equation for n can therefore only be solved algebraically for large n, when t becomes approximately constant and nearly equal to the corresponding point of the normal distribution, or for situations where the standard deviation is known from other information.

For large n, for an intended confidence interval no wider than $\pm C$, the following equation should be used:

$$n_{min} \geq \left(\frac{sz_{1-\alpha/2}}{C}\right)^2 \qquad (8.2)$$

where $z_{1-\alpha/2}$ is the quantile of the normal distribution for probability $1 - \alpha/2$. For a 95% confidence interval, $1 - \alpha/2 = 0.975$ and $z = 1.96$; for 99% confidence, $1 - \alpha/2 = 0.995$ and $z = 2.65$. Equation (8.2) is reasonably accurate for $n > 50$; the accuracy can be improved near $n = 30$ (or $C < 0.4s$) by using $z = 2.0$ for 95% confidence. Like equation (8.1), equation (8.2) is also valid if both s and C are expressed as relative standard deviations.

For small n and larger C, equation (8.2) underestimates the required number of observations, often considerably. Alternative estimates are then necessary. There are two practical methods available:

1. Use tabulated values. A table of confidence intervals for smaller values of n is given in Table 8.1. To use the table, calculate C/s for the desired confidence interval width C and expected standard deviation s, look up the result in the table and read off the corresponding value of n.
2. Use iterative methods; that is, make an initial guess at n and adjust until the desired confidence interval is found. This is relatively easy using a spreadsheet or statistics application and is applicable when tables are not available for the desired level of confidence or confidence interval width.

Example

In developing a validation protocol, an analyst wishes to ensure that confidence intervals at the 95% level are no larger than ±1 standard deviation. How many observations are required?

1. *Using tabulated figures.* Looking down the 95% confidence column in Table 8.1, C/s first drops below 1 at $n = 7$. The validation protocol should require a minimum of seven replicates.
2. *Iteration using a spreadsheet.* Using a spreadsheet, an example using OpenOffice or MS Excel syntax would be similar to that shown in Figure 8.2, which shows a guess of $n = 4$. For this case, since the requirement is in terms of the standard deviation, the standard deviation in column A is set to 1.0; all that is necessary is to adjust the value of n in column B until the value in column C falls just below the required width of 1.0. For a value of 4, the value of cell

Designing Effective Experiments

Table 8.1 Number of observations for desired confidence intervals for $n \leq 50$[a].

n	Confidence 95%	Confidence 99%	n	Confidence 95%	Confidence 99%
2	8.985	45.012	15	0.554	0.769
3	2.484	5.730	16	0.533	0.737
4	1.591	2.920	17	0.514	0.708
5	1.242	2.059	18	0.497	0.683
6	1.049	1.646	19	0.482	0.660
7	0.925	1.401	20	0.468	0.640
8	0.836	1.237	25	0.413	0.559
9	0.769	1.118	30	0.373	0.503
10	0.715	1.028	35	0.344	0.461
11	0.672	0.956	40	0.320	0.428
12	0.635	0.897	45	0.300	0.401
13	0.604	0.847	50	0.284	0.379
14	0.577	0.805			

[a] The table shows the calculated value of C/s for a confidence interval $\bar{x} \pm C$ where $C = t_{\alpha/2, n-1} s/\sqrt{n}$; that is, C is the multiplier that would be used to calculate a confidence interval from a standard deviation s with $n-1$ degrees of freedom. To use the table to find the number of samples required to obtain a particular confidence interval, calculate C/s from the desired confidence interval and expected standard deviation s, look down the column for the relevant level of confidence to find the nearest tabulated value and read off the corresponding value of n, using the next larger value of n where the calculated value of C/s falls between tabulated values.

		A	B	C
	1	s	n	st/sqrt(n)
Formula:	2	1.0		= A2*TINV(0.05, B2−1)/SQRT(B2)
1st guess	2	1.0	4	1.591
2nd guess	2	1.0	8	0.836
3rd guess	2	1.0	7	0.925
Check	2	1.0	6	1.049

Figure 8.2 Example spreadsheet layout and guesses for estimating number of samples for desired confidence interval width. The figure shows the formula required for estimating the number of samples n for a desired two-sided 95% confidence interval, and the effect of successive guesses as described in Section 8.5.2. For different levels of confidence, the value 0.05 in the formula of cell C2 should be adjusted; for 99% confidence, the value is 0.01, and so on.

C2 would be 1.591 (too large). Guessing higher, for $n=8$, C2 displays 0.836 (smaller than necessary); checking to see if a lower number would suffice, at $n=7$ the value becomes 0.925. As above, seven replicates are necessary.

Alternatively, the iterative method can often be implemented using the optimisation, solving or 'goal seeking' facilities available in some spreadsheets and all good statistics software. In Figure 8.2, requesting a 'goal seek' tool to set the value of cell C2 to 1.0 by adjusting cell B2 leads to a value of 6.6 in cell B2. Rounding up gives the minimum number of observations as seven as above.

8.5.3 Number of Samples for a Desired *t*-Test Power

Test *power* is an important consideration when it is necessary to be confident of finding an effect of a particular size. In the example above, an observed bias of $0.925s$ would be (just) detected as significant at the 95% level with seven replicates. However, if the true bias were exactly $0.925s$, about half of the observed biases in a series of experiments would fall on each side of the critical value of 0.925 because of random variation and would be classified as not significant (recall that this is called a *Type II error* as it is the incorrect acceptance of the null hypothesis). At this value for the true bias – with the bias equal to the critical value – there is therefore only a 50% chance of detecting the bias. In statistical terms, the *power* of the test is only 50%. How can the analyst ensure a higher probability of detecting a particular value of bias? Specifically, how many observations are required for a particular (usually higher) test power?

Generally, the chances of the correct conclusion increase as the true effect under study increases. For example, it is much easier to detect a 50% recovery error than a 5% recovery error when the precision is constant at, say, 3% RSD. Hence the test power depends on the size of the effect under study. In Section 8.5.2, the number of samples took account of the required Type I error probability (by selecting the level of confidence), and this remains relevant. To decide the minimum number of observations for a particular test power, therefore, the analyst must know:

1. the type (*t*-test, *F*-test, *etc.*) and details (one- or two-tailed?; one-sample or two-sample?) of the proposed statistical test;
2. the size of the effect that is of interest;
3. the typical standard deviation s;
4. the required level of significance for the test (the Type I error probability α);
5. the desired test power, usually expressed in terms of the probability β of a Type II error.

The first four of these are known or have already been discussed. The required test power will invariably depend on the nature and scale of the risks associated with false negatives and a full discussion is well beyond the scope of this book. However, some pointers can be suggested.

- In investigations such as preliminary method development studies, it is rarely possible to set a criterion for the size of an effect which would be considered important. Under these conditions, test power cannot be calculated. It is, however, usual to set the number of observations such that the uncertainty in estimated effects is less than about one-third of the standard deviation of single results.
- A test power of 80% is commonly used in studies where variation is very large and a false negative is not especially critical. An example might be a trial of a new product, where a weak effect would not lead to a marketable product in any case, but detecting a positive effect is important. Method validation criteria for routine methods might also fall into this category.
- A power of 95% or higher should be used where a false negative is considered particularly important. For example, the rationale for detection limits in environmental work is based on an

implicit assumption of 95% power at 95% confidence, that is, the Type I and Type II error probabilities are set equal. (Detection capability is considered further in Chapter 9, Section 9.5.1.)

Given all this information, there are several methods of estimating n. As with the confidence interval, however, all involve tables or software.

1. *Using tabulated values*

    Table 8.2 gives the number of observations required for a 95% probability of detecting a particular size of bias δ using a t-test at the 95% level of confidence. To use the table, divide the bias δ by an estimate of the likely standard deviation s and look up the ratio in the top row of the table. The required number of observations is given in the second row.

2. *Iteration using a spreadsheet*

    For a one-sample t-test (that is, a t-test for a difference from a reference or hypothesised true value) at a level of confidence $1 - \alpha$, the power $(1 - \beta)$ of the test to detect a true bias δ is approximately:

$$1 - \beta = 1 - T\left(\frac{\delta\sqrt{n}}{s} - t_{\alpha/2, n-1}, n - 1\right) \quad (8.3)$$

where $T(x,v)$ is the cumulative probability of the Student's t-distribution for a value x and v degrees of freedom and $t_{\alpha/2, n-1}$ is the relevant quantile of the t-distribution for a one-tailed probability $\alpha/2$. For example, $t_{\alpha/2, n-1}$ for 95% confidence and large degrees of freedom is approximately 1.96. This approximation is very good when $s/\sqrt{n}$ is smaller than or equal to δ. In MS Excel and OpenOffice spreadsheet software, $T(x,v)$ is implemented using the TDIST function, which takes a positive value x, number of degrees of freedom $n - 1$ and number of tails; t is implemented by the TINV function which returns the quantile for a two-tailed probability. Hence equation (8.3) would become, for 95% confidence:

$$= 1 - \text{TDIST}(\delta * \text{SQRT}(n)/s - \text{TINV}(0.05, n - 1), n - 1, 1) \quad (8.4)$$

(note that the spreadsheet function TINV provides two-tailed values by default, hence the figure 0.05 and not $\alpha/2 = 0.025$). In the above, δ, n and s would be replaced by appropriate cell references. With equation (8.4) inserted in a spreadsheet, adjusting n until the desired power is achieved is straightforward.

Power for a t-test can also be calculated from the non-central t-distribution. This is not generally available in spreadsheet software. Implemented in statistical software, it usually takes the same parameters as t, with the addition of a *non-centrality parameter*, which should be set to $\delta\sqrt{n}/s$ to obtain the probability β.

3. *Using power calculation software*

    Many statistics software suites include power calculations for a variety of tests and experimental designs. Most can return the required number of observations when provided

Table 8.2 Number of observations for 95% power at 95% confidence.[a]

δ/s	0.5	0.6	0.7	0.8	0.9	1.0	1.5	2.0	2.5	3.0
n	55	39	29	23	19	16	9	6	5	4

[a] Based on NIST Special Publication 829, *Use of NIST Standard Reference Materials for Decisions on Performance of Analytical Chemical Methods and Laboratories*.

with the required power, confidence level for the test, the size of the effect and the estimated standard deviation. The implementation is, however, specific to the software. An example is included in Section 8.5.4 for illustration.

8.5.4 Number of Observations for Other Applications and Tests

Similar general principles apply for applications other than uncertainties and tests on mean values. For a minimum number of observations for a given uncertainty in a parameter, n is calculated using the expression for the variance (or standard deviation) of the parameter. For confidence intervals, adjusting n to provide the desired confidence interval based on the appropriate calculation is effective. For power calculations, recourse to specialist software is almost always necessary.

Example

An environmental quality standard for soil testing requires that bias for determination of trace chromium should not be more than 10% of the certified concentration or known spiking level for a validation test material. It further requires that the precision should be no more than 5% (as a relative standard deviation). How many observations are required to ensure a 95% probability of detecting a 10% bias, using a *t*-test at the 95% level of confidence?

1. *Using tabulated values*

    The permitted bias δ is 10% and the precision s is 5%, so $\delta/s=2.0$. Referring to Table 8.2, which gives n for 95% power, a value of δ/s of 2.0 requires a minimum of six observations.

2. *Using power calculation software*

    The free statistics software R (see Reference 2) provides power calculations for *t* tests, via a function power.t.test( ) which takes four of the five parameters (n, power, δ, s and significance level α) and calculates the missing parameter. This example has $\delta = 10$, $s=5$, $\alpha=0.05$ and power=0.95. With these parameters and the additional information about the test type and number of tails, the following output is generated:

    ```
    >   power.t.test(sig.level=0.05, power=0.95, sd=5, delta=10,
        type='one.sample', alternative='two.sided')
    One-sample t test power calculation
    n = 5.544223
    delta = 10
    sd = 5
    sig.level = 0.05
    power = 0.95
    alternative = two.sided
    ```

The result shows a minimum number of $n = 5.54$ observations, which, rounded up to the nearest integer, is six, as found from Table 8.2.

8.6 Controlling Nuisance Effects

Most practical experimental results are subject to 'nuisance factors' – unwanted effects that influence the results and cannot easily be removed. In this section, we describe the main

Designing Effective Experiments

experimental design methods for reducing the impact of these nuisance effects (also sometimes called *uncontrolled* factors because they cannot be removed or stabilised by experimental means).

8.6.1 Randomisation

8.6.1.1 The Advantage of Randomisation

Randomisation is perhaps the most important tool for reducing nuisance effects. It can reduce bias caused by both within-run and between-run effects. An illustration will explain how.

Consider an experiment to assess whether two digestion methods are providing the same elemental analysis result. We label these methods A and B. With due attention to the required power of the test, the experimenter decides to analyse a single reference material (chiefly for its known homogeneity) using six replicates of each digestion method. The intended analysis is a simple *t*-test (as in Chapter 4, Section 4.2.3.2) for significant between-digestion effects.

Ideally, the concentrations in each of the 12 digests would be determined in a single analytical run. A tidy-looking systematic ordering for analysis of the extracts is

A A A A A A B B B B B B,

making it very simple to record observations. However, practical experience shows that instruments tend to drift steadily. Clearly, if the instrument drifts steadily by 10% – not unusual in trace analysis – results near the beginning can be up to 10% higher (or lower) than results near the end. Placing the digests systematically as above would risk creating an entirely spurious difference between the mean result for treatment A and the mean result for treatment B. Figure 8.3a shows the outcome for a marked trend with otherwise good precision and no true difference in digestion methods. The result is, as can be seen from the means and confidence intervals, an apparent strongly significant difference between methods.

If the instrument drift were known to be smooth and linear, systematic alternation, reversed at the midpoint, would lead to the same change in mean value for both treatments. In practice, instrument drift and other uncontrolled effects are rarely so obliging; uncontrolled effects need not be monotonic and may change more or less randomly, making almost any systematic ordering unsafe.

The solution is to randomise the run order. This has two effects. First, if the run order is randomised, the A and B samples all have the same chance of encountering every level of drift, so *in the long term* the difference due to drift will disappear. With a finite number of observations n, the effect of randomisation is to reduce the effect by a factor proportional to $1/\sqrt{n}$. In our particular example, assuming a linear drift of maximum size D and with six observations in each of two groups, the standard deviation of the mean drift effect over many experiments would be about $0.17D$. Although not perfect, this is much better than the $0.5D$ expected for the naïve systematic ordering.

The second effect is on the standard deviation for each treatment. Because the observations are spread over the full range of the drift, the standard deviation for each is expected to increase. This reduces the test power. However, it has the advantage that the *t*-test will be less likely to show an erroneously significant difference. Although the power of the test is poorer, the difference in means is correctly compared with the full standard deviation including drift effects and not a smaller standard deviation that fails to take account of the entire effect of the drift.

Both effects are clearly visible in Figure 8.3b. The spurious difference in means has almost disappeared and the confidence intervals are substantially larger than in Figure 8.3a. The difference in mean values is (correctly) no longer significant. Randomisation therefore leads to a reduction in the nuisance effect and a more realistic estimate of the standard deviation for the experiment.

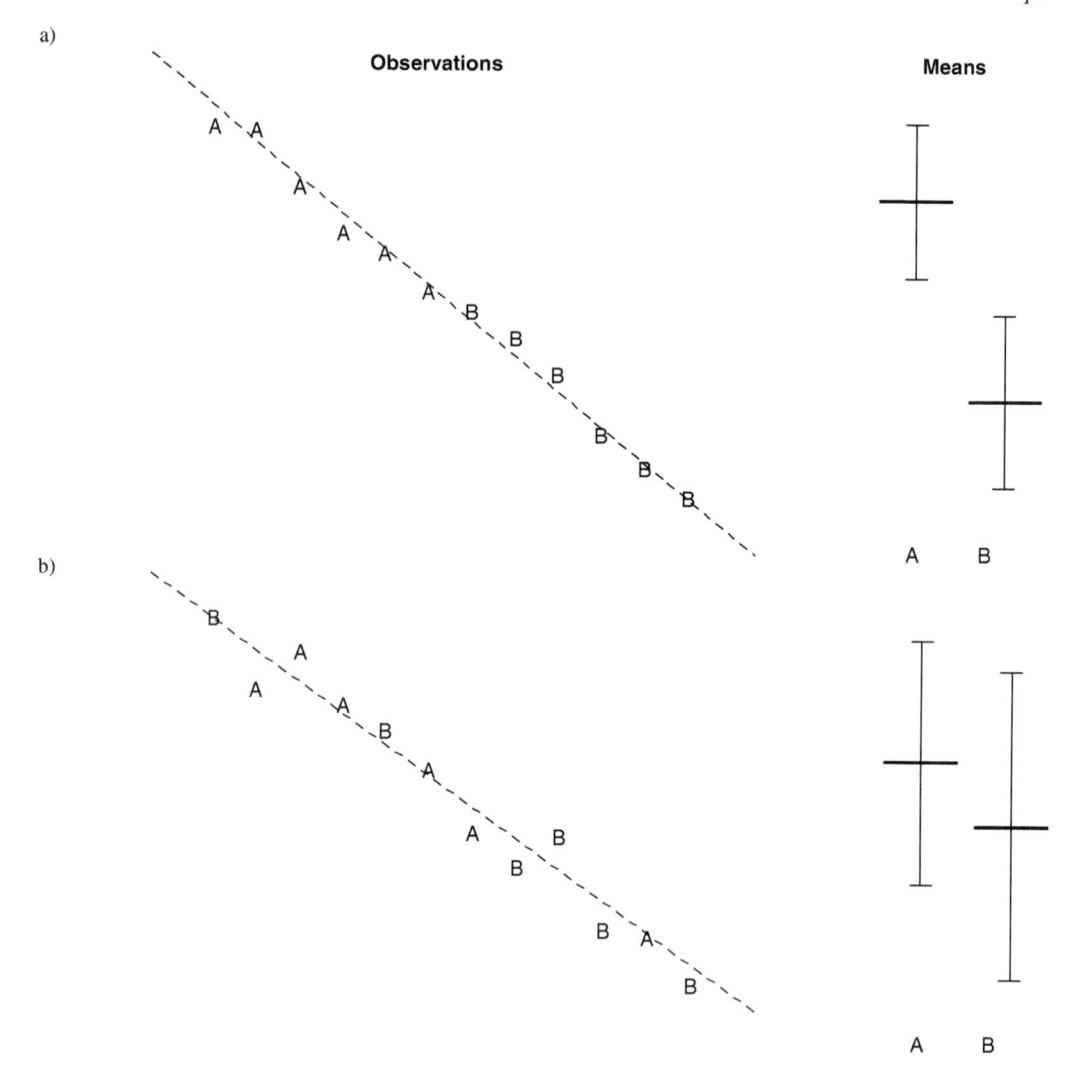

Figure 8.3 Effect of trends and randomisation. The figure shows the effect of a trend (dashed line) during a run with (a) ordered data and (b) randomised data. The observations are shown in run order. Means and 95% confidence intervals are shown to the right of each plot. The data were generated from the trend line $y = 11.0 - 0.1i$, i being the index of the observation, with added random error from a normal distribution with standard deviation 0.05. Units are arbitrary.

8.6.1.2 Implementing Randomisation

Randomisation depends on a source of random numbers. Random numbers can be generated in a number of different ways:

1. *Using random number tables*

    Some textbooks include random number tables. To use such tables, start at any point in the table and read out numbers from that point onward. Allocate the random numbers to the planned list of observations and on completion, re-order the observations in random number order.

Designing Effective Experiments

2. *Using spreadsheet software*

Most spreadsheets include random number generators. In OpenOffice and Excel, the relevant function is RAND( ). These are usually adequate for small experiments, though they do not perform as well as the best random number generators found in good statistical software. Sorting the planned observation list in the order of an adjacent column of random numbers is a simple randomisation method.

Combining RAND( ) with another commonly provided function, RANK( ), provides another convenient way of generating experimental run order. To use them together:

(i) Create a columnar list of the planned observations.
(ii) Next to that list, place a column of random numbers using RAND( ).
(iii) In the next column, place a list of calls to RANK( ), requesting the rank of the random number on that row among the complete list of random numbers.
(iv) Since these are 'live' formulae, it is usually sensible to copy either the random numbers, the ranks or both and paste the *values* back into place, overwriting the formulae, to prevent the order changing on reloading the spreadsheet or on entering other data.
(v) The process is illustrated in Figure 8.4.

a) Formulae

	A	B	C
	Observation	Random number	Run order
2	A(1)	=RAND()	=RANK(B2,$B2:$B13)
3	A(2)	=RAND()	=RANK(B3,$B2:$B13)
12	B(5)	=RAND()	=RANK(B12,$B2:$B13)
13	B(6)	=RAND()	=RANK(B13,$B2:$B13)
14			

b) Results

	A	B	C
	Observation	Random number	Run order
2	A(1)	0.211094	11
3	A(2)	0.364708	9
12	B(5)	0.717843	4
13	B(6)	0.73241	3
14			

Figure 8.4 Implementing randomisation using spreadsheets. The figure illustrates randomisation for 12 observations, six of treatment A and six of treatment B, initially numbered A(1)–A(6) and B(1)–B(6), respectively. Part (a) shows the formulae required; part (b) shows typical results.

3. *Using statistical or experimental design software*
All good statistics software includes a range of functions for generating random numbers, ranking and ordering values and often for generating randomly ordered samples from existing lists. As usual, the implementation depends on the software. A simple example from R is
> sample(X)
which generates a randomly ordered list of the sample numbers stored in variable X.
Experimental design software also invariably includes options for randomising run order.

8.6.1.3 Limitations of Randomisation

Randomisation is a useful general strategy which is expected to reduce uncontrolled effects by a factor proportional to $1/\sqrt{n}$. Use of randomisation is very strongly recommended in essentially all cases to reduce the impact of unknown factors.

Randomisation does, however, suffer from some shortcomings. First, the reduction in the uncontrolled effect is modest and depends on the number of replicates. Second, randomisation is, by its nature, still capable of producing accidental confounding. To see how, consider a new experiment with three treatments A, B and C – perhaps three different sample types treated by the best of the digestion methods studied previously. Three replicates of each are proposed. Clearly, it is not wise to place all three replicates of each treatment in a single run, as in Figure 8.5a. The run effect (shown by the horizontal lines under 'Runs') is confounded with the treatments and leads to a spurious between-treatment effect; notice that the mean for treatment C is high and, because of good within-run precision, the small confidence intervals strongly suggest a between-treatment effect on comparing means. Randomisation is better (Figure 8.5b); the run effect is partly averaged and again the run effect increases the spread of values within each treatment. Even so, the random ordering has resulted in two A samples analysed in the second run and two B samples in the third, so the A and B means are noticeably biased by the run effects. In a particularly bad case, the effect might be almost as serious as a bad systematic choice. This does not make randomisation a bad strategy; it is certainly among the good strategies for minimising uncontrolled effects and remains important for other reasons. However, if more is known about the uncontrolled effects, even more effective strategies exist. These are discussed in the following sections.

8.6.2 Pairing

In Chapter 4, Section 4.2.3.3, the idea of a paired comparison was introduced. In the example, two methods were applied to each of eight different test samples. In terms of experimental design, the (true) concentration in each sample was not of interest; it was a nuisance effect. The paired test eliminated the nuisance effect by considering only the between-method differences for each different sample.

Pairing can be used in other circumstances for comparisons between two groups. For example, in the extraction example in Section 8.6.1, it might be convenient to include a small number of the analyses in each of several routine sample runs. Any between-run effect can be eliminated by including one of each treatment pair in each of several runs (in random order within the run, of course!) and carrying out a paired test.

Pairing is, however, rather limiting; it is not usually convenient to carry out observations two at a time to remove run effects. Fortunately, pairing is only a special case of a more powerful and general method, *blocking*.

Designing Effective Experiments

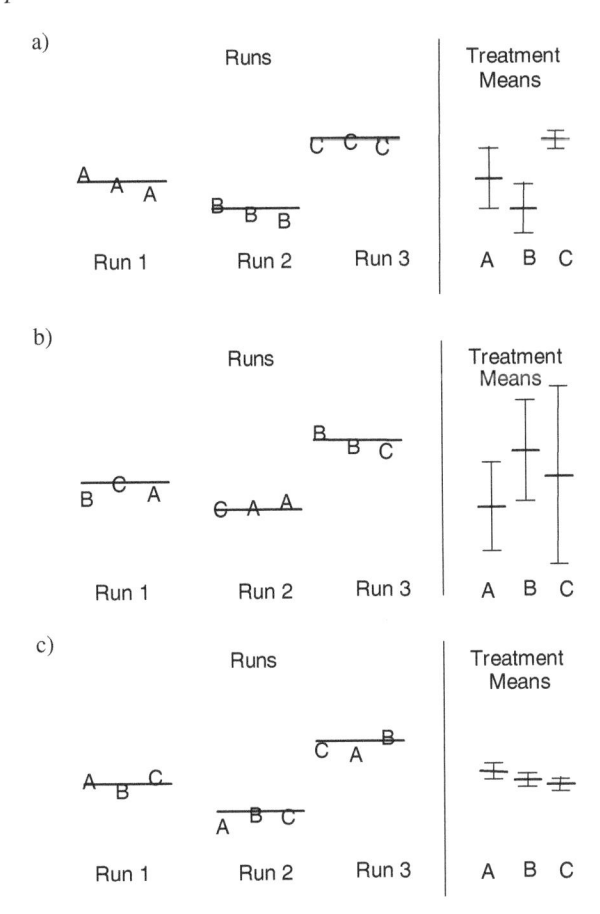

Figure 8.5 Blocked experiments. The figure shows the effect of three strategies for allocating observations to runs: (a) a poor systematic choice of allocation to runs; (b) random allocation to runs; (c) a randomised block design. Individual observations are shown with the underlying run effects (horizontal lines) on the left; the resulting treatment means and 95% confidence intervals are shown on the right. For (c) the confidence intervals are based on the residual error, as is appropriate for comparisons between the treatment means. The data were generated artificially using run effects drawn from a normal distribution with standard deviation of 0.3 and within-run errors drawn from a normal distribution with standard deviation of 0.05.

8.6.3 Blocked Designs

8.6.3.1 Introduction to Blocking

A *block* is a group of test materials subjected to a single level of some nuisance effect or effects. For example, run effects are probably the most common effect addressed by blocking. Returning to the extraction example above, we saw that randomisation partially removed between-run effects and that pairing could in principle remove them entirely if both samples from a pair were analysed in the same run. The run effect disappears because each comparison of interest is made within the same run. An obvious question is, 'If including one treatment comparison in each run eliminates the run effect entirely, why not include more than one comparison in each run?'. That is the principle of blocking. If every treatment comparison occurs the same number of times in each of several runs, the comparisons between treatment effects are unaffected by between-run effects.

The same statement applies to any other nuisance effect; if the experiment is divided into blocks in which every effect of interest appears the same number of times and the nuisance effect only once, the nuisance effect can be eliminated entirely.

For the simple three-level example above, then, we can improve on the randomised layout by placing equal numbers of A, B and C samples in each run. To reduce any residual within-run effects, we randomise the run order within each run, leading to the simple *randomised block design*:

Run		
1	2	3
A B C	A B C	C A B

This is shown in Figure 8.5c. Note that one of the most important effects is on the comparison among the mean results for the samples; the averages of three observations for each sample will *all* 'see' exactly the average of the three run effects, so differences between A, B and C will be unaffected by the run effects. That is, for the three A samples,

$$A_i = A_0 + \delta_i + e_{Ai} \tag{8.5}$$

where A_i is the observed value for sample A in run i, A_0 is the true mean for sample A, δ_i the run effect for the same run and e_{Ai} is the random error affecting sample A in run i. The average $\bar{A}$ of the three A observations is then

$$\bar{A} = [(A_0 + \delta_1 + e_{A1}) + (A_0 + \delta_2 + e_{A2}) + (A_0 + \delta_3 + e_{A3})]/3 = A_0 + \bar{\delta} + \bar{e}_A \tag{8.6}$$

where $\bar{\delta}$ is the mean of the run effects. Similarly, $\bar{B} = B_0 + \bar{\delta} + \bar{e}_B$, and so on. If we then compare the means for, say, groups A and B, by taking the difference in the means, we find that the estimated mean difference is

$$\bar{A} - \bar{B} = (A_0 + \bar{\delta} + \bar{e}_A) - (B_0 + \bar{\delta} + \bar{e}_B) = (A_0 - B_0) + (\bar{e}_A - \bar{e}_B) \tag{8.7}$$

The estimated difference, $(A_0-B_0)+(\bar{e}_A-\bar{e}_B)$, does not include any contribution from the run effects. There is one caveat: the calculation in equation (8.5) assumes that the effects are simply additive. This is often a good approximation, but clearly is unlikely to be exactly true in all cases; the block effects cancel exactly only if the effects are strictly additive. Aside from this caveat, however, the randomised block design has made it possible to compare the treatments with no adverse effects from the run-to-run variation at all.

8.6.3.2 *Analysing Blocked Experiments*

Data from a blocked experiment involving one treatment and one 'nuisance' factor can be analysed using two-factor analysis of variance (Chapter 6, Section 6.4), with the effect of interest as one factor and the blocking ('nuisance') factor as the second factor. In the extraction example in Section 8.6.1.3 and onwards, the treatment factor is the sample type (with levels A–C) and run is the second factor. If treatments appear only once per run (likely in a comparison of recoveries across several sample matrices, for example), two-factor ANOVA without replication is used. If each treatment occurs more than once in each run, two-factor ANOVA with replication is possible and valid (Chapter 6, Section 6.5.2). Analysis for 'main effects' only (that is, two-factor ANOVA with replications but with no interaction term) is also valid if the interaction term is not significant and is possible using specialist statistical software.

Note that simple analytical run effects (as opposed to real changes in conditions between blocks) are not expected to influence treatment effects, so a significant run–treatment interaction term is often attributable to an analytical problem with a specific sample or group of samples in a particular run. Close inspection is recommended before concluding that there is a real run–treatment interaction.

For two-factor experiments with one additional blocking effect, manual subtraction of the block effect and subsequent adjustment of the residual degrees of freedom allow two-factor ANOVA to be used if that is the only approach available. The Appendix to this chapter illustrates the principle. Manual methods are available for particular designs; for example, Box, Hunter and Hunter provide methods of analysis for most of the designs presented.[3] In general, however, the analyst is best advised to use specialist experimental design or statistics software to process data from any but the simplest designs.

Once appropriate software is available, the principle is relatively simple. Although the method of computation may be more complex than the manual methods in this book, the basic tool is ANOVA. The actual treatment of nuisance effects depends on whether the blocking strategy involved deliberate confounding of higher order terms (see Section 8.7.5.2).

Where the blocking strategy does not involve confounding, the statistical analysis can include all effects of interest and should additionally include the nuisance effect. Comparisons of the important effects with the residual error terms (or relevant interaction terms if appropriate) can then proceed as usual. The separated nuisance effect may be estimated or ignored. A simple example is shown in the Appendix to this chapter, which shows how a simple two-factor ANOVA without replication can estimate both an undesirable run effect and test for a between-unit effect of interest in an animal feed example.

Where the blocking strategy uses confounding, the statistical analysis will not generally provide separate estimates and tests for both nuisance effect and confounded effect, although it is sometimes possible to estimate one or other with reduced precision if confounding is not complete. In general, however, the confounded interaction term will be left out of the analysis and only the remaining effects studied.

8.6.4 Latin Square and Related Designs

Blocking is effective when there is one nuisance factor. But what if there is more than one? For example, what if operator and time (like day-to-day variation) both affect the experiment, but are essentially nuisance factors? Or what if a microbiological culture plate suffers from a systematic nutrient gradient in an unknown direction across the plate, so that we need to allow for both vertical and horizontal gradients?

Blocking is still possible; including every treatment in every combination of nuisance factors will work, but it is not very economical: we may be running far more replicates than necessary if there are many nuisance factor combinations. In the microbiological example, of course, we are probably restricted to only one observation at a given plate location, so blocking is not available. Are there designs that properly balance out the nuisance effects while keeping the number of observations manageable?

A traditional class of design for this situation is a *Latin square* (Leonhard Euler studied the mathematical properties of Latin and Graeco-Latin squares in the late 18th century and they saw early application in experimental design). A simple example would include three treatments and two nuisance factors. The Latin square is a simple method of arranging treatments so that every treatment 'sees' all levels of each nuisance effect. The name follows from the layout, which is traditionally drawn as a square table with the number of rows and columns chosen to be equal to the number of treatments. Figure 8.6 shows two possible arrangements of treatments A–C among nuisance effects with three levels each. The basic layout is formed by simply shifting successive rows to the right in the first table and to the left in the second. The resulting set of combinations allows

Nuisance Effect 2 \ Nuisance Effect 1	1	2	3
1	A	B	C
2	C	A	B
3	B	C	A

Nuisance Effect 2 \ Nuisance Effect 1	1·	2	3
1	a	b	c
2	b	c	a
3	c	a	b

Figure 8.6 Two Latin squares for $n = 3$.

Nuisance Effect 2 \ Nuisance Effect 1	1	2	3
1	Aa	Bb	Cc
2	Cb	Ac	Ba
3	Bc	Ca	Ab

Figure 8.7 A Graeco-Latin square.

cancellation of (additive) nuisance effects by analysing the design with three-factor ANOVA without interactions.

The concept can be extended to include another factor of interest or third nuisance factor. Combining the two parts of Figure 8.6 gives the layout in Figure 8.7, termed a *Graeco-Latin square* (named from the common practice of marking one Latin square with Roman characters and one with Greek, instead of upper- and lower-case as here).

The layout in the Figures is convenient for illustration, but would normally be tabulated for experimental instructions and data recording. The tabulated layout of factor combinations for the Graeco-Latin square in Figure 8.7 is shown in Table 8.3. This makes it clearer that although a spatial layout could be implied by the nuisance factor labelling in the figures, any kind of nuisance factor can be accommodated. It is also possible that one or other of the factors labelled 'Study effect' in the table could equally well be a third nuisance factor, giving an efficient design for eliminating three nuisance factors while studying a single experimental factor of interest.

Although not common, yet another factor of interest can be accommodated using so-called *Hyper-Graeco-Latin square* designs, which add a fifth factor. They are thoroughly described elsewhere (see, for example, references 1 and 3) and will not be described here.

Table 8.3 Tabulated layout for a Graeco-Latin square design.

	Factor levels			
Run	Nuisance 1	Nuisance 2	Study effect 1	Study effect 2
1	1	1	A	a
2	2	1	B	b
3	3	1	C	c
4	1	2	C	b
5	2	2	A	c
6	3	2	B	a
7	1	3	B	c
8	2	3	C	a
9	3	3	A	b

Latin square and related designs are simple and relatively economical. The chief disadvantage is that they are restricted to effects that share (or can be coerced to) the same number of levels. Another purely mathematical restriction is that some sizes (that is, some numbers of factor levels) are not available; for example, there are no Graeco-Latin squares of size $n = 2, 6, 10$, *etc.*

Finally, note that although the layouts and tabulation here are given in systematic order, it is just as important with Latin square designs as with any other to randomise the observation order after allocating the treatments. For example, the runs in Table 8.3 should be undertaken in random order as far as possible given the constraints on the experiment.

8.6.5 Validating Experimental Designs

A cautionary note is essential here. Randomisation is simple, fairly effective and universally applicable; it only rarely leads to significant misinterpretation. Blocking, however, can become complex for multi-factor and multi-level experiments and it is disturbingly easy to confound a nuisance effect accidentally with an effect of interest – even a 'main effect' – and thereby generate an experiment incapable of meeting its objectives. It is vital to check that blocked designs actually deliver the effects of interest. Several tests are possible; for example, generating and processing simulated data with and without added nuisance effects will quickly show which effects are confounded with nuisance effects. Box, Hunter and Hunter provide a simple check for confounding in multi-factor two-level designs[3] and some statistical software includes specific checks for confounding or 'aliasing'.

8.7 Advanced Experimental Designs

The following subsections provide a brief introduction to some additional classes of experimental design that improve on or supplement the basic designs described in Section 8.4. Unlike the designs in Section 8.4, data from most of these designs cannot be processed using the statistical methods in this book; most require appropriate software. These descriptions therefore provide only an outline of the general form of each design, its advantages and disadvantages and a brief indication of the type of software or methods used for processing the resulting data. More complete discussions can be found in a range of textbooks and Web-based resources, such as reference 4.

8.7.1 Fractional Factorial Designs

Factorial designs (Section 8.4.4) are excellent tools for discovering the significance of changes in experimental or other parameters and interactions between parameters, but they can become costly

Table 8.4 Factor levels for an example 2^3 factorial.

Treatment combination	Time t	Temperature T	Reagent c	$t \times T \times c$
1	+	+	+	+
2	−	+	+	−
3	+	−	+	−
4	−	−	+	+
5	+	+	−	−
6	−	+	−	+
7	+	−	−	+
8	−	−	−	−

for larger numbers of factors and levels. This disadvantage can be offset if the experimenter is willing to sacrifice some information or precision. This is usually achieved by leaving out factor combinations that are important only for estimating higher order interactions. The idea is simplest to describe by example.

A full three-factor, two-level design (a 2^3 factorial) to study the effects of time, temperature and a reagent concentration on, say, the efficiency of a digestion process, requires eight runs. The factor levels are listed in the second to fourth columns of Table 8.4, where '+' indicates a high level of the relevant factor and '−' a low level. The layout of a 2^3 factorial is also shown schematically in Figure 8.1e. Without further replication, ANOVA with appropriate software would give the significance of the three main effects and the three two-way interaction terms; with duplication to give 16 observations, the three-way interaction also becomes available. However, interactions are rarely as important as main effects and perhaps it would be wise to run a shorter experiment first to see which main effects are important (this is often termed 'screening').

The fifth column of the table shows the sign of the three-way interaction term. Simply leaving out all those combinations with (say) a negative three-way interaction term gives half the number of runs; the remaining runs are shown schematically as dark spots in Figure 8.8a (compare this with the 2^3 factorial in Figure 8.1e). However, it is still possible to calculate all the main effects. This is referred to as a 2^{3-1} fractional factorial, the indices indicating that we started with a 2^3 design and reduced it by half, that is, 2^{-1}.

The cost of this reduction in a three-factor experiment is that we can no longer estimate any interaction terms. Further, it can be shown that the reason for this is that the interaction terms are confounded with main effects. For example, the temperature term $(+,-,+,-)$ has the same sign as the product of the time and reagent levels, indicating that the temperature term is confounded with the time–reagent interaction. Since main effects are usually (but not always) larger than interaction effects, and in particular are unlikely to be completely cancelled out by the confounding, this is often a price worth paying. It is particularly worthwhile for highly fractional factorials such as the ruggedness testing design described in Chapter 9, Table 9.4, where a seven-factor, two-level design has been reduced from 128 to 8 runs. Table 9.4 is actually a 2^{7-4} fractional factorial.

Fractional factorial designs, like full factorial designs, require some replication in order to allow significance tests for the effects. For smaller designs, simply replicating the whole design is practical and retains balance. As designs increase in size, this may become uneconomical and it is common to introduce an additional, replicated, point (a 'centre point') at intermediate levels of ordered or continuous factors to provide an estimate of precision.

In general, fractional factorial designs confound some, but not necessarily all, interaction terms with main effects. The extent of confounding interaction terms retained in a design is often described in terms of design *resolution*. If the smallest order interaction that some main effect is confounded with is p, the resolution is usually $p+1$. In the 2^{3-1} design above, all the two-way effects are confounded with main effects, so the resolution is $2+1=3$ (usually written in Roman numerals, as in 'resolution III'). The term is also used by some statistical software to determine (or limit) the order of interaction terms included in ANOVA. Full factorial designs have no confounding and are said to have 'infinite' resolution, although the issue is largely academic.

It is rarely necessary to use resolutions higher than resolution V, and resolution IV designs may well be sufficient in many cases, so fractional factorial designs have wide application. Resolution III designs (which confound at least some two-way interactions with main effects) with all factors restricted to two levels can be particularly recommended as economical screening designs.

Fractional factorial designs can be analysed manually; for example, Chapter 9, Section 9.7.2, shows how the 2^{7-4} design can be assessed. Generally, however, it is most convenient to analyse the data using ANOVA or regression methods built into proper statistical or experimental design software.

8.7.2 Optimisation Designs

8.7.2.1 Optimisation and Response Surface Modelling

Method development and process improvement share a frequent need for optimisation, that is choosing a set of operating conditions that minimise uncertainty or maximise some measure of efficiency. This usually involves seeking a maximum or minimum response condition among a set of continuous process variables, such as times, temperatures and reagent concentrations. This can be assisted by fitting a mathematical equation through the observed responses and choosing the set of parameters that maximises the response. Because the mathematical equation often involves several parameters, the equation defines a surface, not just a line, so the resulting equation is a *response surface model*. Such models usually require quadratic or higher terms and fitting to quadratics always requires at least three levels of a factor. The following experimental designs provide data with at least three factor levels and are suitable for response surface modelling.

8.7.2.2 Star Design

A star design (Figure 8.8b) varies each factor individually about some central point. It is usually conducted as a three-level experiment for economy. The central point is usually replicated to provide an indication of precision and also because this provides the best precision for second-order terms in the resulting model (the second-order terms are important for defining curvature). It is a relatively ineffective design because it cannot estimate interaction terms. Its chief importance is that it is a component of another important general class, the so-called *central composite design*.

8.7.2.3 Central Composite Designs

A central composite design (Figure 8.8c) combines a star design with a factorial design to provide an experiment that effectively estimates main effects and multiple interaction terms simultaneously. The variant shown in Figure 8.8c is deliberately constructed for a property known as *rotatability*; note that the outer points fall on a circle (or, as in Figure 8.8e, a sphere for a three-factor design, and so on). Rotatable designs such as this have desirable variance properties. However, this variant uses five levels of each factor and the levels are not equally spaced, which can make experiments more costly. The face-centred variant in Figure 8.8d, which restricts the number of factor levels to three, can simplify the experimental procedure at the potential cost of losing some precision.

A potentially important difference between the face-centred and rotatable central composite design is that while the experiment can be run in blocks in either case, only the rotatable central composite with appropriately chosen levels for the 'star' points will allow complete removal of the nuisance effect.[4] If blocking is necessary, therefore, it is probably better to use the rotatable design.

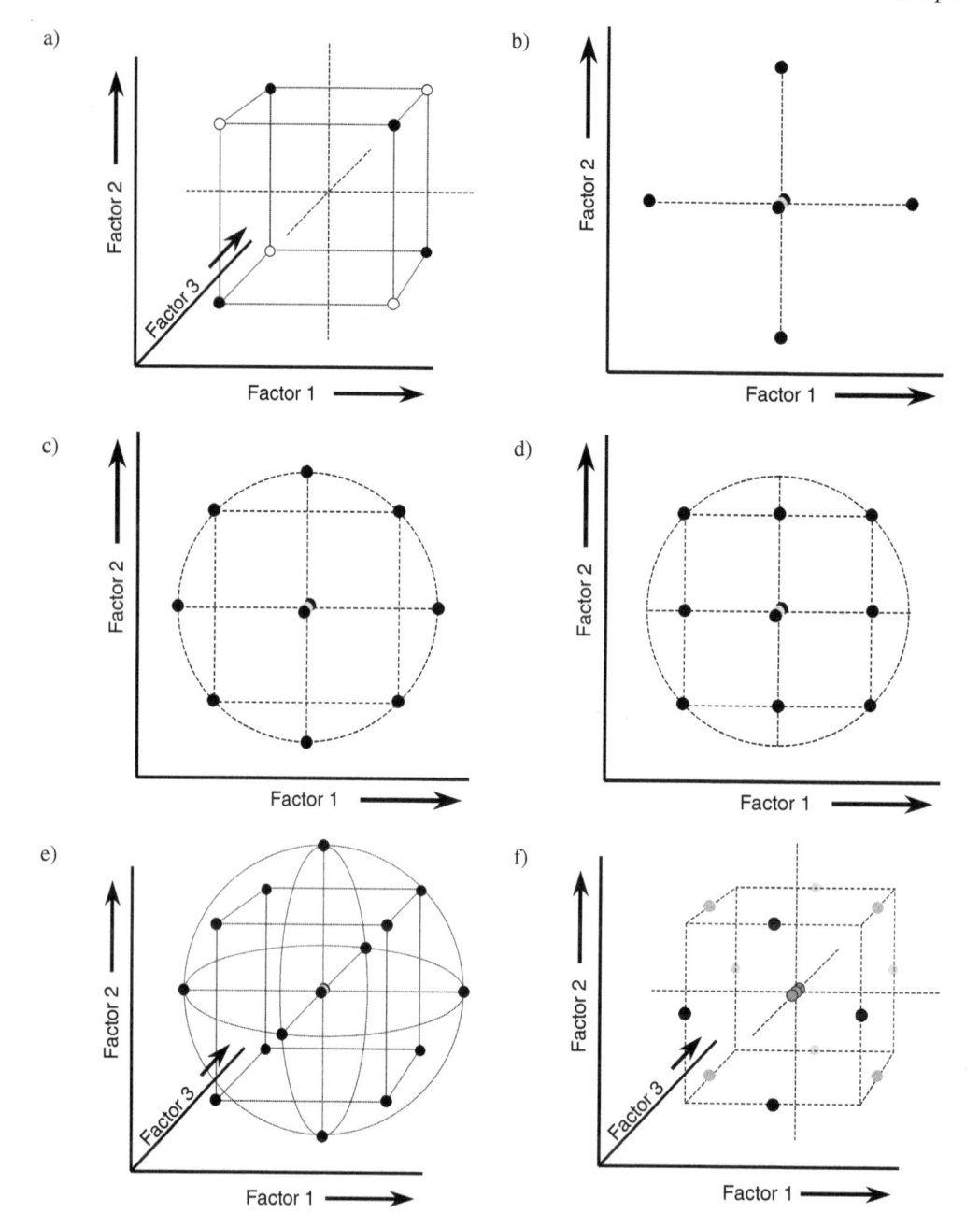

Figure 8.8 Further experimental designs. (a) Two 2^{3-1} fractional factorials (light and dark points, respectively); (b) star design; (c) central composite design; (d) face-centred central composite design; (e) three-factor central composite design; (f) three-factor Box–Behnken design.

8.7.2.4 Box–Behnken Design

An alternative class of optimisation design is shown in Figure 8.8f. This Box–Behnken design (named for its inventors) is a three-level design with design points at the centres of the *edges* of the

Designing Effective Experiments

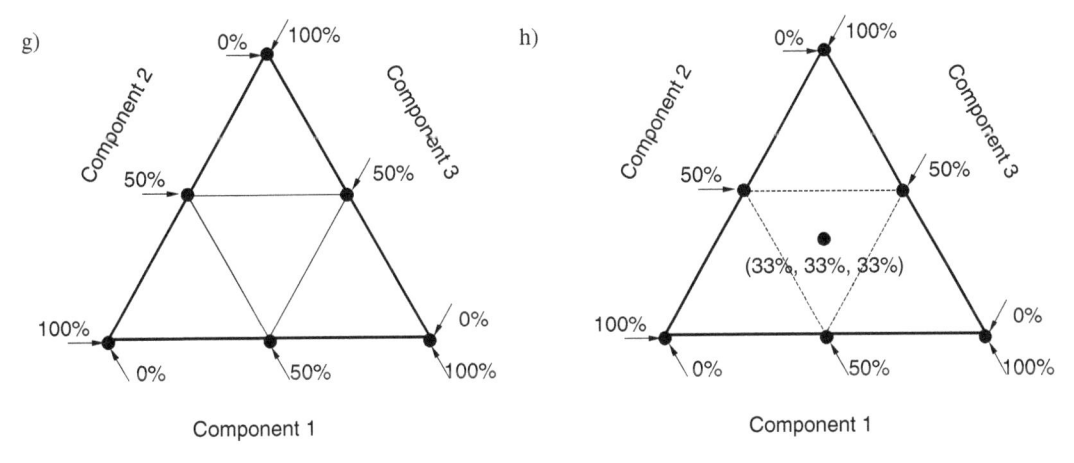

Figure 8.8 (Continued). (g), (h) two three-component mixture designs: (g) simplex-lattice and (h) simplex-centroid.

cubic design space; contrast this with the factorial and central composite designs which use corner and face centre points. The Box–Behnken design is a rotatable design with fewer factor levels than the central composite. It is particularly efficient for three-factor designs, when it requires fewer runs than the central composite.

8.7.2.5 Data Processing for Response Surface Designs

Response surfaces are defined by equations. The most common are quadratic; an example for a two-factor experiment might be

$$Y = b_0 + b_1 X_1 + b_2 X_2 + b_{12} X_1 X_2 \tag{8.8}$$

Equations such as this include a list of coefficients b_i, which are usually estimated using multiple regression software. Multiple regression is a standard feature of statistics and experimental design software, which also includes regression diagnostics. Multiple regression is also available in some spreadsheet packages (including MS Excel, via the Regression tool found in the Data Analysis add-in).

Once the model equation has been fitted, optimisation software is used to find the maximum (or minimum, as appropriate). General-purpose optimisation software is available as part of most statistical software and reasonably robust optimisation functions can be found in some business spreadsheets (for example, the 'solver' in MS Excel). However, response surface modelling benefits considerably from rapid visualisation tools, including contour mapping and response surface plotting, and purpose-designed software is strongly recommended.

8.7.3 Mixture Designs

Many industrial processes produce formulated products made from a mixture of a few components. In analytical science, derivatisation reagents and chromatographic liquid phases also involve mixtures. Mixtures have the property that the sum of the amounts present always equals 100%, introducing a constraint on experimental designs. For example, a three-component mixture cannot be subjected to a full factorial design which requires high and low levels of even, say, 30% and 70% for all components, because all but one of the required factor combinations sums to over 100%.

This kind of problem is treated using a class of designs called *mixture designs*. There are two well-known classes of mixture design: the so-called *simplex-lattice* and *simplex-centroid* designs. These are illustrated for three-component mixtures in Figure 8.8g and h, respectively. The principal difference is that the centroid design includes points at the mixture centroid, whereas the basic lattice design does not.

Mixture designs require special processing because of the constraint that the sum of the components is equal to 1 (or 100%). This effectively removes one parameter from the usual polynomial model. A typical mixture model equation would be, for a three-component mixture:

$$Y = b_1 X_1 + b_2 X_2 + b_3 X_3 + b_{12} X_1 X_2 + b_{13} X_1 X_3 + b_{23} X_2 X_3 \tag{8.9}$$

Note the lack of the separate intercept term b_0 found in most polynomials.

As before, data processing involves multiple regression. Although possible in most statistics and some spreadsheet software, it is worth noting that response surface plots for mixture designs are typically triangular (as in Figure 8.8g and h), and it is important to ensure that the chosen packages provide the relevant features. Again, purpose-designed software has considerable advantages.

8.7.4 D-optimal Designs

It is not unusual to find that the neat, systematic allocation of factor levels implied by the designs described so far is simply not possible. For example, the 'star' points in a central composite design might require reaction temperatures above a solvent's boiling point. One way of accommodating this problem would be to reset all the temperatures so that the maximum was at the reflux temperature, but that may result in most of the experimental observations being far from the nominal conditions under study. Another common problem is in allocating observations to experimental runs; it is simply not always possible to find blocking strategies that make good use of the available equipment and satisfy the basic experimental design.

One reasonably good compromise in such situations is to choose the factor combinations from the available ranges in such a way as to maximise the information (or minimise the uncertainties) in the experiment. Among the most common criteria for this job is the so-called *D-optimal* criterion. This uses a mathematical measure of the information content and chooses combinations of possible experimental runs that maximise the criterion. For the mathematically inclined, the criterion is the determinant of the Fisher information matrix $(\mathbf{X}^T\mathbf{X})$, where $\mathbf{X}$ is the 'design matrix' for the experiment. Both the calculation and the method of choice usually require software which adapts to particular constraints set by the user, so general D-optimal designs are very much the province of specialist software.

Given appropriate software, D-optimal designs represent a useful compromise when experimental constraints prevent adoption of the symmetric designs already discussed above. They are guaranteed to be feasible, but will not generally provide as good a precision as might be available without constraints. Another feature of such optimal choices is that they are not usually unique; several different subsets of possible experimental runs may provide the same value for the criterion and the software will generally choose one of these randomly (or at least based on its own natural ordering). The result is that successive software runs or different software may lead to different experiments for the same conditions. Since all are optimal within the criteria chosen, this raises no substantial issues for analysis; but it can be puzzling for the experimenter using D-optimal designs for the first time.

Data from D-optimal designs are, as for most of the optimisation designs above, processed using multiple regression methods, again requiring specialist software.

Designing Effective Experiments 139

8.7.5 Advanced Blocking Strategies

8.7.5.1 Simple Blocking Extended to Multi-factor Experiments

The simple example in Section 8.6.3 shows one basic blocking strategy; ensure that every level of an effect of interest is included in every block and allow replication to extend over several blocks. Because all effects (the *complete* set) are included the same number of times (for *balance*) in every block, this is called a *balanced complete block design*. With this type of design, it is possible to carry out the comparisons of interest with no influence at all from the blocking factor. This strategy works well for a single factor with any number of levels that can fit conveniently in a single run. Perhaps surprisingly, even if only one instance of every level can be included in each run, it is still possible to carry out the comparison and also possible to determine the analytical within-run precision (see below).

The same strategy can be extended to multi-factor experiments. For example, we may want to study the effects of small changes in derivatisation time and temperature on a chromatographic determination and wish to eliminate interference from between-run effects in the chromatographic analysis. In such a case, we would usually be interested in any time–temperature interaction and would analyse a particular test material at all combinations of high and low temperature and long and short derivatisation time (this is a *full factorial design*, albeit a small one). Because we are interested in the interaction term, replication is essential; perhaps three replicates for each treatment combination might be chosen. If all the experiments cannot be performed within the same chromatographic run, what is the best allocation of samples to runs?

The basic experiment involves four treatment combinations:

Treatment combination	Time	Temperature
ab	Short	Low
Ab	Long	Low
aB	Short	High
AB	Long	High

Following the strategy of placing every treatment combination in every run and replicating across runs, the simplest design includes each of the four combinations in each of three runs, with the run order randomised:

Run	Treatments (in run order)
1	ab, AB, aB, Ab
2	ab, Ab, AB, aB
3	AB, ab, Ab, aB

Using appropriate statistical software, this experiment will provide tests of the effects of time and temperature and of their interaction term, with no interference from the run effect.

8.7.5.2 Confounding with Higher-order Interaction

Often, particularly for multi-factor experiments, it is impossible to include every combination of treatments in every run. Two important strategies then become important. The first is to confound the blocking factor with an effect that is of comparatively little interest. A relatively simple and realistic example is a three-factor, two-level factorial design; for example, the time–temperature experiment above might additionally include a derivatisation reagent concentration to give the full factorial design listed in Table 8.4.

The statistical analysis would normally be a three-factor ANOVA, using statistical software. With no blocking or run effects, a replicated experiment (using, say, duplicate observations at each factor level) would provide tests on three main effects, three two-way interaction terms (time–temperature, time–reagent and temperature–reagent) and a three-way interaction term. Higher-order interactions are often of only moderate interest and we might be prepared to sacrifice this three-way term in the interests of more sensitive tests for the main effects. The final column in the table represents the sign of this three-way interaction term, formed from the product of the three previous columns. Simply assigning all the combinations with a positive three-way interaction term to one run and all those with a negative three-way interaction to another run generates a blocked design with four observations per run and with the three-way interaction confounded with the run effect. Statistical analysis of such an experiment with added replication would allow the estimation of all the main effects and all the two-way interactions without interference from the run effect, at the expense of being unable to test for a three-way interaction independently of run effects.

8.7.5.3 Incomplete Block Designs

A second useful strategy when runs cannot include all levels of an effect in a multi-level experiment is to ensure that, if each run can include m levels of n, then all possible groups of m are represented an equal number of times among the runs. This *combinatoric design* is among a family of designs called *balanced incomplete block* designs ('incomplete' because not all treatments are included in every block). For example, for a three-level experiment, in which some factor takes one of three levels A, B and C but each run can study only two levels at a time, a simple balanced incomplete block design contains three runs, with factor levels (A, B), (A, C) and (B, C), respectively. Replication within each run is permitted, as is duplicating the entire set of runs if more replication is desired. Again, appropriate statistical analysis will generate an accurate test for differences between the levels A, B and C with no interference from the run effect. Balanced incomplete block designs include the combinatoric designs in which all combinations of m levels are represented, and also some considerably more economical designs applicable for particular numbers of factors and levels; various textbooks provide extensive detail.[3]

Appendix: Calculations for a Simple Blocked Experiment

It is always best to use the right statistical software to analyse more complex experimental designs. However, this is not always possible. Where such software is not available, it is often possible to reduce the problem to a simpler case by removing the block effect manually and adjusting the calculated degrees of freedom and mean squares accordingly. The procedure is as follows (assuming each run is one block):

1. Calculate the run mean for each run.
2. Subtract the run mean from all the observations in the run.
3. Carry out an ANOVA on the corrected data in the usual way to obtain the necessary sums of squares.
4. Correct the residual (or error) degrees of freedom by subtracting the number of degrees of freedom for the runs (number of runs − 1) and recalculate the corresponding residual (error) mean square from the new degrees of freedom.
5. Recalculate the F values and critical values using the new error mean square and degrees of freedom.

Designing Effective Experiments

The following example illustrates the procedure for a simple case which can also be analysed using two-factor ANOVA. The principle can, however, be applied to other simple experiments in which replication crosses runs. For example, a two-factor experiment in which one replicate for each factor combination is run in each of two experimental runs (a total of two replicates per treatment), can be reduced to a simple two-factor ANOVA with replication in exactly the same way.

Example

Six tins of animal food have been analysed to determine whether fibre content is consistent among the tins. Three replicates per tin were desired. Because the analytical equipment could handle only six samples per run, subsamples from each tin were run once in each run. This is a simple blocked design with replication across blocks. The results for fibre are given in Table 8.5. These results can be analysed perfectly well using two-factor ANOVA without replication, but to illustrate the procedure above, we will first use one-way analysis with the correction above.

Table 8.6 shows the simple one-way ANOVA table, ignoring the runs for the moment and treating all results for the same tin as simple replicates. Note that there is apparently no significant difference between groups.

The experiment actually includes three separate runs; we now apply the procedure above to eliminate any run effects.

Steps 1 and 2 above calculate run means and subtract them from the data. The observations with the run means subtracted are given in Table 8.7.

Step 3 is to calculate the initial sums of squares and degrees of freedom for the run-corrected data, neglecting the run effect. This initial ANOVA table is shown as Table 8.8.

Table 8.5 Fibre in animal food[a].

			Tin				
Run	A	B	C	D	E	F	Run mean
Run 1	1.857	2.233	1.109	1.627	1.854	1.481	1.693
Run 2	1.773	1.884	1.138	1.733	2.154	1.698	1.730
Run 3	1.645	1.164	0.837	1.260	1.312	1.279	1.250

[a]Fibre is expressed as mass% throughout this example.

Table 8.6 ANOVA neglecting the run effect.

Source of variation	SS	df	MS	F	p-Value	F_{crit}
Between groups	1.242	5	0.248	2.389	0.1005	3.106
Within groups	1.248	12	0.104			
Total	2.490	17				

Table 8.7 Fibre observations with run effect subtracted.

			Tin			
Run	A	B	C	D	E	F
Run 1	0.163	0.540	−0.585	−0.066	0.161	−0.212
Run 2	0.043	0.154	−0.592	0.003	0.424	−0.032
Run 3	0.396	−0.085	−0.412	0.010	0.062	0.029

Table 8.8 Simple one-way ANOVA for the run-corrected data.

Source of variation	SS	df	MS	F	p-Value	F_{crit}
Between groups	1.242	5	0.248	7.665	0.0019	3.106
Within groups	0.389	12	0.032			
Total	1.631	17				

Table 8.9 ANOVA corrected for run effect and for run degrees of freedom[a].

Source of variation	SS	df	MS	F	p-Value	F_{crit}
Between groups	1.242	5	0.248	**6.388**	**0.0065**	**3.326**
Within groups	0.389	**10**	**0.039**			
Total	1.631	**15**				

[a]Figures in bold are affected by the degrees of freedom adjustment.

Table 8.10 Two-factor ANOVA for animal feed data.

Source of variation	SS	df	MS	F	p-Value	F_{crit}
Rows (run)	0.859	2	0.429	11.039	0.0029	4.103
Columns (tins)	1.242	5	0.248	6.388	0.0065	3.326
Error	0.389	10	0.039			
Total	2.490	17				

Step 4 corrects the residual degrees of freedom. In this experiment, there are three runs and therefore $3 - 1 = 2$ degrees of freedom for the runs. Subtracting this from the residual (within-groups) degrees of freedom and recalculating the new within-group mean square gives the figures in the second row of Table 8.9 (in which the final row has also been corrected to show the new total degrees of freedom).

Step 5 is the calculation of the revised value for F ($0.248/0.039 = 6.388$) and the new critical value and p-value; these are also given in Table 8.9.

Note that the between-groups term is now statistically significant at the 95% level. Why has the conclusion changed?

Comparing Table 8.9 with the naïve one-way table in Table 8.6, note that the between-groups sum of squares, degrees of freedom and mean square are unchanged in Table 8.9. The change in significance comes entirely from the corrected within-group mean square and degrees of freedom; neglecting the run effect has apparently inflated the within-group mean square – in this case by so much that it concealed a real between-tin effect.

To confirm that this general approach gives the same results as including the run effect in the analysis and to gain a little more insight into the reason for the changed conclusion, this particular example can also be run as a two-factor ANOVA without replication. The results are given in Table 8.10. Comparing with Table 8.9, note that the row for the effect of the tins is identical in both tables, as is the 'Error' row, confirming that the manual procedure gives results identical with those of the complete formal analysis. We can also see why the naïve one-way analysis in Table 8.6 gave an insignificant value; the run effect is clearly significant. The effect of this is to increase the apparent within-group sum of squares. The within-group sum of squares in Table 8.6 (1.248) includes the run effect and is exactly equal to the sum of the run and error sums of squares ($0.859 + 0.389 = 1.248$) in Table 8.10.

References

1. ISO 3534-3:1999, *Statistics – Vocabulary and Symbols – Part 3: Design of Experiments*, International Organization for Standardization, Geneva, 1999.
2. R Development Core Team, *R: A Language and Environment for Statistical Computing*. R Foundation for Statistical Computing, Vienna, ISBN 3-900051-07-0; URL http://www.R-project.org.
3. G. E. P. Box, W. G. Hunter and J. S. Hunter, *Statistics for Experimenters: an Introduction to Design, Data Analysis and Model Building*, Wiley, New York, 1978, ISBN 0-471-09315-7; see also the revised second edition, *Statistics for Experimenters: Design, Innovation and Discovery*, Wiley, New York, 2005, ISBN 0-471-71813-0.
4. NIST/SEMATECH, *e-Handbook of Statistical Methods*, http://www.itl.nist.gov/div898/handbook/ (accessed 22 April 2008).

CHAPTER 9
Validation and Method Performance

9.1 Introduction

Method validation is the process of evaluating the performance of a test method to determine whether it is capable of producing test results that are fit for a particular purpose. This chapter gives an overview of the key method performance parameters that are frequently studied in method validation and illustrates how some of the statistical techniques described in earlier chapters can be used. More detailed information on method validation can be found in references 1–5.

The main stages in obtaining a suitable method for a particular analytical requirement can be summarised as specification, selection/development of a method and validation. These are discussed in more detail below:

- *Specification of the measurement requirement:* Measurements are always made for a reason, generally to help answer a particular question. In addition to knowing the analyte(s) to be measured and the nature of the test samples, it is important to have a clear understanding about why the analysis needs to be carried out and what the data will be used for. This will enable criteria for the performance of the method to be set (for example, the acceptable uncertainty in the measurement results). Once the specification is complete, a candidate test method can be selected.
- *Selection of a test method:* In some cases, a published method which meets the measurement requirement may already be available. However, it remains necessary to carry out an evaluation of the performance of the method to demonstrate that its performance is acceptable when used in the laboratory for the analysis of test samples. If a published method is not available, then development of a new method, or adaptation of an existing method, will be required.
- *Method validation:* Validation is the process of demonstrating that the chosen test method is capable of producing results that are fit for purpose.

The validation process can be broken down into three stages:
- *Assessment of method performance:* This stage involves planning and carrying out a range of experiments to evaluate different aspects of method performance. The main performance parameters and their evaluation are discussed in the following sections. It may not be necessary to study every aspect of method performance. This will depend on the scope and intended use of the method and on its history (for example, whether there is any existing validation data available).

Validation and Method Performance

- *Comparison of method performance data with the measurement requirement:* The data on method performance obtained from the experimental studies are compared with the performance criteria set at the measurement specification stage. Statistical tests, such as *t*- and *F*-tests, can be used to assist with the comparison.
- *Decision on whether the method is fit for purpose:* If the performance of the method meets the requirements, then it can be considered to be valid and used for the analysis of test samples. If one or more parameters do not meet the method specification, further method development, followed by re-validation, will be required. It is important to remember that method validation is not complete until this decision on fitness for purpose has been made.

When developing a new method, it is often difficult to identify the point at which development stops and validation starts. During the development process, data will be generated which will start to give a picture of the method performance. However, data should be generated which confirm that the performance of the final version of the method is fit for purpose before the method is used for the analysis of test samples.

Table 9.1 shows the key method performance parameters that are studied during method validation. The amount of effort required to validate a method will depend on the criticality of the measurement, the scope of the method and the level of information and experience already available. Not all of the performance parameters described in this chapter will be relevant for every test method. Validating a new method for trace analysis would require a detailed study of all of the parameters highlighted in Table 9.1. However, if the intention is to implement a method for the measurement of a major component which had been previously validated by an interlaboratory study, a limited study of precision and bias to confirm that the method is being applied correctly may be all that is required. The process of checking the performance of a previously validated method against established method performance targets is sometimes referred to as *verification* rather than validation.

9.2 Assessing Precision

Precision is defined as 'The closeness of agreement between independent test/measurement results obtained under stipulated conditions'.[6] Precision is usually expressed as a standard deviation (or relative standard deviation) (see Chapter 4, Section 4.1.4) obtained from replicate measurements of an appropriate sample. The sample should be stable, sufficiently homogeneous and representative of test samples in terms of the matrix and analyte concentration.

Table 9.1 Key method performance parameters required for the validation of different types of analysis.

Parameter	Type of analysis			
	Qualitative	Quantitative		Physical property
		Major component[a]	Trace analysis[b]	
Precision		✔	✔	✔
Bias		✔	✔	✔
Specificity/selectivity	✔	✔	✔	✔
Limit of detection	✔		✔	
Limit of quantitation			✔	
Linearity/working range		✔	✔	✔
Ruggedness	✔	✔	✔	✔

[a]Major component: analyte concentration in the range 1–100% by mass or where detection of presence or absence is not an issue.
[b]Trace analysis: analyte concentration less than 100 mg kg^{-1} or where detection capability is important.

Evaluation of the precision of the analytical instrument used to make the determination of analyte concentration may provide some useful information, but it is insufficient for method validation purposes. In a method precision study, each replicate result should be obtained from the application of the entire test method, including any sample preparation and/or pretreatment. If the method being validated is to be applied to the analysis of a range of analyte concentrations and/or sample types, the precision should be studied for a representative range of samples.

When studying method precision, there are two factors that need to be considered: the type of precision estimate required and the design of the precision experiment.

9.2.1 Types of Precision Estimate

The conditions under which the replicate measurements are made determine the nature of the precision estimate obtained.

9.2.1.1 *Repeatability*

'Repeatability' refers to a precision estimate obtained from replicate measurements made in one laboratory by a single analyst using the same equipment over a short time scale (these are called 'repeatability conditions'). It gives an indication of the short-term variation in measurement results and is typically used to estimate the likely difference between replicate measurement results obtained in a single batch of analysis. Precision under repeatability conditions, often called simply 'repeatability', is also referred to as within-batch or intra-assay precision.

9.2.1.2 *Intermediate Precision*

Intermediate precision refers to a precision estimate obtained from replicate measurements made in a single laboratory under more variable conditions than repeatability conditions. Ideally, intermediate precision conditions should mirror, as far as possible, the conditions of routine use of the method (for example, measurements made on different days by different analysts using different sets of equipment within the same laboratory). It is not always possible to vary all of these factors within a laboratory; the conditions under which an estimate of intermediate precision is obtained should therefore always be recorded. ISO 5725-3 suggests nomenclature for recording the conditions.[9] 'Intermediate precision' is also known as 'within-laboratory reproducibility'.

9.2.1.3 *Reproducibility*

Reproducibility as defined by ISO 3534[6] refers to a precision estimate obtained from replicate measurements carried out in different laboratories by different analysts using different pieces of equipment. It therefore has to be evaluated by carrying out an interlaboratory study. Note that the term 'reproducibility' has also been used more generally to describe any conditions of measurement other than repeatability; it is therefore often useful to qualify the term (for example, 'interlaboratory reproducibility').

9.2.2 Experimental Designs for Evaluating Precision

9.2.2.1 *Simple Replication Studies*

A simple precision study involves making repeated measurements on a suitable sample under the required measurement conditions. The precision is expressed as the standard deviation (or relative

Validation and Method Performance

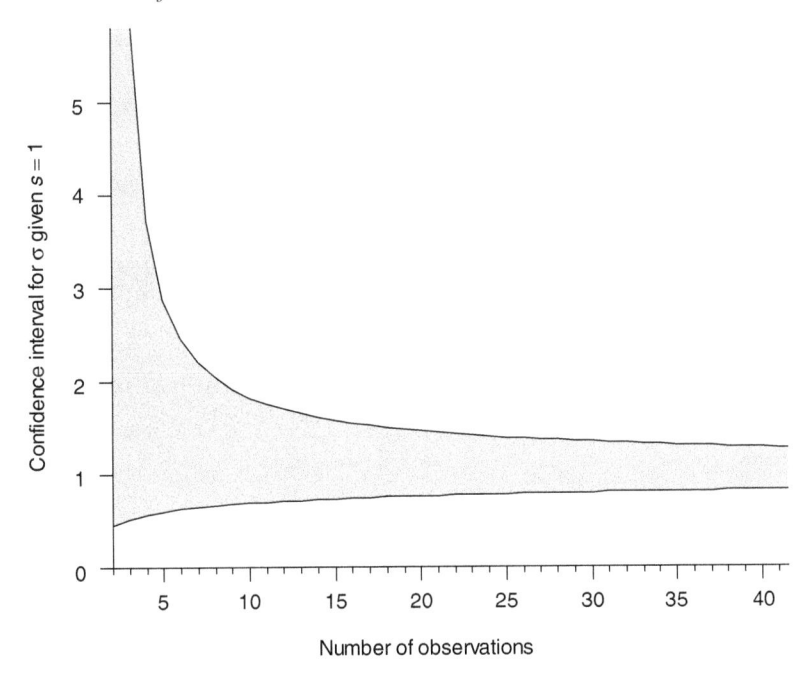

Figure 9.1 Confidence interval for the population standard deviation. The 95% confidence interval for σ given $s=1$ is shown.

standard deviation) of the results. The number of replicate measurements should be sufficient to provide a reliable estimate of the standard deviation. Figure 9.1 shows the confidence interval for the population standard deviation (σ) plotted against the number of independent replicates (n).

With fewer than six replicates, the confidence interval becomes very wide. This illustrates that sample standard deviations based on small data sets will give an unreliable estimate of the population standard deviation. As the number of replicates increases, the confidence interval narrows. However, beyond around 15 replicates there is little change in the confidence interval.

9.2.2.2 Nested Design

A nested design (see Chapter 8, Section 8.4.3) can be used to obtain an estimate of repeatability when it is not possible to generate sufficient data under repeatability conditions using a simple replication study (for example, because the method run time is very long). A nested design is also an efficient way of generating data when studying the effect of changing various parameters such as analyst, instrument or laboratory and additionally provides reproducibility data (see below). The results from this type of study are evaluated using analysis of variance (see Chapter 6).

A nested design requires groups of measurement data which are each generated under repeatability conditions, but conditions can be varied from one group of measurements to the next as required. For example, a study of intermediate precision may involve different runs/batches of analyses carried out on different days (possibly by different analysts) as shown in Figure 9.2. Interlaboratory method validation studies typically involve a number of laboratories, each making repeat measurements on portions of the same bulk sample under repeatability conditions.

A one-way ANOVA of the data from a nested design of the type shown in Figure 9.2 will yield a table similar to that shown in Table 9.2.

If each of p groups of data contains the same number of replicates (n) the total number of results, N, is equal to pn. This leads to the degrees of freedom shown in Table 9.2.

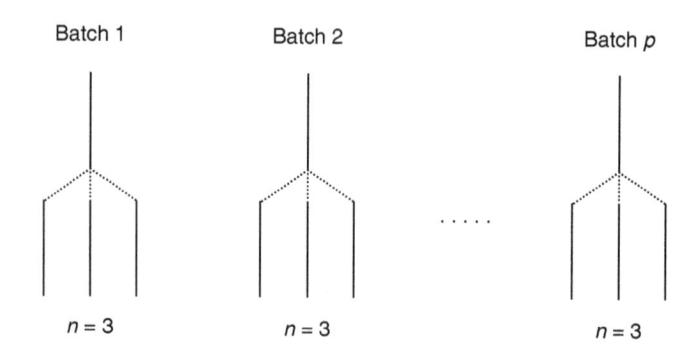

Figure 9.2 Study of intermediate precision in which portions of the same bulk sample are analysed in triplicate under repeatability conditions in p batches of analysis.

Table 9.2 ANOVA table for precision study with p groups of data each containing n replicates.

Source of variation	Sum of squares	Degrees of freedom v	Mean square	F
Between groups	S_b	$p - 1$	$M_b = S_b/(p-1)$	M_b/M_w
Within groups	S_w	$p(n-1)$	$M_w = S_w/p(n-1)$	
Total	$S_{tot} = S_b + S_w$	$pn - 1$		

If the groups contain different numbers of replicates, the degrees of freedom associated with the within-group sum of squares is the sum of the degrees of freedom of the individual groups. It is also equal to $N - p$.

An estimate of the repeatability standard deviation, s_r, is obtained using the equation

$$s_r = \sqrt{M_w} \tag{9.1}$$

this is identical with the calculation of the within-group variance in Chapter 6, Table 6.5.

An estimate of the between-group standard deviation, s_b, is calculated using the equation

$$s_b = \sqrt{\frac{M_b - M_w}{n}} \tag{9.2}$$

Note that if n is not the same for each group of data then a reasonable approximation of s_b can be obtained by replacing n in equation (9.2) with the average number of replicates per group; most statistical software also provides exact calculations of s_b with different n.

An estimate of the intermediate precision standard deviation, s_I, is obtained by combining s_r and s_b

$$s_I = \sqrt{s_r^2 + s_b^2} \tag{9.3}$$

If the groups of data were produced by different laboratories participating in a collaborative study, equation (9.3) will give an estimate of the reproducibility standard deviation, s_R. Detailed information on the organisation of collaborative studies for the evaluation of method precision can be found in ISO 5725.[7–10] Note that ISO 5725 recommends applying Cochran's test (see Chapter 5, Section 5.2.5) to check for unusually high 'within-group' variation before carrying out ANOVA to obtain an estimate of reproducibility. The standard also recommends using Grubbs' tests to

Validation and Method Performance

identify group means that are possible outliers (see Chapter 5, Section 5.2.4), as extreme values will influence the estimate of the reproducibility standard deviation, s_R.

9.2.3 Precision Limits

A repeatability limit is calculated using the equation

$$r = t \times \sqrt{2} \times s_r \quad (9.4)$$

This is the confidence interval (typically 95% confidence) for the difference between two results obtained under repeatability conditions.

A reproducibility limit is calculated using the equation

$$R = t \times \sqrt{2} \times s_R \quad (9.5)$$

This is the confidence interval for the difference between two results obtained under reproducibility conditions (that is, results from different operators, working in different laboratories using different equipment).

In the above equations, t is the two-tailed Student's t value for the required level of confidence and the appropriate number of degrees of freedom. A common approximation is to use a value of $t = 2$ (approximately the Student's t value for a large number of degrees of freedom). This leads to limits of $2.8s$.[10]

9.2.4 Statistical Evaluation of Precision Estimates

Precision estimates can be evaluated statistically using the F-test (see Chapter 4, Section 4.2.4).

9.2.4.1 Comparison with the Precision of an Existing Method

If the validation study is being carried out with the aim of replacing an existing method, it is useful to be able to test whether the precision of the new method is significantly different from (or perhaps better than) the precision of the existing method. The test statistic, F, is calculated using the equations given in Table 4.7 in Chapter 4.

9.2.4.2 Comparison with a Specified Precision Value

If the standard deviation obtained from the precision study marginally exceeds the value set during the specification of the measurement requirement, an F-test can be used to determine whether the precision estimate is *significantly greater than* the target value:

$$F = \frac{s^2}{\sigma_0^2} \quad (9.6)$$

where s is the precision estimate obtained from the validation study and σ_0 is the target precision (expressed as a standard deviation). To obtain the appropriate critical value for F, σ_0 is assumed to have infinite degrees of freedom. See Chapter 4, Section 4.2.5, for further information on this type of test.

9.3 Assessing Bias

Precision describes the variation in results obtained from the replicate analysis of a sample but it gives no information as to how close the results are to the true concentration of the analyte in the sample. It is possible for a test method to produce results that are in very close agreement with one another but that are consistently lower (or higher) than they should be. Trueness is defined as 'The closeness of agreement between the expectation of a test result or a measurement result and a true value'.[6] Trueness is normally expressed in terms of bias. The definition includes a note that in practice, an 'accepted reference value' is substituted for the true value. Bias can be evaluated by comparing the mean of measurement results ($\bar{x}$) and an accepted reference value (μ_0), as shown in Figure 9.3.

Evaluating bias involves carrying out repeat analysis of a suitable material containing a known amount of the analyte (this is the reference value, μ_0). The bias is the difference between the average of the test results and the reference value:

$$\text{Bias} = \bar{x} - \mu_0 \tag{9.7}$$

Bias is also frequently expressed as a percentage:

$$\text{Bias}(\%) = \frac{\bar{x} - \mu_0}{\mu_0} \times 100 \tag{9.8}$$

and as a ratio (particularly when assessing 'recovery'):

$$\text{Recovery}(\%) = \frac{\bar{x}}{\mu_0} \times 100 \tag{9.9}$$

One of the problems when planning a study of method bias is the selection of a suitable reference value. There are a number of options:

- a certified reference material (CRM)
- spiked test samples
- a reference method.

A certified reference material is a material that has been produced and characterised to high standards and that is accompanied by a certificate stating the value of the property of interest (usually the concentration of a particular analyte) and the uncertainty associated with the value. If a suitable CRM is available (that is, one that is similar to test samples in terms of sample form, matrix

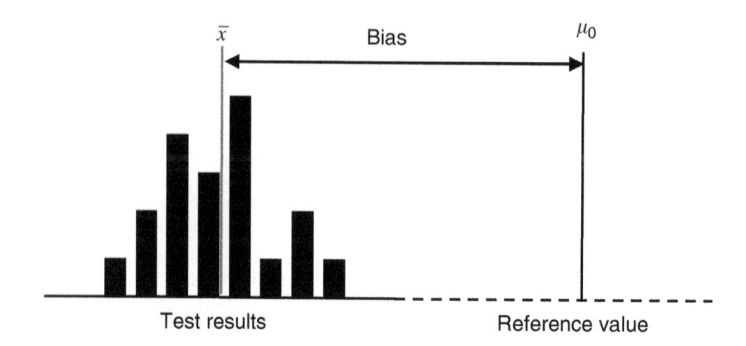

Figure 9.3 Evaluating bias.

composition and analyte concentration), then it should be the first choice material for a bias study. However, compared with the very large number of possible analyte–matrix combinations, the number of CRMs available is relatively limited, so a suitable material may not be available. An alternative is to prepare a reference sample in the laboratory by spiking a previously analysed sample with an appropriate amount of the analyte of interest. With this type of study, care must be taken to ensure that the spike is in equilibrium with the sample matrix before any measurements are made.

Bias can also be evaluated by comparing results obtained from a reference method with those obtained using the method being validated. If this approach is taken, the evaluation can be carried out using test samples rather than a CRM or spiked sample. If a single test sample is used, the means of the results obtained from the two methods are compared using a two-sample t-test (Chapter 4, Section 4.2.3.2). Alternatively, a range of different samples can be analysed using both methods and the differences between the pairs of results obtained evaluated statistically to determine whether there is a significant difference between the results produced by the two methods (see Chapter 4, Section 4.2.3.3).

The number of replicate measurements that needs to be carried out for a bias study depends on the bias that needs to be detected and the precision of the method. This is discussed in Chapter 8, Section 8.5.3. Table 8.2 indicates the number of replicates required to detect a given bias (δ) reliably (specifically, with 95% power) if the precision (s) is known. The table shows the number of replicates for different values of the ratio (δ/s). For example, if it has been agreed in the measurement specification that a maximum bias of $\pm 5\%$ is acceptable, Table 8.2 can be used to determine how many replicate measurements would be required to enable such a bias to be detected reliably. If the precision of the method (repeatability) is 5% then $\delta/s=1$ and Table 8.2 shows that at least 16 replicates are required.

As in the case of precision studies, a bias study should aim to cover a representative range of sample types.

9.3.1 Statistical Evaluation of Bias Data

Bias estimates are evaluated statistically using Student's t-tests (see Chapter 4, Section 4.2.3).

9.3.1.1 Comparison with a Reference Value

Once a bias estimate has been obtained, for example using equation (9.7), a Student's t-test can be carried out to determine whether the observed bias is statistically significant. The test statistic is calculated using the equation

$$t = \frac{|\bar{x} - \mu_0|}{s/\sqrt{n}} \tag{9.10}$$

where n is the number of replicate measurements used to obtain $\bar{x}$ and s is the standard deviation of the n measurement results. See Chapter 4, Section 4.2.3.1, for further information on this type of t-test.

9.3.1.2 Comparison with an Existing Method

When planning to replace one method by another, it is necessary to establish that the new and existing methods give comparable results. Two approaches are commonly used. The first approach involves using the two methods to make a number of replicate measurements on the same sample. This will result in two independent sets of data (one from each method) for which the mean and standard deviation are calculated. Following the procedure in Chapter 4, Section 4.2.3.2 the test statistic is calculated using the equation

$$t = \frac{|\bar{x}_1 - \bar{x}_2|}{\sqrt{\left(\frac{1}{n_1}+\frac{1}{n_2}\right)\left[\frac{s_1^2(n_1-1)+s_2^2(n_2-1)}{n_1+n_2-2}\right]}} \qquad (9.11)$$

Remember that this approach assumes that the standard deviations of the two datasets (s_1 and s_2) are not significantly different. This can be confirmed using an F-test as described in Chapter 4, Section 4.2.4. If the standard deviations are significantly different, then a modified test is required (see Chapter 4, Section 4.2.3.2).

In some cases, it is not possible to make a large number of replicate measurements on a single sample. An alternative approach is to carry out a paired comparison, which requires a single measurement using each method on each of a number of different samples (see Chapter 4, Figure 4.7). The difference between each pair of results is calculated. If there is no difference between the results obtained using the two methods then, on average, the differences should be equal to zero. The t-test is therefore carried out using the mean and standard deviation of the differences ($\bar{d}$ and s_d, respectively) to determine whether $\bar{d}$ is significantly different from zero, as shown in the equation

$$t = \frac{|\bar{d}|}{s_d/\sqrt{n}} \qquad (9.12)$$

where n is the number of pairs of data. See Chapter 4, Section 4.2.3.3, for further information on paired comparisons.

9.4 Accuracy

Accuracy is defined as 'The closeness of agreement between a test result or measurement result and the true value'.[6] Accuracy is the property of a *single* measurement result. It describes how close a single measurement result is to the true value and therefore includes the effect of both precision and bias, as illustrated in Figure 9.4.

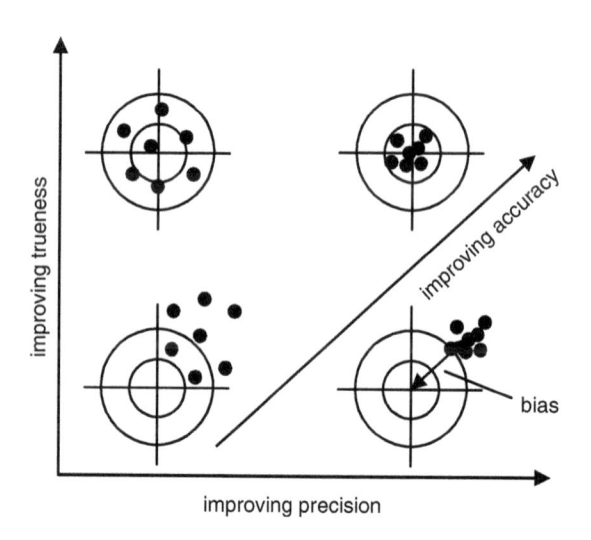

Figure 9.4 Precision, bias and accuracy.

Validation and Method Performance 153

9.5 Capability of Detection

9.5.1 Limit of Detection

The limit of detection (LOD) is the minimum concentration of the analyte that can reliably be detected with a specified level of confidence. There are a number of approaches which can be used to evaluate the LOD. When quoting LOD values, it is important to state the approach used.

The LOD can be evaluated by obtaining the standard deviation of results obtained from replicate analysis of a blank sample (containing none of the analyte of interest) or a sample containing only a small amount of the analyte. Ideally, 6–10 replicate results for this sample, taken through the whole analytical procedure, should be obtained. The limit is calculated by multiplying the standard deviation by a suitable factor. The multiplying factor is based on statistical reasoning and is specified so that the risk of false positives (wrongly declaring the analyte to be present) and false negatives (wrongly declaring the analyte to be absent) is kept to an acceptable level (a 5% probability is usually specified for both types of error).

Figure 9.5 shows the expected distribution of results for a sample which contains zero concentration of the analyte. After correction for any blank reading, the results are distributed approximately normally about zero. About half of the results would (incorrectly) indicate that there is some analyte present (that is, the instrument response would lead to a calculated analyte concentration greater than zero). This is clearly unacceptable, so a limit is set such that the probability of false positive results is reduced to an acceptable level. This probability is represented by α in Figure 9.5. The limit (identified in the figure as the 'critical value') is usually obtained by multiplying the standard deviation of observations from a blank sample, σ_0, by the one-tailed Student's t value for infinite degrees of freedom and the appropriate value of α, then adding this to the mean blank response if the blank response is significantly different from zero. A false positive rate of 5% (that is, $\alpha = 0.05$) is typically used, which gives a critical value of $1.65\sigma_0$ if the mean blank response is zero.

A result above the critical value is equivalent to a positive result in a significance test for exceeding zero; a result above the critical value should therefore be treated as a positive finding that the true concentration is above zero.

This has set a decision criterion: if the result is above the critical value, the sample is declared 'positive'.

Figure 9.6 shows the distribution of calculated concentrations expected for a true analyte concentration corresponding exactly to the critical value. Clearly, half of the observations would be

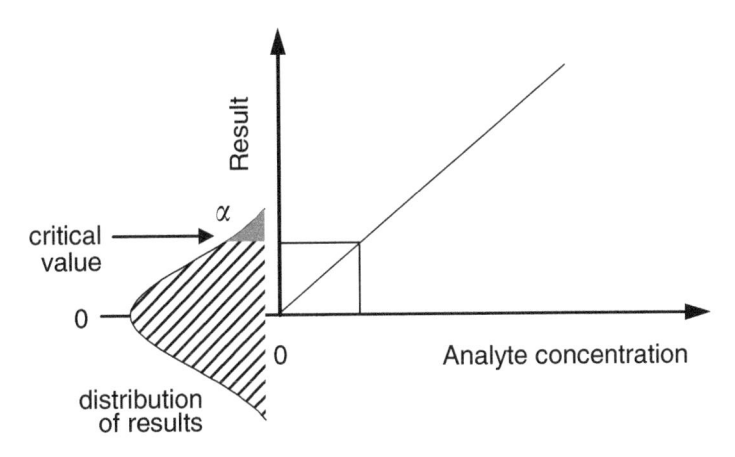

Figure 9.5 Distribution of results for a sample containing zero concentration of the analyte (after correction for any blank reading).

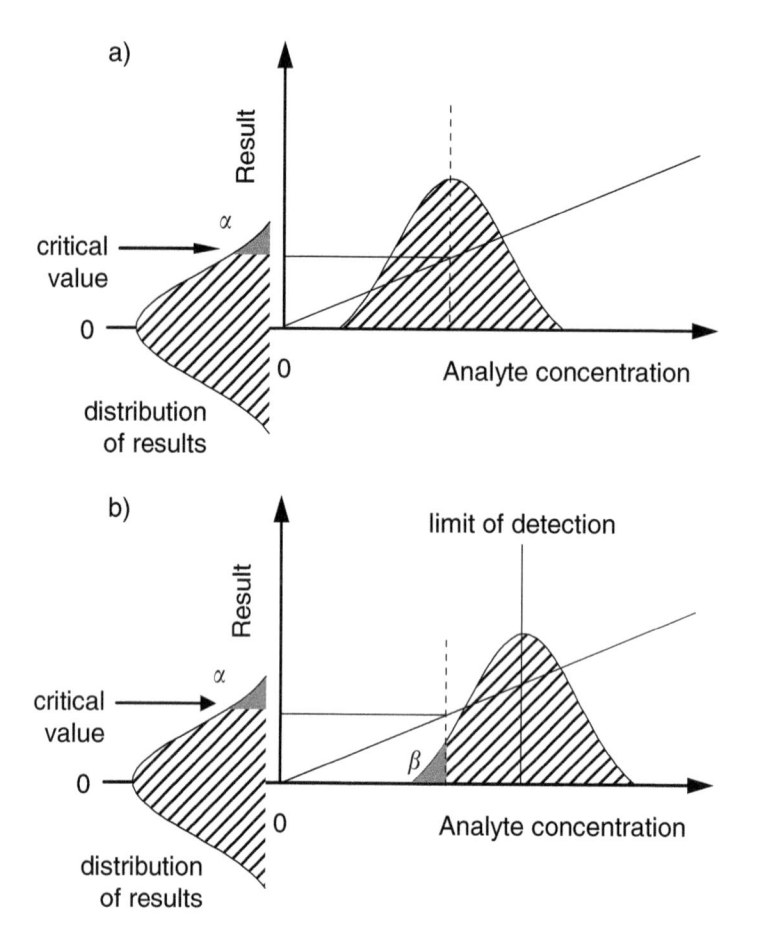

Figure 9.6 Statistical basis of limit of detection calculations.

expected to fall below the critical value, leading to a negative finding – a 50% false negative rate. This seems very high. What true level of analyte needs to be present for the false negative rate to be low, say 5% or less? Remember that this was the question posed by the definition of 'limit of detection' – the minimum concentration of the analyte that can be reliably detected with a specified level of confidence.

Figure 9.6 shows how this level is estimated; by calculating a true concentration such that the critical value cuts off only an area β of the expected distribution of calculated concentrations. It is this new, higher, level that is properly called the limit of detection. Typically, β is set equal to α, that is, 0.05, to give a 5% false negative rate. It follows that the limit of detection is (assuming large degrees of freedom) $1.65\sigma_0$ above the concentration corresponding to the critical value.

Note that this calculation depends on having *and using* a critical value above which samples are declared 'positive'. If another critical value is chosen, the limit of detection must also change.

The general form of the equation for calculating the limit of detection from experimental data is

$$x_0 + k_l s t_{(\nu,\alpha)} + k_l s t_{(\nu,\beta)} \tag{9.13}$$

where x_0 is a correction term to take account of any baseline offset or reading obtained from the blank sample, s is the standard deviation of the results obtained from the limit of detection study, k_l is a constant used to correct the estimate of s and t is the one-tailed Student's t value for ν degrees of freedom and a significance level of α or β. The value of k_l will depend on whether the

Validation and Method Performance

Table 9.3 Values of x_0 and k_l for use in the calculation of the limit of detection.

Validation study of LOD s based on:	Routine use of method[a]	
	Will test results be baseline corrected?	
	Yes	No
Observations with independent baseline corrections	$x_0 = 0$ $k_l = 1.0$	NA
Observations without independent baseline corrections	$x_0 = 0$ $k_l = \sqrt{1 + 1/n_B}$	$x_0 = x_{blank}$ $k_l = 1.0$

[a] n_B is the number of observations averaged to obtain the value of the blank correction. x_{blank} is either the mean blank response or the observed baseline offset, as appropriate. NA, not applicable.

LOD study uses independent blank corrections, on whether such corrections are applied in normal use of the method and on the number of observations averaged to calculate the blank correction. For most purposes, it is sufficient to assume $\alpha = \beta = 0.05$ and take t for a large number of degrees of freedom, that is, $t = 1.65$. The limit of detection is then approximately $x_0 + 3.3k_l s$. Table 9.3 shows the appropriate values of x_0 and k_l for a number of scenarios.

'Independent baseline correction' requires every result to be baseline corrected by an independent determination of the blank sample. As this approach requires a significant number of additional analyses when the method is used routinely, it is more common to use one or two blank samples per batch of test samples and use the results to correct all the results from the test samples in the batch. In this case, the results are being corrected, but as the corrections are not independent, the last row in Table 9.3 applies.

The limit of detection is often specified as $3.3s$ (sometimes rounded to $3s$) or $4.65s$. Using equation (9.13) and the values in Table 9.3, the origin of these limits can be explained. If independent baseline corrections are used in the validation study, $k_l = 1$ and for baseline-corrected test results, $x_0 = 0$. If a value of $t = 1.65$ is used then the LOD is

$$\text{LOD} = (1 \times s \times 1.65) + (1 \times s \times 1.65) = 3.3s$$

If results in the validation study are corrected by a single baseline reading ($n_B = 1$), then $k_l = \sqrt{2}$. The limit of detection in this case is

$$\text{LOD} = \left(\sqrt{2} \times s \times 1.65\right) + \left(\sqrt{2} \times s \times 1.65\right) = 4.67s$$

If test results are not baseline corrected, then $k_l = 1$ and the limit of detection is

$$\text{LOD} = x_{blank} + (1 \times s \times 1.65) + (1 \times s \times 1.65) = x_{blank} + 3.3s$$

It is also possible to estimate the critical value and detection limit directly from a calibration curve; ISO 11843-2 gives the relevant calculations in full.[11]

The LOD determined during method validation should be taken only as an indicative value. If sample concentrations are expected to be close to the limit of detection, then the LOD should be monitored regularly after validation.

9.5.2 Limit of Quantitation

The limit of quantitation (LOQ) is the lowest concentration of analyte that can be determined with an acceptable level of uncertainty. A value of $10s$ is frequently used (where s is the standard deviation of the results from replicate measurements of a blank or low-concentration sample). Unlike the calculations used for obtaining an estimate of the LOD, there is no statistical basis to the conventions used for estimating the LOQ. The aim is to identify the concentration below which the measurement uncertainty becomes unacceptable. The value of 10s provides a reasonable estimate for many test methods.

9.6 Linearity and Working Range

The working range of a method is the concentration range over which the method has been demonstrated to produce results that are fit for purpose. The lower end of the working range is defined by the LOD or LOQ as appropriate for the application (see Section 9.5). The upper end is usually signified by a change in sensitivity, for example a 'tailing-off' or 'plateauing' in the response, as illustrated in Figure 9.7.

Figure 9.7 shows a schematic response curve. The concentration range corresponding to the working range is shown, together with the range in which change in response is directly proportional to the change in concentration – the linear range. The linear range and working range need not coincide exactly. The determination of linearity and working range is important at the method validation stage, because it allows the analyst to establish the suitability of the method over the range required by the measurement specification.

In order to assess the working range and confirm its fitness for purpose, standards whose concentration range extends beyond the required concentration range by ±10% or even ±20% of the required range should be studied. As discussed in Chapter 7, the standards should be evenly spaced across the concentration range studied. Establishing linearity during method validation will normally require more standards (and more replication at each concentration) than is typical for calibration of a validated method in regular use. It is recommended that at least seven different concentration levels are studied during method validation. This approach provides sufficient information for a useful visual check of the linearity of the method. There are other designs which allow for a more powerful statistical test of non-linearity. For example, if it is suspected that the data follow a quadratic function, a design with three concentration levels (low, middle and high), with replication at each level will allow a powerful statistical test for non-linearity. The main disadvantage of this approach is that, with only three concentration levels, a plot of the data will not be very informative. Therefore, if the statistical evaluation indicates non-linearity it may not be

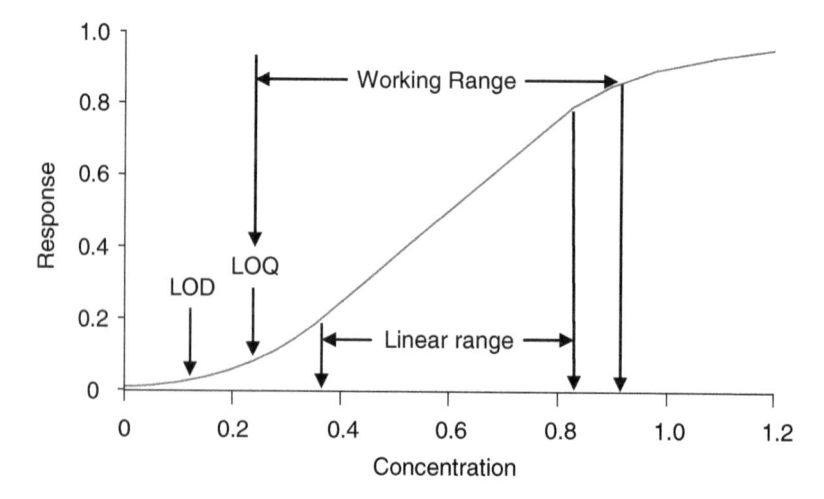

Figure 9.7 Linearity and working range.

Validation and Method Performance

possible to determine the cause of the non-linearity. It is common in analytical chemistry for the method to follow a linear function up to a certain concentration, above which the instrument detector may become saturated and the instrument response plateaus. With only three concentration levels it will not be possible to determine where the response becomes non-linear. See Chapters 7 and 8 for further information on the planning and interpretation of linear regression studies.

Many chemical test methods require the test sample to be treated in some way so as to get the analyte into a form suitable for measurement. During method validation, it is sensible to carry out an initial study to evaluate the response of the instrument to the analyte across the required concentration range. This can be done by analysing standards containing known concentrations of the analyte in a suitable solvent. This study will enable the calibration function for the instrument to be established. Once the instrument performance has been demonstrated to be satisfactory, the linearity of the whole method should be studied. This requires the analysis of CRMs, spiked samples or matrix-matched standard solutions (that is, solutions containing a known amount of the analyte plus the sample matrix). If the instrument response has been demonstrated to be linear, then any non-linearity observed in the second study may indicate problems such as the presence of interfering compounds or incomplete extraction of the analyte from the sample matrix.

Linearity should be assessed initially by constructing a scatter plot of response *versus* concentration (Chapter 2, Section 2.9). The data can then be evaluated by carrying out linear regression to establish the gradient and intercept of the line (see Chapter 7, Section 7.1.4) and the correlation coefficient (r) (see Chapter 7, Section 7.1.6). Note that the correlation coefficient should not be used in isolation to make a judgement on the linearity of a method; it should always be interpreted in conjunction with a plot of the data. When assessing linearity, it is also useful to obtain a plot of the residual values (see Chapter 7, Section 7.1.5). Chapter 7 also includes objective tests for linearity.

9.7 Ruggedness

The ruggedness of a method describes its ability to withstand small changes in its operating conditions.[12] A method which is unaffected by changes in experimental parameters, within the control limits defined in the method protocol, is said to be rugged (the term 'robustness' is also used to describe the same concept[13]).

Ruggedness testing is carried out by a single laboratory and evaluates how small changes in the method conditions affect the measurement result, for example, small changes in temperature, pH, flow rate or composition of extraction solvents. The aim is to identify method conditions that might lead to variations in measurement results when measurements are carried out at different times or in different laboratories. Ruggedness testing is often carried out as part of method development to help identify the optimum method conditions and controls.

Ruggedness testing can be carried out by considering each effect separately. However, this can be labour intensive as a large number of effects may need to be considered. Since most of the effects can be expected to be small for a well-developed method, it is possible to vary several parameters at the same time. Any stable and homogeneous sample within the scope of the method can be used for ruggedness testing experiments. Ruggedness tests typically use fractional factorial designs as described in Chapter 8, Section 8.7.1; Youden and Steiner describe a particularly simple experimental design which allows seven independent factors to be examined in eight experiments.[12] Factors identified as having a significant effect on measurement results will require further study. If it is not possible to reduce the impact of the parameters by employing tighter control limits, then further method development will be required.

9.7.1 Planning a Ruggedness Study

Youden and Steiner[12] make use of a Plackett–Burman experimental design.[14] The first step is to identify up to seven method parameters to be studied. Each parameter is assigned two experimental values or levels. The first levels for the parameters are identified by the letters A–G. The alternative levels are denoted by the letters a–g. The levels can be chosen in one of two ways. The approach given by the AOAC is to set parameters A–G at the 'normal' level used in the method and parameters a–g at a different level, which may be higher or lower. This is useful if information is already available about how the parameter might affect the method.

For example, consider a method that has been developed using an extraction time of 30 min. It is suspected that using a shorter extraction time would reduce the recovery of the analyte, leading to a lower result. The normal level would therefore be 30 min and the alternative level would be a shorter time, say 25 min. This approach is also used for non-continuous variables, such as when investigating the effect of using different types of HPLC columns. For example, the normal level might be a C_{18} column whilst the alternative is a C_8 column.

An alternative approach is to take the extremes of a range about the normal value. For example, to investigate the effect of a 5 °C change in extraction temperature from the 40 °C normally used, one level would be 35 °C and the alternative level would be 45 °C. A combination of both approaches can be used when assigning the levels for parameters in a ruggedness study.

The Plackett–Burman experimental design for seven parameters is shown in Table 9.4. The sample chosen for the study is analysed eight times, according to the experimental conditions laid out in the experimental design. For example, in experiment 1 all of the experimental parameters will be at the level identified by an upper-case letter. However, for experiment 2, parameters c, e, f and g will be at the alternative level. The order in which the eight experiments are carried out should be randomised where possible. (See Chapter 8 for detailed information on experimental design and randomisation.) The eight experimental results produced are identified by the letters s–z. These results are used to determine whether the changes in any of the parameters studied have a significant effect on the method performance.

9.7.2 Evaluating Data from a Ruggedness Study

The experimental design shown in Table 9.4 is balanced so that whenever the results of the eight experiments are split into two groups of four on the basis of one of the parameters, the effects of all the other parameters cancel. To investigate the effect of a parameter, the difference between the average of the results obtained with the parameter at the normal level and the average of the results with the parameter at the alternative level is calculated. For example, the difference for parameter

Table 9.4 Plackett–Burman experimental design for studying seven experimental parameters.

Experimental parameter	Experiment number							
	1	2	3	4	5	6	7	8
A or a	A	A	A	A	a	a	a	a
B or b	B	B	b	b	B	B	b	b
C or c	C	c	C	c	C	c	C	c
D or d	D	D	d	d	d	d	D	D
E or e	E	e	E	e	e	E	e	E
F or f	F	f	f	F	F	f	f	F
G or g	G	g	g	G	g	G	G	g
Observed result	s	t	u	v	w	x	y	z

B, Δ_B, is given by

$$\Delta_B = \frac{s+t+w+x}{4} - \frac{u+v+y+z}{4} \qquad (9.14)$$

The next step is to arrange the seven differences, Δ_A–Δ_G, in order of magnitude (ignoring the sign). If one or two parameters have a greater effect on the method than the others, their calculated differences will be substantially larger than the others. However, we need to decide whether any of the parameters is having a significant effect on the results in relation to the required method performance. One parameter showing a greater effect on the method does not necessarily mean the method is not rugged. The differences can be evaluated statistically by calculating a critical difference:

$$\Delta_{\text{crit}} > \frac{ts}{\sqrt{2}} \qquad (9.15)$$

where s is the expected precision of the test method (expressed as a standard deviation) and t is the two-tailed Student's t value for the required level of confidence (usually 95%) with appropriate degrees of freedom. The number degrees of freedom used for t is that associated with the estimate of the standard deviation s. If any of the values $|\Delta_A|$ to $|\Delta_G|$ are greater than Δ_{crit}, the effects of those parameters are statistically significant.

References

1. LGC, *In-House Method Validation: a Guide for Chemical Laboratories*, LGC, Teddington, 2003, ISBN 0-948926-18-X.
2. C. Burgess, *Valid Analytical Methods and Procedures*, Royal Society of Chemistry, Cambridge, 2000, ISBN 978-0-85404-482-5.
3. E. Mullins, *Statistics for the Quality Control Chemistry Laboratory*, Royal Society of Chemistry, Cambridge, 2003, ISBN 978-0-85404-671-3.
4. C. C. Chan, Y. C. Lee, H. Lam and X.-M. Zhang, *Analytical Method Validation and Instrument Performance Verification*, Wiley, Hoboken, NJ, 2004, ISBN 0-471-25953-5.
5. J. Ermer and J. H. McB. Miller (Eds), *Method Validation in Pharmaceutical Analysis*, Wiley-VCH, Weinheim, 2005, ISBN 978-3-527-31255-9.
6. ISO 3534-2:2006, *Statistics – Vocabulary and Symbols – Part 2: Applied Statistics*, International Organization for Standardization, Geneva, 2006.
7. ISO 5725-1: 1994, *Accuracy (Trueness and Precision) of Measurement Methods and Results – Part 1: General Principles and Definitions*, International Organization for Standardization, Geneva, 1994.
8. ISO 5725-2: 1994, *Accuracy (Trueness and Precision) of Measurement Methods and Results – Part 2: Basic Method for the Determination of Repeatability and Reproducibility of a Standard Measurement Method*, International Organization for Standardization, Geneva, 1994.
9. ISO 5725-3: 1994, *Accuracy (Trueness and Precision) of Measurement Methods and Results – Part 3: Intermediate Measures of the Precision of a Standard Measurement Method*, International Organization for Standardization, Geneva, 1994.
10. ISO 5725-6: 1994, *Accuracy (Trueness and Precision) of Measurement Methods and Results – Part 6: Use in Practice of Accuracy Values*, International Organization for Standardization, Geneva, 1994.

11. ISO 11843-2: 2000, *Capability of Detection – Part 2: Methodology in the Linear Calibration Case*, International Organization for Standardization, Geneva, 2000.
12. W. J. Youden and E. H. Steiner, *Statistical Manual of the Association of Official Analytical Chemists*, AOAC, Arlington, VA, 1975, ISBN 0-935584-15-3.
13. ICH Harmonised Tripartite Guideline, *Validation of Analytical Procedures: Text and Methodology*, Q2(R1), ICH, 2005, http://www.ich.org (accessed 2 April 2009).
14. R. L. Plackett and J. P. Burman, *Biometrika*, 1946, **33**, 305.

CHAPTER 10
Measurement Uncertainty

10.1 Definitions and Terminology

The definition of the term uncertainty (of measurement) is provided by the ISO *Guide to the Expression of Uncertainty in Measurement* ('the GUM'):[1]

'A parameter associated with the result of a measurement, that characterises the dispersion of the values that could reasonably be attributed to the measurand.'
(Note: the 3rd edition of the *International Vocabulary of Metrology* extends the definition slightly.[2])

Notes to the GUM definition additionally indicate that the 'parameter' may be, for example, a standard deviation (or a given multiple of it) or the width of a confidence interval and also that uncertainty of measurement comprises, in general, many components. An interpretation of the ISO Guide for analytical chemistry is also available from Eurachem.[3]

When expressed as a standard deviation, an uncertainty is known as a *standard uncertainty*, usually denoted u. For reporting, an *expanded uncertainty*, U, is used. Strictly, this is defined as an interval containing a high proportion of the distribution of values that could reasonably be attributed to the measurand. For most practical purposes in analytical measurement, however, it may be thought of as an interval believed to include the 'true value' with a high level of confidence.

The ISO Guide defines two different methods of *estimating* uncertainty. Type A estimation uses statistical analysis of repeated observations; Type B uses any other method. For example, the standard deviation of the mean of the observations on a test item would be classed as Type A estimation of part of the uncertainty (that part associated with variation during the course of the measurement). Using the stated uncertainty in the certified value of a reference material used for calibration would be classed as Type B. The most important point about Type A and Type B estimation, however, is that uncertainties estimated by either method are treated in the same way when combining uncertainties. In either case, they are converted to standard uncertainties in the result and combined using the formulae used to combine variances from more than one cause.

Finally, we will use four other terms in this chapter. A *source of uncertainty* is some factor or effect which can change the analytical result by an amount which is not exactly known; for example, imperfect analytical matrix match or the temperature used in an extraction process. The term *influence quantity*, as it is called in the GUM, is a source of uncertainty associated with a quantitative factor (any factor that takes a measurable value); in practice, 'influence quantity' covers practically all sources of uncertainty so the terms are nearly synonymous. We will use the term '*component* of uncertainty' for the uncertainty associated with the value of an influence quantity; for example, an uncertainty in our extraction temperature (note that this will have the units of temperature). Finally, we will use the term '*contribution* to uncertainty' for the uncertainty that each source of uncertainty contributes to the combined uncertainty in the analytical result. Note that this will have the same units as the result itself, usually a concentration.

10.2 Principles of the ISO Guide to the Expression of Uncertainty in Measurement

10.2.1 Steps in Uncertainty Assessment

Uncertainty can arise from more than one source. Two mentioned above are good examples; it is common to see random variation in analytical results and most analysts use reference materials for calibration, so both the random error and the reference material uncertainty contribute to the uncertainty in the result. To use and combine separate uncertainties, several steps are necessary:

1. Write down a clear 'specification of the measurand'; that is, write down clearly what is being measured.
2. Identify all the relevant sources of uncertainty and, as far as possible, write an equation that relates them all to the final result. Usually, this will be the calculation used to obtain the result itself, extended if necessary to include any additional terms required for uncertainty estimation. A precision term is a good example of such an additional term.
3. Obtain uncertainty information on each source of uncertainty and, where necessary, convert it to a standard uncertainty.
4. If necessary, convert the standard uncertainty in each input quantity to a standard uncertainty u in the final analytical result.
5. Combine all the separate contributions.
6. If required, multiply by a suitable coverage factor to obtain the expanded uncertainty U.

The following sections describe each step. Later sections give some practical suggestions which can simplify the work considerably.

10.2.2 Specifying the Measurand

The measurand is usually the analyte concentration of interest. It is usually important to be clear about whether the quantity of interest is the total analyte present or just the amount extracted using a specified procedure (for example, there is a difference between 'total cadmium' and 'leachable cadmium'). It is also important to be clear about whether the result is intended to relate to the sample received in the laboratory or to a larger batch or area from which the samples were taken; in the latter cases, sampling effects should be considered in estimating the uncertainty.

10.2.3 Identifying Sources of Uncertainty – the Measurement Equation

Most analytical results are calculated from other quantities, such as weights, volumes, concentrations and peak area ratios. Start by writing down the equation used to calculate the result. If other effects are important, add a term for each. For example, for precision (either associated with a single input quantity or with the measurement process as a whole), add a term for random variation; if an extraction time is specified and could affect the result, add a term for 'difference from nominal

Measurement Uncertainty

extraction time'. These terms can be additive or, if the equation involves only multiplication and division, could be included as multiplicative factors. In practice, these terms are allocated a nominal value of zero with an uncertainty if additive or 1.0 with a relative uncertainty if multiplicative. This complete equation is usually called the 'measurement equation' and written generally as

$$y = f(x_1, x_2, \ldots, x_n) \quad (10.1)$$

where x_1, x_2, *etc.* are simply the quantities which affect the result y and $f(\ )$ denotes the equation or function of these variables.

Example

Imagine we wish to estimate the density of ethanol, d_{EtOH}, at 25 °C by weighing a known volume V in a suitable volumetric container held at 25 °C. For simplicity, assume that we calculate the density from the observed net weight of ethanol, m_{EtOH} (this will normally involve two weighings, but for extra simplicity we will treat it as the observed weight by difference). Imagine also that we wish to use the precision of the entire procedure as part of the uncertainty estimate, by carrying out several filling, temperature conditioning and weighing runs. The density is then calculated from

$$d_{EtOH} = m_{EtOH}/V \quad (10.2)$$

For clarity, only three effects will be considered: equipment calibration, the temperature of the liquid and the precision of the determination.

Equipment calibration for mass and volumetric equipment in principle implies a correction to the nominal volume or indicated mass, that is, for mass, a small additive correction δ_m and for volume a small correction δ_V. In practice, analysts rarely make such corrections; they are usually assumed to be negligible within the uncertainty on the calibration certificate. Where this is the case, there is no need to change equation (10.2); we need only associate the calibration uncertainties with the mass and volume in equation (10.2).

We know, however, that temperature can affect the result. We could include a simple multiplication factor to take account of this. In this instance, however, we know the approximate form of density change with temperature and we know that it involves the volume coefficient of expansion for ethanol, α_{EtOH}. The temperature-corrected form of equation (10.2) is

$$d_{EtOH} = (m_{EtOH}/V)/[1 - (T - 25)\alpha_{EtOH}] \quad (10.3)$$

where T is the actual temperature of the liquid in °C. T is expected to be 25 °C, so there is no change in the calculated result, but there will be some uncertainty about the temperature measurement (certainly involving temperature calibration uncertainty), which can now be associated with a variable in the extended measurement equation.

So far, we have not considered the overall precision of the measurement. We could, in principle, include the precision of mass, volume and temperature measurements in the estimated uncertainty for each of these terms. Since we wish to use the observed precision of the measurement, however, this is not useful. Instead we extend equation (10.3) by adding a term, e_d, for the random variation in observed density:

$$d_{EtOH} = (m_{EtOH}/V)/[1 - (T - 25)\alpha_{EtOH}] + e_d \quad (10.4)$$

e_d has expected value zero, but will have a standard uncertainty equal to the observed standard deviation of the replicate measurements performed.

This extended equation now shows how the different uncertainty components can affect the final result; it is a measurement equation we can use for uncertainty estimation.

10.2.4 Obtaining Standard Uncertainties for Each Source of Uncertainty

Repeated observations can be used to calculate standard deviations, which can usually be used directly in uncertainty estimation. However, Type B estimates of uncertainty are often based on information in different forms. For example, uncertainty information on calibration certificates or manufacturer's tolerances are often expressed in the form of limits or confidence intervals. To convert these to an estimated standard uncertainty u, the following rules are applied:

- The standard deviation s of n observations which are averaged to give an intermediate value or a result should be converted to the standard uncertainty in the mean value, which is simply the standard deviation of the mean, by dividing by $\sqrt{n}$ (see Chapter 4, Section 4.1.4.3); that is, $u = s\sqrt{n}$.
- The uncertainty due to random variation when using only a single observation must be based on the standard deviation for a similar material or the precision estimated during validation. For example, if validation gives a relative standard deviation of s', the uncertainty in a single value x, due to random variation, is xs'.
- For a tolerance in the form '$x \pm a$': $u = a/\sqrt{3}$
  Note: this arises as a consequence of an assumed rectangular distribution (see Chapter 3, Section 3.5.1) of width $2a$, symmetrical about x, that is, it describes the assumption that all values in the interval $x \pm a$ are equally probable. If there is reason to believe that values close to x are much more likely, the GUM recommends that a triangular distribution (Chapter 3, Section 3.5.2) is assumed. This has standard uncertainty $u = a/\sqrt{6}$.
- For a confidence interval in the form '$x \pm d$ at $p\%$ confidence': $u = d/t$
  where t is the two-tailed value of Student's t for the relevant level of confidence and number of degrees of freedom (see Appendix A, Table A.4). Where the number of degrees of freedom for the confidence interval is not given, assume t for infinite degrees of freedom. For 95% confidence, $t = 1.96$ for large degrees of freedom; 2.0 is a more than adequate approximation for this purpose.
- Given an expanded uncertainty in the form '$x \pm U$', with a coverage factor k: $u = U/k$
  If the coverage factor is not given, assume $k = 2$, so that $u = U/2$.

These rules are summarised in Table 10.1 for quick reference.

Examples

1. The following four aflatoxin results (in mg kg^{-1}) are averaged to give a mean aflatoxin B1 content for a foodstuff: 2.1, 1.9, 2.2, 2.2. The mean is 2.10 mg kg^{-1}. The standard deviation is 0.141 mg kg^{-1}. However, because the results are averaged, the standard uncertainty in the mean value is $0.141/\sqrt{4} = 0.071$ mg kg^{-1}.
2. A thermometer has a scale marked in whole degrees Celsius, so a reading to the nearest degree will always be correct to only ± 0.5 °C. This describes a rectangular distribution of width 1 °C or half-width $a = 0.5$ °C. The standard uncertainty associated with the thermometer scale reading is therefore $0.5/\sqrt{3} = 0.29$ °C.
3. A certified calibration solution has a lead content of 1000 mg L^{-1} with a 95% confidence interval of 1000 ± 0.6 mg L^{-1}. The number of degrees of freedom is not stated. The standard uncertainty is calculated assuming large degrees of freedom and is $0.6/1.96 = 0.31$ mg L^{-1}.

Measurement Uncertainty

Table 10.1 Conversion factors for standard uncertainty.

Information	Assumed distribution	Standard uncertainty
Standard deviation of observed values s	Normal	$s/\sqrt{n}$, where n is the number of observations averaged to obtain the result. For a single observation, $n=1$; the standard deviation may be used unchanged.
Tolerance in the form $x \pm a$; no reason to expect values near x more likely	Rectangular	$a/\sqrt{3}$
Tolerance in the form $x \pm a$; good reason to expect values near x more likely	Triangular	$a/\sqrt{6}$
Confidence interval in the form $x \pm d$ with only a stated level of confidence $1 - \alpha$[a]	Normal	$d/z(\alpha)$, where $z(\alpha)$ is the quantile of the normal distribution for a two-tailed significance level α. $z(\alpha) = 1.96$ for 95% confidence.
Confidence interval in the form $x \pm d$ with stated level of confidence $1 - \alpha$ and degrees of freedom v[a]	Student's t	$d/t(\alpha,v)$, where $t(\alpha,v)$ is the critical value of Student's t for a two-tailed test with significance level α and degrees of freedom v. For example, $t(\alpha,v) = 2.57$ for 95% confidence and 5 degrees of freedom.
Expanded uncertainty U with stated coverage factor k	Normal	U/k

[a]Remember that a p% level of confidence is the same as $100(1 - \alpha)$% confidence.

10.2.5 Converting Uncertainties in Influence Quantities to Uncertainties in the Analytical Result

There are two simple cases:

- If a quantity x_i is simply added to or subtracted from all the others to obtain the result y, the contribution to the uncertainty in y is simply the uncertainty $u(x_i)$ in x_i.
- If a quantity is multiplied by, or divides, the rest of the expression for y, the contribution to the *relative* uncertainty in y, $u(y)/y$, is the *relative* uncertainty $u(x_i)/x_i$ in x_i.

These are special cases of more general rules. If the measurement equation involves a mixture of different algebraic operations, these simple rules do not apply and the uncertainty contribution must be obtained by other means. The basic question is 'how much does the result change if the input quantity changes by its uncertainty?'. The general answer is 'the uncertainty in x_i times the rate of change of y with x_i'. This can be obtained by one of the following methods:

1. *Experiment:* Set up an experiment in which the variable of interest is varied deliberately and measure the effect on the result. Ideally, plot a graph of the observed result y against the value of the variable x_i and obtain the gradient (which we shall call c_i) of the line. Given this gradient, the uncertainty $u_i(y)$ in y due to an uncertainty $u(x_i)$ in x_i is given by

$$u_i(y) = c_i u(x_i) \tag{10.5}$$

An estimate of the gradient based on a single change in the variable may well be sufficient for many uncertainties. It is important, however, to obtain a good estimate of the gradient, which in turn usually involves changing the variable x_i by much more than its standard uncertainty.

2. *Calculation:* One of the simpler methods is to change the input quantity by its standard uncertainty and recalculate the result. If $y(x_i)$ is the result found for the nominal or observed value of the quantity x_i and $y[x_i + u(x_i)]$ the result with the value plus its standard uncertainty, then the standard uncertainty $u_i(y)$ in the result is given approximately by

$$u_i(y) = y[x_i + u(x_i)] - y(x_i) \qquad (10.6)$$

If the gradient term c_i is required, this too can be obtained approximately as

$$c_i \approx \frac{y[x_i + u(x_i)] - y(x_i)}{u(x_i)} \qquad (10.7)$$

This can be carried out simply and automatically using a spreadsheet, using a method first described by Kragten[4] and later elaborated on;[5] see Section 10.3.1 for additional details.

3. *Algebraic differentiation of the measurement equation:* The GUM uses algebraic differentiation of the measurement equation. Specifically, the gradient term (often called the sensitivity coefficient in the GUM) is given by

$$c_i = \partial y / \partial x_i \qquad (10.8)$$

where $\partial y / \partial x_i$ is the partial differential of y with respect to x_i. The uncertainty $u_i(y)$ is then

$$u_i(y) = c_i u(x_i) = (\partial y / \partial x_i) u(x_i) \qquad (10.9)$$

10.2.6 Combining Standard Uncertainties and 'Propagation of Uncertainty'

The GUM combines standard uncertainties using the same rules as for combining random variation due to several causes.

Given the contributions to uncertainty – the terms $u_i(y) = c_i u(x_i)$ above – the rule is very simple: independent *contributions* combine as the root sum of their squares. Diagrammatically, this looks like Pythagoras' theorem (Figure 10.1); mathematically:

$$u(y) = \sqrt{\sum_{i=1,n} c_i^2 u(x_i)^2} \qquad (10.10)$$

where y, x_i and c_i have the meanings given above. The general form is sometimes called the 'law of propagation of uncertainty'.

Equation (10.10) leads to the two very simple cases at the beginning of Section 10.2.5 which are so common they are worth remembering: if variables are added or subtracted, the uncertainties combine as simple standard uncertainties; if variables are multiplied or divided, their uncertainties combine as *relative* uncertainties (Table 10.2).

Examples

1. A method requires 100 mg of an internal standard to be weighed out on a four-figure balance. The balance calibration certificate gives a standard uncertainty of ±0.0002 g, excluding weighing-to-weighing precision. Replicate weighings of a 100 mg check weight on the four-figure balance had a standard deviation of 0.000071 g, which can be used directly as an

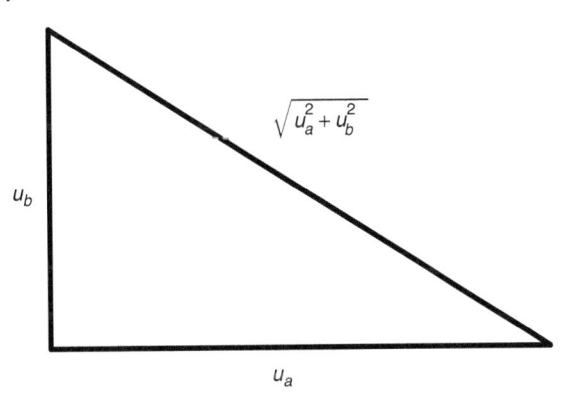

Figure 10.1 Simple rule for combining independent uncertainty contributions.

Table 10.2 Simple rules for uncertainty combination.

Result y calculated from:	Uncertainty u_y in y
$y = a + b$ or $y = a - b$	$u_y = \sqrt{u_a^2 + u_b^2}$
$y = a \times b$ or $y = a/b$	$\dfrac{u_y}{y} = \sqrt{\left(\dfrac{u_a}{a}\right)^2 + \left(\dfrac{u_b}{b}\right)^2}$

estimate of the standard uncertainty associated with weighing precision. There are therefore two uncertainties, both directly affecting the observed weight and arising from additive effects. The standard uncertainty in the result of the weighing is therefore

$$\sqrt{0.0002^2 + 0.000071^2} = 0.00021 \text{ g}$$

2. The concentration c of a (nominally) 1000 mg L^{-1} solution of an organic analyte is calculated from the mass m of the analyte and the volume v of the volumetric flask used to prepare the standard solution. The concentration $c = 1000\, m/v$. The mass was 100.4 mg with standard uncertainty 0.21 mg; the volumetric flask has a volume of 100 mL with standard uncertainty 0.16 mL. The equation is entirely multiplicative, so the uncertainties are combined as relative uncertainties, giving

$$\frac{u(c)}{c} = \sqrt{\left(\frac{0.21}{100.4}\right)^2 + \left(\frac{0.16}{100}\right)^2} = 0.0026$$

and, since $c = 1000 \times 100.4/100 = 1004$ mg L^{-1}, the standard uncertainty is $1004 \times 0.0026 = 2.6$ mg L^{-1}.

Finally, note that equation (10.10) holds only where the uncertainties lead to independent effects on the result. Where they are not independent, the relationship is more complex:

$$u[y(x_{i,j,\ldots})] = \sqrt{\sum_{i=1,n} c_i^2 u(x_i)^2 + \sum_{\substack{i,k=1,n \\ i \neq k}} c_i c_k u(x_i, x_k)} \qquad (10.11)$$

where $u(x_i,x_k)$ is the covariance between x_i and x_k. This is dealt with in more detail in the GUM and also in reference 5.

10.2.7 Reporting Measurement Uncertainty

For most purposes, an expanded uncertainty, U, should be reported. U is obtained by multiplying u, the standard uncertainty, by a *coverage factor*, k. The choice of the factor k is based on the level of confidence desired. For an approximate level of confidence of 95%, k is often taken as 2. For exacting requirements, k is calculated using Student's t for the *effective degrees of freedom*, v_{eff}, for the uncertainty estimate. Detailed procedures for calculating v_{eff} are provided in the GUM.

In practice, it is rarely useful to quote uncertainties to more than two significant figures and one significant figure often suffices. Results should also be rounded to be consistent with the reported uncertainty.

Example

The concentration of 1004 mg L^{-1} calculated above has a standard uncertainty of 2.6 mg L^{-1}. For an approximate level of confidence of 95%, the coverage factor is 2. The expanded uncertainty is therefore $2 \times 2.6 = 5.2$ mg L^{-1}. Rounding to 5 mg L^{-1}, the concentration can be reported as 1004 ± 5 mg L^{-1}.

Note that rounding reported uncertainties upwards is often considered prudent; here, this would give a reported value of 1004 ± 6 mg L^{-1}.

10.3 Practical Implementation

10.3.1 Using a Spreadsheet to Calculate Combined Uncertainty

Section 10.2.5 referred to the possibility of calculating a combined uncertainty using ordinary spreadsheet software, following the approach described by Kragten.[4] This saves a great deal of time and is particularly useful for checking, for example, that a reference material is up to the job or that temperature control arrangements lead to negligible uncertainty. It is achieved as described below. The description follows that in the Eurachem Guide.[3]

The basic spreadsheet (Figure 10.2) is set up as follows, assuming that the result y is a function of three parameters p, q and r and using the spreadsheet formula format used by MS Excel and OpenOffice:

1. Enter the values of p, q, *etc.* and the formula for calculating y in column **A** of the spreadsheet and the values of the uncertainties $u(p)$ and so on in column **B** as shown in Figure 10.2a. Copy column **A** across the following columns (**C**, *etc.*) once for each variable in y (see Figure 10.2a).
2. Add $u(p)$ to p in cell **C2**, $u(q)$ to q in cell **D3**, *etc.*, as in Figure 10.2b. On recalculating the spreadsheet, cell **C6** then becomes $f(p+u(p), q, r, \ldots)$ [denoted by f(p', q, ...) in Figure 10.2], cell **D6** becomes $f(p, q+u(q), r, \ldots)$, *etc.*
3. In row 7 enter row 6 minus **A6** (for example, cell **C7** becomes = C6-A6) (Figure 10.2b). Row 7 then contains the values of $u_i(y)$, calculated as

$$u_i(y) = f(p+u(p), q, r, \ldots) - f(p, q, r, \ldots), \text{ etc.} \tag{10.12}$$

Measurement Uncertainty   169

a)

	A	B	C	D	E
1					
2	p	u(p)	p	p	p
3	q	u(q)	q	q	q
4	r	u(r)	r	r	r
5					
6	y = f(p,q,..)		y = f(p,q,..)	y = f(p,q,..)	y = f(p,q,..)
7					
8					

b)

	A	B	C	D	E
1					
2	p	u(p)	p+u(p)	p	p
3	q	u(q)	q	q+u(q)	q
4	r	u(r)	r	r	r+u(r)
5					
6	y = f(p,q,..)		y = f(p',...)	y = f(..q',..)	y = f(..r',..)
7			= C6-A6	= D6-A6	= E6-A6
8					

c)

	A	B	C	D	E
1					
2	p	u(p)	p+u(p)	p	p
	q	u(q)	q	q+u(q)	q
4	r	u(r)	r	r	r+u(r)
5					
6	y = f(p,q,..)	u(y)[a]	y = f(p',...)	y = f(..q',..)	y = f(..r',..)
7			= C6-A6	= D6-A6	= E6-A6
8					

[a] $u(y)$: The formula is =SQRT(SUMSQ(C7:E7)).

Figure 10.2 Spreadsheet calculation of combined uncertainty.

4. To obtain the combined standard uncertainty in y, these individual contributions are squared, added together and then the square root taken. The shortest method is to set cell B6 to the formula

=SQRT(SUMSQ(C7:E7))

which gives the standard uncertainty in y (Figure 10.2c).

The contents of the cells C7, D7, *etc.* show the contributions $u_i(y) = c_i u(x_i)$ of the individual uncertainty components to the uncertainty in y, making it easy to see which components are significant.

The principle can be extended to cope with correlated uncertainty contributions if necessary.[4,5]

10.3.2 Alternative Approaches to Uncertainty Evaluation – Using Reproducibility Data

There are two broad approaches to uncertainty evaluation:

1. Study of individual contributions to the overall uncertainty and combination using an appropriate mathematical model.
2. Studies involving replication of the whole measurement procedure to give a direct estimate of the uncertainty for the final result of the measurement.

The first is the procedure given in the GUM and described in Section 10.2. It is sometimes called a 'bottom up' approach. It is most applicable where the major effects on the result can be described by theoretical models, although the general principles of combining standard uncertainties apply fairly widely.

The second is typified by use of data from interlaboratory study, in-house validation or ongoing quality control. Because it uses values describing the performance of the whole method, it is sometimes referred to as a 'top down' approach. This approach is particularly appropriate where individual effects are poorly understood. 'Poorly understood' in this context does not imply lack of knowledge or experience on the analyst's part; only a lack of quantitative theoretical models capable of predicting the behaviour of analytical results for particular sample types. A good, detailed procedure applicable where interlaboratory study data are available is provided by the document ISO/TS 21748:2004.[6] Briefly, this approach takes the reproducibility standard deviation, or some other overall precision estimate, as the basis for an uncertainty estimate and adds additional contributions as required using the principles of the GUM. In many cases, it is found that no additional terms are required, as few are significant compared with reproducibility from interlaboratory data or long-term within-laboratory precision.

These two approaches should generally be seen as extremes of a range of approaches. In practice, even interlaboratory study may be unable to explore all the possible effects operating and in-house studies or theoretical considerations are needed to at least check some possible effects. At the other extreme, even well-understood systems often need experimental data on, for example, random effects, which can only be obtained by replication experiments.

10.4 A Basic Methodology for Uncertainty Estimation in Analytical Science

For many routine analytical measurements, method development and validation proceed until the well-understood and controllable effects are tested and shown to have essentially negligible contributions to the dispersion of results. Because of this, analytical measurement uncertainty is very often dominated by poorly understood effects which appear only in the dispersion of results over extended times or in interlaboratory comparisons. This suggests that a simplified methodology should be applicable to many routine analytical measurements. This methodology can be summarised as follows:

1. Write the specification of the measurand and the complete equation used to calculate the analytical results from all intermediate measurements and equipment (this information will normally be found in the Operating Procedure for the method).
2. Obtain a good estimate of the long-term precision of the analytical process. This should be, for example, the precision for a typical test material used as a QC material; an estimate of within-laboratory reproducibility obtained from one-way ANOVA applied to the data from a multi-day experiment (see Chapter 6, Section 6.3, and Chapter 9, Section 9.2.2); or a published reproducibility standard deviation for the method in use. In the last case, the performance of the laboratory should be checked to ensure that it is performing with precision and bias consistent with the published reproducibility standard deviation or RSD. Repeatability data are *not* sufficient for uncertainty estimation.
3. Using the spreadsheet method in Section 10.3.1, check that uncertainties for parameters in the measurement equation are small compared with the available precision estimate. In general, an uncertainty less than 20% of the largest uncertainty contribution may be neglected.
4. Identify any other suspected effects (for example, change of LC column, effect of flow rates, effect of reagent concentrations) and, using validation data or new experimental data, test that they are indeed negligible. Ruggedness studies (see Chapter 9, Section 9.7) are particularly well suited to this task.
5. If all the effects studied in steps 3 and 4 are negligible, use the precision estimate in step 2 as the standard uncertainty. If any other effects prove significant, make appropriate allowances and combine them with the precision data using the usual ISO procedure.

Finally, it is always important to check that uncertainty estimates are consistent with actual performance. If a model-based approach estimates uncertainties at 10% and the observed performance shows 30–50% variation, the model is almost certainly incomplete; in general, uncertainty calculated using a modelling approach should not be less than the observed dispersion of results over a period of time. Monitoring proficiency testing results can also provide a good indication of whether the uncertainty estimate is realistic.[7]

References

1. ISO Guide 98:1995, *Guide to the Expression of Uncertainty in Measurement*, International Organization for Standardization, Geneva, 1995, ISBN 92-67-10188-9.
2. ISO Guide 99:2007, *International Vocabulary of Metrology – Basic and General Concepts and Associated Terms (VIM)*, International Organization for Standardization, Geneva, 2007. Also available for free download as JCGM 200:2008 at the BIPM website, http://www.bipm.fr.
3. S. L. R. Ellison, M. Roesslein and A. Williams (Eds), *Eurachem/CITAC Guide: Quantifying Uncertainty in Analytical Measurement*, 2nd edn, 2000, ISBN 0-948926-15-5. Available via http://www.eurachem.org (accessed 11 December 2008).
4. J. Kragten, *Analyst*, 1994, **119**, 2161.
5. S. L. R. Ellison, *Accred. Qual. Assur.*, 2005, **10**, 338.

6. ISO/TS 21748:2004, *Guidance for the Use of Repeatability, Reproducibility and Trueness Estimates in Measurement Uncertainty Estimation*, International Organization for Standardization, Geneva, 2004.
7. AMC Technical Brief No. 15, *Is My Uncertainty Estimate Realistic?*, Analytical Methods Committee, Royal Society of Chemistry, Cambridge, 2003; Available from http://www.rsc.org/amc/(accessed 2 April 2009).

CHAPTER 11
Analytical Quality Control

11.1 Introduction

Analytical quality control (QC) procedures are used to provide evidence of the reliability of analytical results. Statistical control implies that factors affecting the uncertainty of measurement results have not deteriorated since method validation. An important QC activity is the analysis of QC materials. A QC material is a material of known composition which is similar to test materials in terms of the matrix and analyte concentration. A QC material may be prepared by the laboratory for the purpose of quality control or it may be excess test material from a previous batch of analysis. Some reference material suppliers also sell materials for QC purposes. The QC material must be sufficiently homogeneous and stable and available in sufficient quantity to allow it to be analysed regularly over an extended time period. The QC material is analysed alongside test materials and should receive exactly the same treatment as the test materials. The data from the analysis of QC materials are most readily evaluated by plotting on a control chart.

Control charts are used to monitor the performance of a measurement system over time. We need to be able to identify when the variation in measurement results increases or when drift or bias occur. There are a number of different types of control chart but they all involve plotting data in a time-ordered sequence, for example the results obtained from successive analysis of a quality control material.

This chapter introduces Shewhart charts and CuSum charts. Further information on QC in analytical chemistry laboratories can be found in references 1–3.

11.2 Shewhart Charts

11.2.1 Constructing a Shewhart Chart

The Shewhart chart is a sequential plot of the observations obtained from a QC material analysed in successive runs, together with 'warning' and 'action' limits which are used to identify when problems with the measurement results have arisen. The observations plotted are the mean QC result for each run if more than one result is obtained per run or single observations if only one QC result is obtained per run. Figure 11.1 shows an example of a Shewhart chart using single observations.

To construct the action and warning limits, a target value T or mean μ and the standard deviation σ for the QC observations are estimated. These values are used to calculate the warning and action

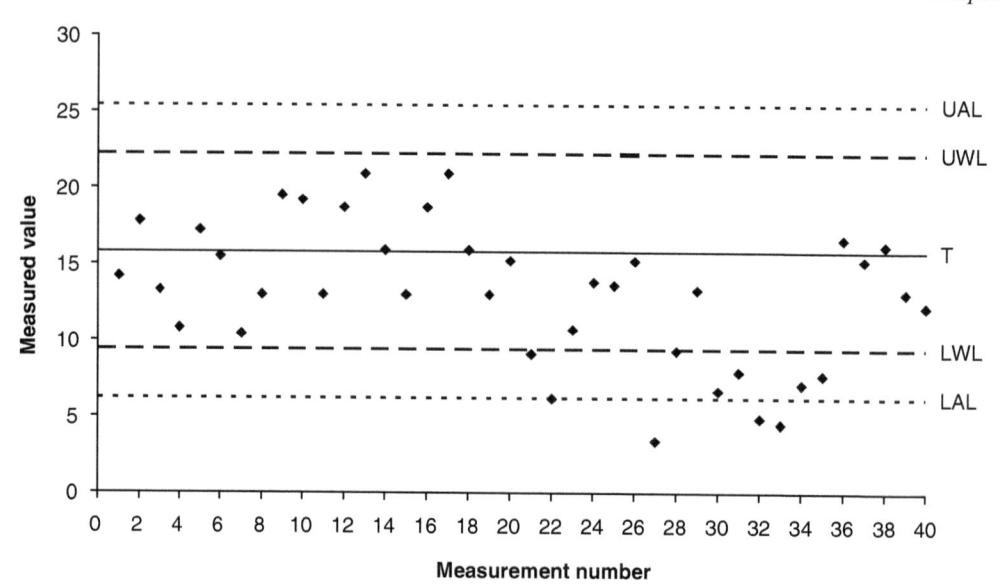

Figure 11.1 Shewhart chart. UAL=upper action limit; UWL=upper warning limit; LWL=lower warning limit; LAL=lower action limit; T=target value.

limits, as in the following equations:

$$\text{Warning limits:} \quad \mu \pm 2\sigma \quad (11.1)$$
$$\text{Action limits:} \quad \mu \pm 3\sigma \quad (11.2)$$

For charts involving only one QC measurement per run, μ and σ are generally obtained by calculating the mean and standard deviation of a number of results obtained from the analysis of the QC material in different runs. In the example shown in Figure 11.1, the target value T was set to the mean of the first 20 results plotted on the chart, and, since only one QC observation was taken for each run, σ was set to the standard deviation of the first 20 observations.

Where more than one QC result is averaged per run, the standard deviation σ must be calculated from separate estimates of within- and between-run variances. These are usually obtained by analysis of variance (ANOVA) applied to the results of several runs as described in Chapter 6, Section 6.3.2. The calculations in Chapter 6, Table 6.5, provide an estimated within-group variance s_w^2 and between-group variance s_b^2. Using these and with n QC results averaged per run, σ is calculated as

$$\sigma = \sqrt{s_b^2 + \frac{s_w^2}{n}} \quad (11.3)$$

Note that where single QC results are taken per run, $n = 1$. σ may then be estimated directly from the standard deviation of single results obtained in different runs.

Shewhart charts depend heavily on good estimates of σ for successful operation. Estimates of standard deviation can, however, be very variable when small numbers of observations are used (Chapter 9, Section 9.2.2). Initial set-up should therefore use as many observations as is practical.

Remember also that the QC material should be treated in exactly the same way as test materials. For example, if test materials are analysed in duplicate and the mean of the results is reported, the QC material should also be analysed in duplicate. The mean of the duplicate results should be plotted on the chart and σ calculated using $n = 2$ in equation (11.3).

Analytical Quality Control

Finally, when setting up a QC chart, it is important to make sure, as far as possible, that only valid results are included in the data used to set up the limits. Outlying values should be inspected carefully and removed if there is reason to suspect a problem. It may also be useful to use robust statistics to estimate μ and σ (Chapter 5 gives details of outlier detection methods and of robust statistics). It is also useful to revise the limits after a modest period of operation; early data may be more (or less) subject to mistakes and out-of-control conditions and it is usually useful to revisit the limits when the method has had time to settle into routine operation.

11.2.2 Shewhart Decision Rules

For a normal distribution, the warning limits calculated in equation (11.1) correspond approximately to the 95% confidence level and the action limits in equation (11.2) would correspond approximately to the 99.7% confidence level. Results outside action limits are therefore very unlikely to occur by chance, while results outside warning limits are expected to occur, on average, about once per 20 observations.

If a result falls outside the action limits, the measurement process should be stopped and an investigation into the cause of the problem carried out. A laboratory should also have defined procedures in place setting out the action to be taken if results fall between the warning limits and the action limits.

In addition to using the action and warning limits, one should look for any trends in the data. If the system is under statistical control, the results should be randomly distributed about the central line. A significant drift in data or a step change could indicate a problem and the cause should be investigated. Patterns of results which should give cause for concern include the following:

- two successive points outside warning limits but inside action limits (e.g. points 21 and 22 in Figure 11.1);
- nine successive points on the same side of the mean (e.g. points 19 to 27 in Figure 11.2);
- six consecutive points steadily increasing or decreasing.

These three conditions should normally be treated as if an action limit were exceeded. ISO 8258:1991 gives additional rules for identifying abnormal patterns in the data.[4]

11.3 CuSum Charts

11.3.1 Constructing a CuSum Chart

Constructing a CuSum chart involves plotting the cumulative sum of the deviation of QC results from a reference or target value. CuSum charts are better at highlighting small changes in the mean value of the measurements (compared with the target value) than Shewhart charts. Another advantage of CuSum charts is that they may be used for '*post mortem*' investigations, that is, the retrospective examination of historical data. By investigating information from past records it may be possible to determine the cause of unexpected changes in results.

As with Shewhart charts, a target value needs to be set. This can be obtained from repeated analysis of the QC material. Once the target value has been set, the difference between each subsequent result from the analysis of the QC material and the target value is calculated and added to a running total. This is the cumulative sum, C_i, that is plotted on the CuSum chart:

$$C_i = \sum_{j=1,i} (\bar{x}_j - \mu) \qquad (11.4)$$

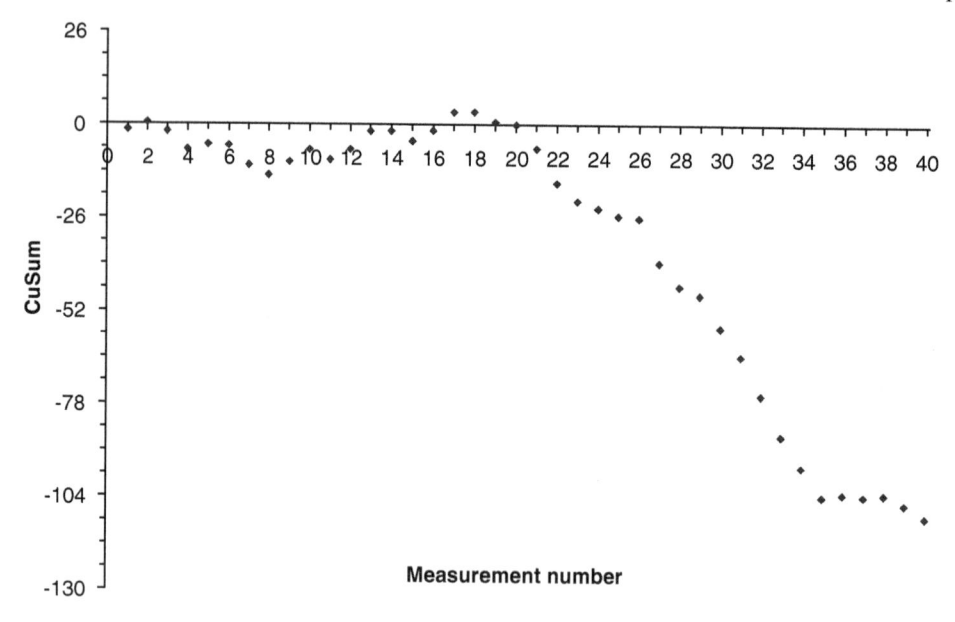

Figure 11.2 CuSum chart.

where $\bar{x}_j$ is a measurement result obtained for the QC material (this may be a single result or the average of n results if the QC material is analysed n times), μ is the target value and i is the measurement number. C_i is therefore simply the sum of all differences so far, up to and including the current observation $\bar{x}_i$. An example of a CuSum chart is shown in Figure 11.2. This chart was constructed from the data plotted on the Shewhart chart shown in Figure 11.1. The calculation of the cumulative sum is shown in Table 11.1. The target value, μ, is 15.8. As an example, note that C_4, corresponding to measurement number 4, is simply the sum of all the differences to that point, that is,

$$C_4 = (\bar{x}_1 - \mu) + (\bar{x}_2 - \mu) + (\bar{x}_3 - \mu) + (\bar{x}_4 - \mu)$$
$$= (-1.6) + (2.0) + (-2.5) + (-5.0) = -7.1.$$

C_4 can also be calculated from C_3:

$$C_4 = C_3 + (\bar{x}_4 - \mu) = (-2.1) + (-5.0) = -7.1$$

which is a particularly simple form when using a spreadsheet.

If the measurement system is operating such that the mean is close to the target value, the gradient of the CuSum plot will be close to zero. A positive gradient indicates that the operating mean is greater than the target value, whereas a negative gradient indicates an operating mean that is less than the target value. A step change in a set of data shows up in a CuSum as a sudden change of gradient. Gradual drift in a system causes small but continuous changes to the mean which translate into a constantly changing gradient, causing a curve in the CuSum plot. In Figure 11.2 there appears to be a sharp change in gradient around measurement number 21. The gradient of the plot becomes negative, indicating that the operating mean is now lower than the target value.

Analytical Quality Control

Table 11.1 Calculation of the cumulative sum.

Measurement number	QC result $\bar{x}_i$	$\bar{x}_i - \mu$	C_i	Measurement number	QC result $\bar{x}_i$	$\bar{x}_i - \mu$	C_i
1	14.2	−1.6	−1.6	21	9.1	−6.7	−6.7
2	17.8	2.0	0.4	22	6.2	−9.6	−16.3
3	13.3	−2.5	−2.1	23	10.7	−5.1	−21.4
4	10.8	−5.0	−7.1	24	13.8	−2.0	−23.4
5	17.2	1.4	−5.7	25	13.6	−2.2	−25.6
6	15.5	−0.3	−6.0	26	15.2	−0.6	−26.2
7	10.4	−5.4	−11.4	27	3.4	−12.4	−38.6
8	13.0	−2.8	−14.2	28	9.3	−6.5	−45.1
9	19.5	3.7	−10.5	29	13.3	−2.5	−47.6
10	19.2	3.4	−7.1	30	6.7	−9.1	−56.7
11	13.0	−2.8	−10.0	31	7.9	−7.9	−64.7
12	18.7	2.9	−7.1	32	4.9	−10.9	−75.6
13	20.9	5.1	−2.0	33	4.5	−11.3	−86.9
14	15.9	0.1	−1.9	34	7.1	−8.7	−95.6
15	13.0	−2.8	−4.7	35	7.7	−8.1	−103.7
16	18.7	2.9	−1.8	36	16.6	0.8	−102.9
17	20.9	5.1	3.3	37	15.2	−0.6	−103.5
18	15.9	0.1	3.4	38	16.2	0.4	−103.1
19	13.0	−2.8	0.6	39	13.1	−2.7	−105.8
20	15.2	−0.6	0.0	40	12.2	−3.6	−109.4

11.3.2 CuSum Decision Rules

CuSum charts cannot be interpreted using warning and action limits of the type used in the interpretation of Shewhart charts. In a CuSum chart, significant deviations from the target value are indicated by changes in the gradient of the CuSum plot. Changes in gradient can be checked using a 'V-mask'. An example is shown in Figure 11.3.

When using a V-mask, the relationship between the scales used on the *x*- and *y*-axes in a CuSum chart is important. The axes are conventionally scaled so that the divisions on both axes are the same length when plotted. A division on the *x*-axis should represent a single unit, whereas a division on the *y*-axis should be equivalent to $2\sigma/\sqrt{n}$ where σ and n are as defined in equation (11.3). In the example shown in Figure 11.2, σ was equal to 3.2 and n equal to 1.

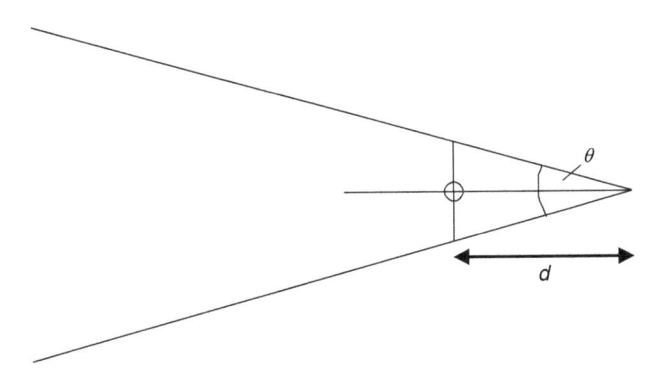

Figure 11.3 Full V-mask.

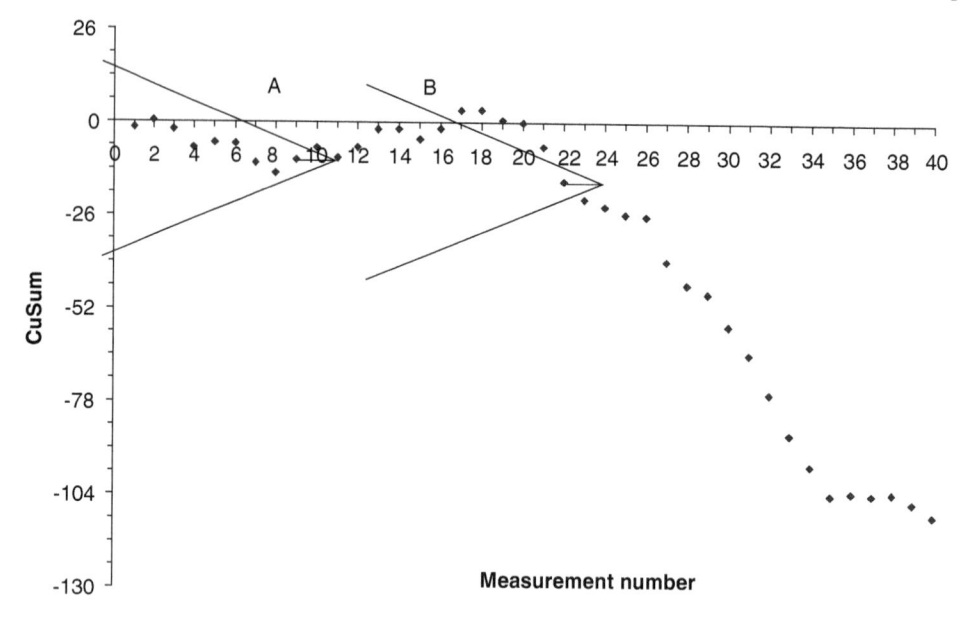

Figure 11.4 Use of a V-mask to interpret a CuSum chart.

The V-mask itself is normally made of a transparent material so that it can be laid over the CuSum chart. The data are evaluated by laying the mask over the chart with the cross (⊕) in Figure 11.3 overlaying each point in turn. The horizontal line in the centre of the V-mask is always kept parallel to the x-axis. If all the preceding data lie within the arms of the mask (or their projections), the system is in control. This situation is shown in scenario A in Figure 11.4, which shows application of the V-mask to point 9. When preceding data points fall outside the arms of the mask, the system has fallen out of control, as shown in scenario B in Figure 11.4.

The control limits for a CuSum chart are therefore defined by the values of d and θ used to construct the V-mask. The aim is to be able to identify quickly when the system has gone out of control (*i.e.* when measurement results deviate significantly from the target value) but to avoid too many 'false alarms'. The dimensions of the V-mask can be chosen by trial and error or by examining historical data. The mask should be designed to give an early warning that the data have gone out of control. If historical data are not available then an initial mask can be constructed using $d=2$ units along the x-axis and $\theta = 22°$ (assuming the conventional axis scaling described above). If necessary, after the chart has been in use for some time, the dimensions of the V-mask can be adjusted to give the required level of discrimination.

A more detailed explanation of the construction and interpretation of CuSum charts is given in references 5–8.

References

1. M. Thompson and R. Wood, *Pure Appl. Chem.*, 1995, **67**, 649.
2. AMC Technical Brief No. 12, *The J-chart: a Simple Plot that Combines the Capabilities of Shewhart and CuSum Charts, for Use in Analytical Quality Control*, Analytical Methods Committee, Royal Society of Chemistry, Cambridge, 2003; Available from http://www.rsc.org/amc/ (accessed 2 April 2009).
3. R. J. Howarth, *Analyst*, 1995, **120**, 1851.

4. ISO 8258:1991, *Shewhart Control Charts*, International Organization for Standardization, Geneva, 1991.
5. ISO/TR 7871:1997, *Cumulative Sum Charts – Guidance on Quality Control and Data Analysis Using CUSUM Techniques*, International Organization for Standardization, Geneva, 1997.
6. BS 5703-1:2003, *Guide to Data Analysis and Quality Control Using CuSum Techniques. Uses and Value of CuSum Charts in Business, Industry, Commerce and Public Service*, British Standards Institution, Chiswick, 2003.
7. BS 5703-2:2003, *Guide to Data Analysis and Quality Control Using CuSum Techniques. Introduction to Decision-Making Using CuSum Techniques*, British Standards Institution, Chiswick, 2003.
8. BS 5703-3:2003, *Guide to Data Analysis and Quality Control Using CuSum Techniques. CuSum Methods for Process/Quality Control Using Measured Data*, British Standards Institution, Chiswick, 2003.

CHAPTER 12
Proficiency Testing

12.1 Introduction

The primary aim of proficiency testing is to allow laboratories to monitor and optimise the quality of their routine analytical measurements. Proficiency testing is described in several international guidelines and standards, including the IUPAC *International Harmonized Protocol for the Proficiency Testing of Analytical Chemistry Laboratories*[1] and ISO/IEC Guide 43-1:1997, *Proficiency Testing by Interlaboratory Comparisons – Part 1: Development and Operation of Proficiency Testing Schemes.*[2] The discussion and nomenclature in this chapter largely follow the IUPAC Protocol.

A proficiency testing round typically involves the simultaneous distribution of sufficiently homogeneous and stable test materials to participants. The participants analyse the samples using a method of their choice and submit their results to the scheme organisers. The organisers carry out statistical analysis of all the data and provide each participant with a 'score' that allows them to judge their performance in that particular round. There are a number of different scoring systems used in proficiency testing schemes; the majority involve comparing the difference between the participant's result (x) and a target or assigned value (x_a) with a quality target. The quality target is usually expressed as a standard deviation, often called the 'standard deviation for proficiency assessment' and denoted σ_p.[1] Each scoring system has acceptability criteria to allow participants to evaluate their performance.

This chapter provides a brief introduction to some of the key statistical aspects of the organisation of proficiency testing schemes and the interpretation of the results. A detailed discussion can be found in references 1–3.

12.2 Calculation of Common Proficiency Testing Scores

12.2.1 Setting the Assigned Value and the Standard Deviation for Proficiency Assessment

Two critical steps in the organisation of a proficiency testing scheme are specifying the assigned value and setting the standard deviation for proficiency assessment. These directly influence the scores that participants receive and therefore how they interpret their performance in the scheme.

Proficiency Testing

12.2.1.1 *Assigned Value*

The assigned value is the value attributed to a particular quantity being measured. It is accepted as having a suitably small uncertainty which is appropriate for a given purpose. There are a number of approaches to obtaining the assigned value.

Assigned value obtained by formulation

A known amount or concentration of the target analyte is added to a base material containing no native analyte (or a very small but well characterised amount).

Advantages
- Useful when the assigned value for the proficiency testing round is the amount added to individual test objects (for example, air filters).
- The analyte can be added to the base material accurately by volumetric or gravimetric methods. The uncertainty in the assigned value should therefore be small and relatively easy to calculate.
- The traceability of the assigned value can usually be established.

Disadvantages
- There may be no suitable blank or well-characterised base material available.
- It may be difficult to achieve sufficient homogeneity in the bulk material.
- Formulated samples may not be truly representative of test materials as the analyte may be in a different form or less strongly bound to the matrix.

Assigned value is a certified reference value

If the test material is a certified reference material (CRM), the assigned value is the certified value quoted on the CRM certificate.

Advantages
- The assigned value and its uncertainty are obtained directly from the CRM certificate.
- Traceability of the assigned value is established by the CRM producer.

Disadvantages
- Limited CRM availability so it may not be possible to obtain the required combination of analyte, matrix and analyte level.
- The certified value uncertainty may be high.
- Generally, CRMs are expensive.

Assigned value is a reference value

The assigned value is determined by a single expert laboratory using a primary method of analysis (for example, gravimetry, titrimetry, isotope dilution mass spectrometry) or a fully validated test method which has been calibrated with CRMs.

Advantages
- Likely to be less expensive than distributing CRMs as the test material.
- The assigned value is traceable via a primary method or to the CRM values used for calibration.

Disadvantages
- A suitable primary method or suitable CRMs may not be available.
- Relies on results from a single laboratory.
- Costs may be too high for routine proficiency testing rounds.

Assigned value from consensus of expert laboratories
The assigned value is obtained from the results reported by a number of expert laboratories that analyse the material using suitable methods. The expert laboratories must be able to demonstrate their proficiency in the measurements of interest. The measurements can be carried out as part of the proficiency testing round or prior to the distribution of the test material. The assigned value is normally a robust estimate of the mean of the expert laboratories' results. Robust statistics are discussed in Chapter 5, Section 5.3.

Advantages
- No additional analysis required if results are obtained as part of the proficiency testing round.

Disadvantages
- There may be an unknown bias in the results produced by the expert laboratories.
- The choice of expert laboratories may be contentious.

Assigned value from consensus of scheme participants
The assigned value is obtained from the results from all the participants in the proficiency testing round. The assigned value is normally based on a robust estimate, to minimise the effect of extreme values in the data set (see Chapter 5, Section 5.3).

Advantages
- No additional analysis of the test material prior to distribution of the samples to participants.
- Useful approach for standardised empirical test methods.
- Low cost.

Disadvantages
- There may be no real consensus in the results produced by participants.
- The consensus may be biased, which may perpetuate poor methodology.
- It may be difficult to establish the uncertainty and traceability of the assigned value.

12.2.1.2 Standard Deviation for Proficiency Assessment

The standard deviation for proficiency assessment is set by the scheme organiser, usually with specialist advice, and is intended to represent the uncertainty regarded as fit for purpose for a particular type of analysis. Ideally, the basis for setting the standard deviation should remain the same over successive rounds of the proficiency testing scheme so that interpretation of performance scores is consistent over different rounds. Due allowance for changes in performance at different analyte concentrations is usually made. This makes it easier for participants to monitor their performance over time. There are a number of different approaches to defining the standard deviation for proficiency assessment.

Standard deviation set by prescription
The standard deviation is chosen to ensure that laboratories that obtain a satisfactory score are producing results that are fit for a particular purpose. This may be related to a legislative requirement.

Advantages
- Standard deviation is directly related to fitness for purpose criteria for a particular application.

Disadvantages
- In some cases the fitness for purpose criteria may not be known and there may be no legislative requirement.

Proficiency Testing

The standard deviation can also be set to reflect the perceived current performance of laboratories or to reflect the performance that the scheme organiser and participants would like to be able to achieve. This is sometimes referred to as setting the standard deviation 'by perception'.

Standard deviation based on the results from a reproducibility experiment

If the analytical method used by the participants in the proficiency testing scheme has been the subject of a formal collaborative study, the standard deviation for proficiency assessment can be calculated from the reproducibility and repeatability estimates obtained from the study. The standard deviation for proficiency assessment, σ_p, is given by

$$\sigma_p = \sqrt{\sigma_L^2 + \left(\frac{\sigma_r^2}{n}\right)} \tag{12.1}$$

where

$$\sigma_L = \sqrt{\sigma_R^2 - \sigma_r^2} \tag{12.2}$$

and σ_R is the reproducibility standard deviation obtained from the collaborative study, σ_r is the repeatability standard deviation obtained from the collaborative study and n is the number of replicate measurements that each participant is to perform on the proficiency test sample. Repeatability and reproducibility are discussed in detail in Chapter 9, Section 9.2.

Advantages
- Standard deviation linked directly to expected method performance.

Disadvantages
- Requires data from a collaborative study.
- Participants might use a range of test methods with different performance characteristics.
- 'Expected method performance' may not meet fitness for purpose criteria.

Standard deviation obtained from a general model

In this case the standard deviation for proficiency assessment is derived from a general model which describes method precision, such as the Horwitz function.[4]

The Horwitz function is an empirical relationship based on statistics from a large number of collaborative studies. It describes how the reproducibility standard deviation varies with analyte level:

$$\%CV = 2^{(1-0.5\log C)} \tag{12.3}$$

where %CV is the predicted reproducibility standard deviation (σ_R) expressed as a percentage of the analyte concentration, C. The concentration must be expressed as a decimal fraction (*e.g.* 1 mg kg^{-1} = 10^{-6}). Therefore, if the analyte concentration in the test sample is known, σ_R can be calculated and used to estimate σ_p.

Note that the original Horwitz relationship loses applicability at concentrations lower than about 10 ppb (ppb=10^{-9} mass fraction) and a modified form of the function has been recommended.[5]

Advantages
- Easy to calculate.
- No additional method performance data required.
- Some sectors routinely use models of this kind as criteria for fitness for purpose.

Disadvantages
- The model may not reflect the true reproducibility of certain test methods.

Standard deviation obtained from participants' results

If none of the other approaches described is appropriate, the standard deviation can be obtained from a robust estimate of the standard deviation of the participants' results. Robust estimates of the standard deviation are discussed in Chapter 5, Section 5.3.3.

Advantages
- Obtained directly from data from the proficiency testing round.

Disadvantages
- Estimates of standard deviation may vary substantially from round to round, making it impossible for participants to use z-scores to monitor performance over a number of rounds.
- Not related to fitness for purpose criteria.

Note that because of the disadvantages, this approach is discouraged in the IUPAC protocol.[1]

12.2.2 Scoring PT Results

12.2.2.1 The z-Score

The most common scoring system is the z-score:

$$z = \frac{x - x_a}{\sigma_p} \tag{12.4}$$

where x is the result submitted by the participant, x_a is the assigned value and σ_p is the standard deviation for proficiency assessment.

z-scores are usually interpreted as follows:

$|z| \leq 2$ satisfactory performance
$2 < |z| \leq 3$ questionable performance
$|z| > 3$ unsatisfactory performance.

The interpretation is based on the properties of the normal distribution (see Chapter 3, Section 3.3.1). If all laboratories perform exactly in accordance with the performance requirement, about 95% of values are expected to lie within two standard deviations of the mean. There is then only a 5% chance that a valid result would fall further than two standard deviations from the mean. The probability of finding a valid result more than three standard deviations away from the mean is very low (approximately 0.3% for a normal distribution). A score of $|z| > 3$ is therefore considered unsatisfactory. These designations are most meaningful if the value of σ_p is based on fitness for purpose criteria.

Proficiency Testing

Table 12.1 shows the data and z-scores from a round of a proficiency testing scheme for the analysis of contaminated soil samples. In this round, participants were required to determine the concentration of nickel obtained from a soil sample using an aqua regia extraction procedure. The assigned value (x_a) was based on consensus of the participants' results and calculated as the median of the results. The standard deviation for proficiency assessment (σ_p) was based on a fitness for purpose criterion and set at 10% of the median value. Figure 12.1 shows a plot of the z-scores shown in Table 12.1.

In this round, the performance of laboratories 17, 22 and 25 was unsatisfactory as they received scores of $|z| > 3$. The performance of laboratories 1, 6 and 21 was questionable as their z-scores were in the range $2 < |z| \leq 3$. The performance of all other participants was considered satisfactory.

12.2.2.2 The Q-Score

An alternative scoring system is the Q-score:

$$Q = \frac{x - x_a}{x_a} \qquad (12.5)$$

This is simply a relative measure of laboratory bias and, unlike the z-score, takes no account of a target standard deviation. In the ideal situation where there is no bias in the participants' measurement results, the distribution of Q-scores will be centred on zero. Because Q-scores are not based on a fitness for purpose criterion, interpretation relies on an external prescription of acceptability. For example, the participants or scheme organiser may set criteria for acceptable percentage deviation from the target value and the Q-score is compared with these criteria.

12.3 Interpreting and Acting on Proficiency Test Results

The statistical interpretation of z-scores was discussed in the previous section; for example, a score of $|z| > 3$ is generally considered unsatisfactory. A laboratory should take action following any unsatisfactory score (this is a requirement for laboratories accredited to ISO/IEC 17025[6]). A laboratory is also advised to take action if two consecutive questionable results for the same measurement are obtained (that is, $2 < |z| \leq 3$) or if nine consecutive z-scores of the same sign are obtained. Many good laboratories also choose to investigate questionable results in case a need for corrective action is found.

Table 12.1 Data and z-scores from a proficiency testing round for the determination of aqua regia extractable nickel from soil: $x_a = 198$ mg kg^{-1}, $\sigma_p = 19.8$ mg kg^{-1}.

Laboratory ID No.	Ni/mg kg^{-1}	z-Score	Laboratory ID No.	Ni/mg kg^{-1}	z-Score
1	140	−2.93	15	207.07	0.46
2	178	−1.01	16	159	−1.97
3	165	−1.67	17	260	3.13
4	211.3	0.67	18	215	0.86
5	224	1.31	19	165	−1.67
6	241	2.17	20	180	−0.91
7	162.33	−1.80	21	147	−2.58
8	213	0.76	22	133.1	−3.28
9	197	−0.05	23	226.4	1.43
10	232	1.72	24	177.2	−1.05
11	202	0.20	25	313.3	5.82
12	196	−0.10	26	195	−0.15
13	198	0.00	27	214.9	0.85
14	208	0.51			

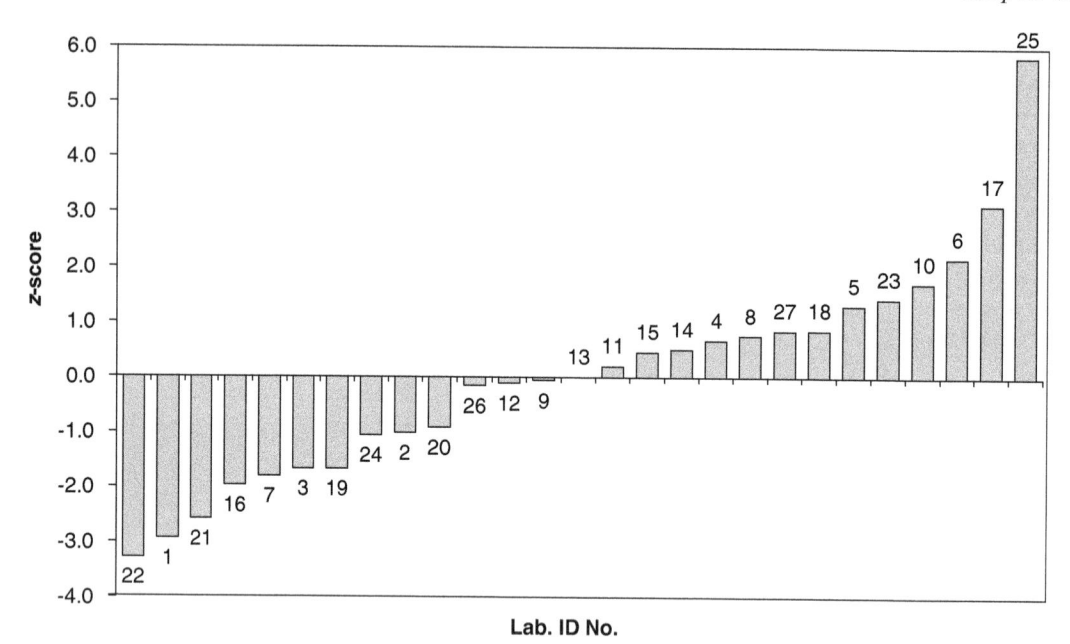

Figure 12.1 Plot of z-scores from proficiency testing round for the determination of aqua regia extractable nickel from soil (data taken from Table 12.1).

If an unsatisfactory score is obtained there are a number of issues for the laboratory to consider:

- The overall standard of performance for the round:
  — Did a large number of participants obtain unsatisfactory results? If so, the problem may not lie within the laboratory.

- Test method performance:
  — Which test methods were used by the other participants in the round? Did other laboratories use methods with very different performance characteristics?
  — How was the standard deviation for proficiency assessment established? Was it appropriate for the laboratory's own needs?

- Test sample factors:
  — Was the material for that round within the scope of the laboratory's normal operations? Proficiency testing schemes often cover a range of materials appropriate to the scheme, but individual laboratories may receive materials that differ in composition from their routine test samples.

- Proficiency testing scheme factors:
  — How many results were submitted? Small numbers of results can make it difficult to establish the assigned value if the consensus approach is used.
  — Were there any problems with the organisation of that particular round? Occasionally there may be problems such as unexpected test material behaviour, data entry or reporting errors, software problems or unsuitable evaluation criteria (for example, choice of assigned value or standard deviation for proficiency assessment).

Proficiency Testing

If none of the above applies, the laboratory should investigate further to try to identify the cause of the unsatisfactory result and implement and document any appropriate corrective actions. There are many possible causes of unsatisfactory performance. These can be categorised as analytical errors or non-analytical errors; both are equally important.

Examples of analytical errors
- incorrect calibration of equipment
- analyst error such as incorrect dilution of samples or standards
- problems with extraction and clean-up of samples, such as incomplete extraction of the analyte from the sample matrix
- interferences
- performance characteristics of chosen test method not fit for purpose
- instrument performance not optimised.

Examples of non-analytical errors
- calculation errors
- transcription errors
- results reported in incorrect units or incorrect format.

12.4 Monitoring Laboratory Performance – Cumulative Scores

Performance over time can be monitored using parameters such as the rescaled sum of z-scores (RSZ) and the sum of squared z-scores (SSZ)

$$\text{RSZ} = \sum_{i=1,n} \frac{z_i}{\sqrt{n}} \qquad (12.6)$$

$$\text{SSZ} = \sum_{i=1,n} z_i^2 \qquad (12.7)$$

where n is the number of scores to be combined. The RSZ can be interpreted in the same way as a z-score. A disadvantage of this approach is that if the majority of the z-scores are acceptable but there are a small number of relatively high scores, the RSZ will 'hide' the poor scores. Further, scores with opposite signs may cancel to give a misleadingly low RSZ.

SSZ is sensitive to single high z-scores and is unaffected by the direction of the deviation. Interpretation of SSZ is complicated by the fact that the SSZ is expected to increase with increasing numbers of z-scores. For normally distributed z, however, the SSZ is expected to be distributed as chi-squared with n degrees of freedom, so, for example, if SSZ exceeds the upper 95th percentile of the chi-squared distribution, the SSZ should be considered to be evidence of 'questionable' performance. For example, for a combination of five z-scores, an SSZ exceeding 11.1 would be considered 'questionable'.

Although these cumulative scores offer an index of performance over time or over several analytes, their tendency to understate important individual aberrations makes them unsuitable for monitoring performance in isolation. Neither of the major proficiency testing guidelines recommends their use.[1,2] Instead, it is recommended that performance should always be monitored primarily on individual scores and that if an overview is required, simple graphical approaches should be used as shown in Figure 12.2. This will show whether unsatisfactory or questionable

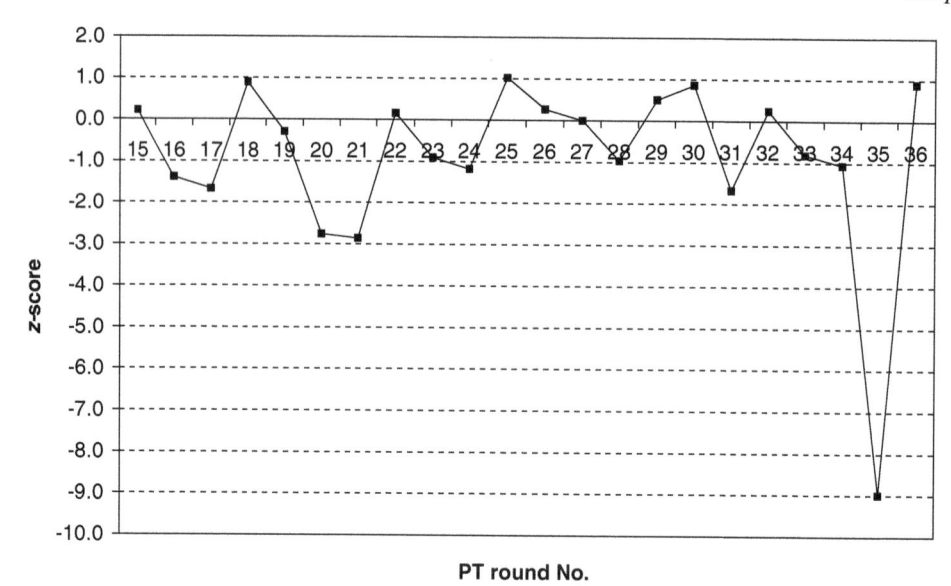

Figure 12.2 Plot of z-scores from a single laboratory for the determination of nickel in soil by aqua regia extraction, last 22 rounds.

results are a one-off or part of a longer term trend. Some useful alternative displays and monitoring methods for summarising performance over several analytes and several rounds have also been described in a freely available AMC Technical Brief.[7]

12.5 Ranking Laboratories in Proficiency Tests

It is often tempting to look at the ranking of laboratories in a proficiency testing round to see which laboratory is 'best'. Unfortunately, ranking is not a reliable indication of performance; even a very imprecise laboratory can approach the assigned value closely by chance and, in a group of equally competent laboratories, ranking is essentially random from round to round. For this reason, most internationally accepted guidelines for proficiency testing strongly discourage ranking as a performance indicator.

References

1. M. Thompson, S. L. R. Ellison and R. Wood, *Pure Appl. Chem.*, 2006, **78**, 145.
2. ISO/IEC Guide 43-1:1997, *Proficiency Testing by Interlaboratory Comparisons – Part 1: Development and Operation of Proficiency Testing Schemes*, International Organization for Standardization, Geneva, 1997.
3. ISO 13528:2005, *Statistical Methods for Use in Proficiency Testing by Interlaboratory Comparisons*, International Organization for Standardization, Geneva, 2005.
4. R. Albert and W. Horwitz, *Anal. Chem.*, 1997, **69**, 789.
5. M. Thompson, *Analyst*, 2000, **125**, 385.
6. ISO/IEC 17025:2005, *General Requirements for the Competence of Testing and Calibration Laboratories*, International Organization for Standardization, Geneva, 2005.
7. AMC Technical Brief No. 16, *Proficiency Testing: Assessing z-Scores in the Longer Term*, Analytical Methods Committee, Royal Society of Chemistry, Cambridge, 2007. Available via http://www.rsc.org/amc/.

CHAPTER 13
Simple Sampling Strategies

13.1 Introduction

Sampling is the process of selecting a portion of material to represent or provide information about a larger body of material. ISO/IEC 17025[1] defines sampling as 'A defined procedure whereby a part of a substance, material or product is taken to provide for testing or calibration a representative sample of the whole. Sampling may also be required by the appropriate specification for which the substance, material or product is to be tested or calibrated'.

The analyst often – and unfortunately so – has little or no involvement with the selection of the samples that are submitted to the laboratory for analysis. However, it is important to remember that if an appropriate sampling strategy has not been used prior to analysis, the results produced in the laboratory will be of little value. It is frequently found that sampling introduces the great majority of the uncertainty associated with measurements. Careful planning and execution of a sampling protocol tailored for the specific application are therefore important.

Planning and implementation of sampling, especially outside the laboratory, are far beyond the scope of this book. Good introductions[2,3] and an advanced treatment[4] can be found elsewhere; reference 2 in particular includes a useful bibliography. This chapter is intended only to give a brief introduction to some of the strategies available for sampling and their strengths and weaknesses. Two of these, simple random sampling and stratified random sampling, are often useful within a laboratory setting and the statistical procedures required to obtain mean values and standard errors from these two strategies are provided.

13.2 Nomenclature

The word 'sample' means different things to analysts and statisticians. To the former, a sample is usually a single quantity of material subjected to analysis; to the latter, a sample is typically a subset of items taken from a larger population. In this chapter, therefore, we adopt the following terms:

Bulk The material which is, ultimately, the subject of the measurement and from which a sample is taken.
Increment Individual portion of material collected by a single operation of a sampling device.[5,6]

Laboratory sample Sample as prepared for sending to the laboratory and intended for inspection or testing.[7]
Note: here, a laboratory sample may comprise several distinct items.
Sub-sample 'A selected part of a sample',[8] usually, here, a selected part of the laboratory sample.
Test sample Sample, prepared from the laboratory sample, from which *test portions* are removed for testing or analysis.[5,6]
Test portion Quantity of material, of proper size for measurement of the concentration or other property of interest, removed from the test sample.[5]

13.3 Principles of Sampling

13.3.1 Randomisation

In general, random sampling is a method of selection whereby each possible member of a population has an equal chance of being selected. As in experimental design (see Chapter 8), randomisation in sampling is one of the most important defences against unintended bias. A random sample provides an unbiased estimate of the population parameters of interest (usually analyte concentrations).

13.3.2 Representative Samples

In ordinary usage, 'representative' means something like 'sufficiently like the population to allow inferences about the population'. A random sample provides unbiased estimates of the mean for a bulk material and in that sense is always representative. However, in statistics, 'unbiased' means that the average over a large number of operations is expected to be equal to the true mean value. This says nothing about how close an *individual* sample's composition will be to the composition of the bulk. It is entirely possible that the composition of a particular sample taken at random will be completely unlike the bulk composition. The larger the sampling variance, the more likely it is that individual samples will be very different from the bulk.

In practice, therefore, representative sampling involves obtaining samples which are not only unbiased, but which also have sufficiently small variance for the task in hand. The various sampling strategies described in Section 13.4 have different variance properties; ideally, the analyst should select those that provide small variance in addition to choosing randomisation procedures that provide unbiased results. Within a laboratory, where the laboratory sample is comparatively accessible and sampling costs are relatively low, variants on stratified random sampling are likely to be the most useful for all but the simplest situations.

13.3.3 Composite Samples

Often, it is possible and useful to combine a collection of sub-samples (then usually called 'increments') into a single homogenised test sample for analysis. The measured value for the composite is then taken as an estimate of the mean value for the bulk material. The principal advantage is a reduction in analytical effort.

Not all sampling strategies deliver a set of increments which provide a composite sample with the same mean as the bulk material. Simple random sampling does so. Stratified random sampling (Section 13.4.2) only does so if the strata are sampled proportionately. It is therefore important to choose sampling strategies carefully when forming a composite sample.

13.4 Sampling Strategies

13.4.1 Simple Random Sampling

13.4.1.1 Description

In simple random sampling, each item in the population (or laboratory sample) has an equal chance of being selected. For example, when the quality of manufactured products is being monitored, a table of random numbers is frequently used to ensure that each item in a batch of the product has an equal chance of being selected for testing.

A variety of simple procedures may be used to provide random sampling in a laboratory setting. For sub-sampling from a set of discrete items, such as ampoules or food packages, the usual method is to number the items sequentially, then use a table of random numbers or random numbers generated from software to select items randomly from the set (Figure 13.1). For particulate materials, sub-samples can be obtained by repeated cone-and-quartering, riffle sampling or rotatory sampling, with due care to avoid bias from size or density selection. The common feature of all these procedures is that they are intended to generate sub-samples in which every member of the population (laboratory sample) has an equal chance of appearing in a sub-sample.

13.4.1.2 Advantages and Disadvantages

Simple random sampling is simple to implement and often sufficient for sub-sampling within a laboratory. Simple random samples can, however, include very variable intervals between successive test items (Figure 13.1), making it a poor choice for monitoring items from a continuous sequence. Further, if the material is inhomogeneous and particularly if it consists of identifiable sub-groups with substantially different properties, the variance in simple random sampling is among the highest of the strategies described here.

13.4.1.3 Statistical Treatment

Statistical treatment for random samples is comparatively straightforward. Where the test samples are measured separately, the simple mean $\bar{x}$ of the measured values x_i for each test sample provides an unbiased estimate of the bulk material composition.

The uncertainty in $\bar{x}$ due to sampling variation (usually, for within-laboratory sampling, arising from inhomogeneity in the material) depends on the number of test samples n taken relative to the number of items N in the population. For example, in taking 10 tins of a product as representative of a laboratory sample of 50 tins, $n = 10$ and $N = 50$. The standard error $s(\bar{x})$ in $\bar{x}$ arising from

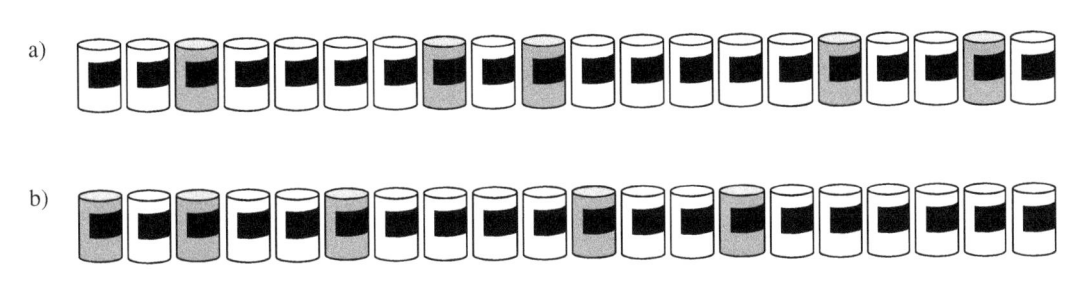

Figure 13.1 Simple random samples. Two simple random samples of five items (shown in dark grey) from a laboratory sample of 20 items.

sampling is given by

$$s(\bar{x}) = s_{\text{sam}}\sqrt{\frac{1-f}{n}} \tag{13.1}$$

where $f = n/N$ and s_{sam} is the standard deviation found for the sampling process. If the analytical uncertainty is small, s_{sam} is simply the standard deviation s of the observations x_i. Otherwise, if the test samples are measured with replication, s_{sam} can be obtained from the between-group component of variance from a one-way ANOVA (s_b in Chapter 6, Section 6.3.2.2).

When N is very large, $s(\bar{x})$ converges to $s_{\text{sam}}/\sqrt{n}$ as usual; for $n < 0.1N$ the correction f can usually be ignored for analytical purposes. n/N should also be considered small, and f ignored, if the test samples to hand represent a random sample from a much wider population. There is also an argument for treating sampling from particulate materials as a large-N situation (because the number of possible different sub-samples is extremely large).

As n approaches N the uncertainty associated with sampling variability approaches zero; if we have sampled the entire population, there is no remaining possibility of variation due to a different choice of test samples.

13.4.2 Stratified Random Sampling

13.4.2.1 Description

The population being sampled is divided into segments (or strata) and a simple random sample is selected from each. Figure 13.2 shows a simple example applied to a sequence of items. The number of items selected from each segment depends on the object of the exercise.

Three particular variants are important:

1. *Equal number of items per stratum:* The same number of items is selected from each identified stratum (Figure 13.3a). This is convenient and common when there is no information about the variance within each stratum. Composite samples may be formed within a stratum, but a composite of the whole does not provide an unbiased estimate of the bulk sample composition unless all the strata are of equal size.
2. *Proportional sampling:* The number of items taken per stratum is proportional to the fraction of each stratum in the bulk (Figure 13.3b). This is ideal for forming a composite sample, as the expected mean for the composite is the same as the mean for the bulk. The variance is smaller than for simple random sampling and often sufficiently close to optimal (see below) to provide a reasonable compromise when the stratum variances are unknown.
3. *Optimal allocation:* Optimal allocation of samples sets the number of samples per stratum in proportion to the size and the sampling standard deviation of each stratum. Although rarely possible without substantial prior information, this strategy provides the smallest sampling

Figure 13.2 Stratified random sampling for sequences. A stratified random sample of five items, shown in dark grey, is formed by dividing the laboratory sample into five strata on the basis of sequence number, and selecting one sample at random from within each stratum. Vertical lines show the divisions between strata. This limits the interval between selected items while still providing a random selection of items.

Simple Sampling Strategies

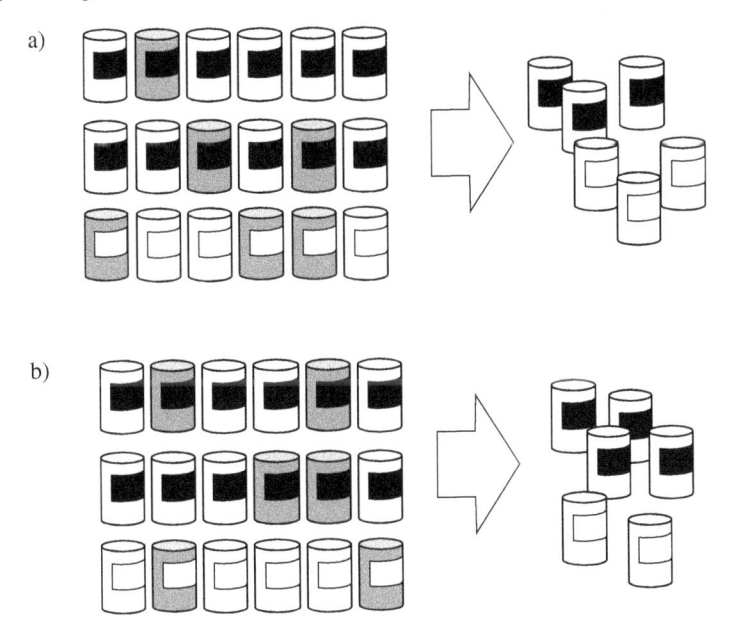

Figure 13.3 Variations on stratified random sampling. Strata are shown by item label colour (black or white). Selected samples are shown in dark grey, and the selected subsample shown to the right of the arrow. (a) Equal number per stratum; (b) proportional sampling. In (b), the number of samples of each type is in the same ratio as in the bulk sample.

variance for a given total number of test samples. If the total number of items to take is n, the optimum number n_i^{opt} taken from the ith stratum is given by

$$n_i^{opt} = \frac{P_i \sigma_i}{\sum P_i \sigma_i} n \qquad (13.2)$$

where P_i is the proportion of stratum i in the population and σ_i the (true) sampling standard deviation of values in that stratum. σ_i can, of course, be approximated by an experimental standard deviation.

13.4.2.2 Advantages and Disadvantages

Where strata differ appreciably, stratified random sampling provides smaller sampling variance than simple random sampling and can be planned to provide the minimum variance available. The principal disadvantage is added complexity in selecting the sample and calculating the mean value.

13.4.2.3 Statistical Treatment

Statistical treatment for stratified random sampling is somewhat more complex than in simple random sampling. Assume that there are n_s strata, each containing N_i items with a total mass m_i and that the measured mean, standard deviation and number of items tested in stratum i are $\bar{x}_i$, s_i and n_i, respectively. The standard deviation s_i is assumed to be equivalent to s_{sam} in Section 13.4.1, that is, the standard deviation of observations for the stratum if the analytical variation is considered negligible or an estimate of the sampling variance alone if not. (Mass proportion can, of course, be replaced by whatever proportion is most appropriate.) The proportion P_i of each stratum in the bulk is

$$P_i = \frac{m_i}{\sum_{i=1,n_s} m_i} \qquad (13.3)$$

If we require the mean value $\bar{X}$ for the bulk:

$$\bar{X} = \sum_{i=1,n_s} P_i \bar{x}_i \qquad (13.4)$$

This has variance $s^2(\bar{X})$ given by

$$s^2(\bar{X}) = \sum_{i=1,n_s} \left[P_i^2 s_i^2 (1 - f_i)/n_i \right] \qquad (13.5)$$

where $f_i = n_i/N_i$. The corresponding standard error, which can be used directly as the standard uncertainty from sampling, is given by

$$s(\bar{X}) = \sqrt{s^2(\bar{X})} \qquad (13.6)$$

If $\bar{X}$ is an estimate of the mean mass fraction of analyte in the bulk and an estimate of the total mass of analyte is required, this can be calculated from the mean mass fraction $\bar{X}$ as $m_{tot}\bar{X}$ with standard uncertainty from sampling equal to $m_{tot}s(\bar{X})$, m_{tot} being the total mass of the bulk material. The same result can also be obtained by calculating the amount per stratum separately and using the uncertainty estimation principles of Chapter 10 to estimate the combined uncertainty in the total mass of analyte.

Example
In a raid on a suspected illegal drug importer, police seize 35 bags each containing 200 g of white powder later found to contain imported heroin with closely similar composition (based on infrared spectroscopy). In addition, 273 smaller bags each containing 5 g of heroin with added 'cutting' agent are found on the premises; again, infrared spectroscopy confirms that the smaller bags are closely similar in composition. An estimate of the total amount of heroin is required. With two distinct types of material and evidence of similarity within each, a stratified random sample is considered appropriate. In the absence of detailed variance information, 10 bags of each size are taken and analysed by HPLC, with the results shown in the upper part of Table 13.1. The subsequent calculations of the mean heroin concentration over all seized material are shown in the bottom part of the table (assuming that variation is predominantly from sampling), allowing the calculation of total heroin in the seizure and providing an uncertainty in the estimate.
Note that this is a simulated example for the purpose of illustrating the calculations in stratified random sampling and not a general recommendation for treating forensic samples.

13.4.3 Systematic Sampling

13.4.3.1 Description

The first test sample is selected at random but subsequent samples are then selected at a fixed interval, such as every fifth item (Figure 13.4a). For example, consider 1000 sequentially numbered items in a batch of a product of which 20 need to be selected for testing. The first is selected at

Simple Sampling Strategies

Table 13.1 Results of sampling drug packages.

	Large bags	Small bags	Whole consignment
Analytical results			
Stratum mass m_i (g)	7000	1365	
Total mass m_{tot} (g)			8365
Total number of items N_i	35	273	
Number in test sample n_i	10	10	
Mean concentration $\bar{x}_i$ of heroin (g per 100 g)	43.2	14.3	
Standard deviation of observations s_i (g per 100 g)	3.5	2.7	
Calculations			
Proportion $P_i = m_i / \sum m_i$	0.837	0.163	
$P_i \bar{x}_i$	36.2	2.3	
$f_i = n_i / N_i$	0.286	0.037	
$P_i^2 s_i^2 (1-f_i)/n_i$	0.613	0.019	
Mean concentration $\bar{X}$ (g per 100 g)			$(36.2 + 2.3) = 38.5$
$s(\bar{X})$ (g per 100 g)			$\sqrt{0.613 + 0.019} = 0.795$
Total heroin $(m_{tot}\bar{X}/100)$ (g)			3221
Uncertainty $[m_{tot} s(\bar{X})/100]$ (g)[a]			67

[a]Due to sampling variation only.

random from items 1–50, after which every 50th item is chosen. If item number 8 was selected at random as the first test sample, the subsequent samples chosen would be 58, 108, ..., 958.

13.4.3.2 Advantages and Disadvantages

Systematic sampling is usually simpler to implement than random or stratified random sampling over a period of time or a stream of material. It also reduces the occurrence of long or very short intervals that may occur in simple random sampling. However, it can only be considered equivalent to random sampling if the variation in the material is reasonably independent of the item numbering or location, as in Figure 13.4a, and in particular shows no systematic variation correlated with the sampling points. This is not always the case. For example, systematic sampling of effluent every Monday might well provide a very biased picture of the average effluent composition. Figure 13.4b illustrates a particularly extreme possibility.

13.4.3.3 Statistical Treatment

Systematic samples are generally treated as if random, so the statistical treatment is as for simple random sampling, in Section 13.4.1.

13.4.4 Cluster and Multi-stage Sampling

13.4.4.1 Description

Cluster sampling can be used to overcome time and cost constraints when a relatively small number of samples are required from a large and widely dispersed population which would be costly to examine in full using simple random sampling. The first stage involves dividing the population into

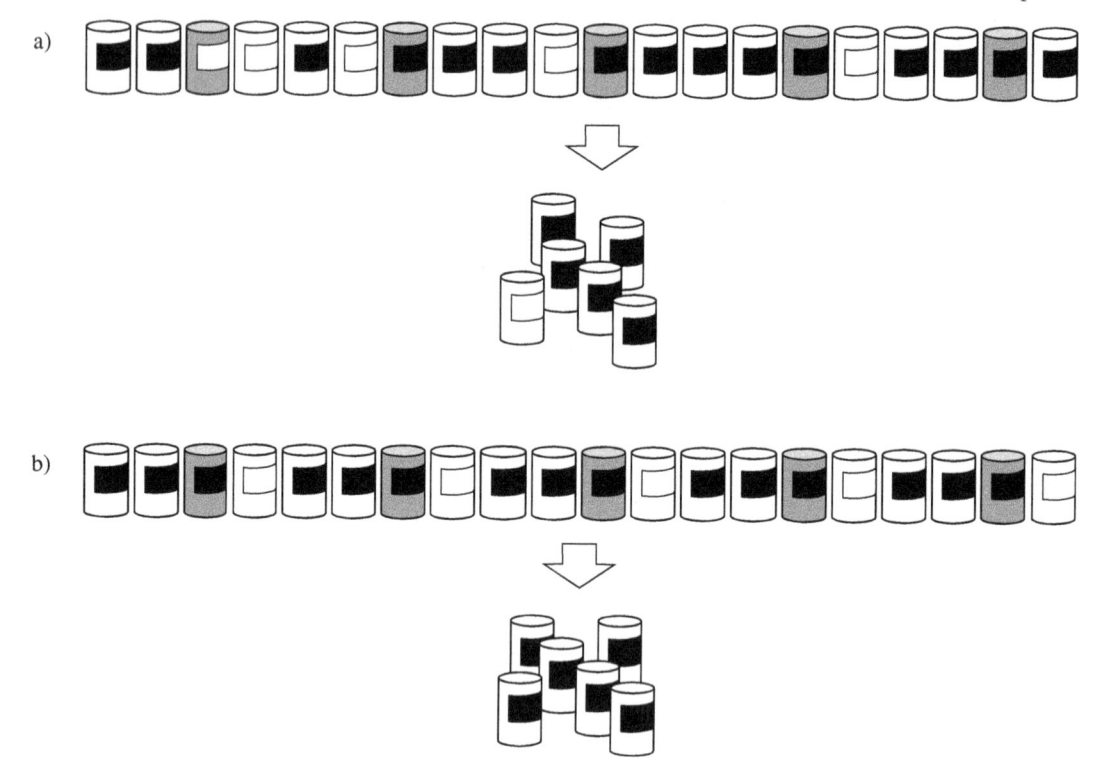

Figure 13.4 Systematic sampling. (a) A systematic sample of five items (shown in dark grey) from an inhomogeneous laboratory sample in which different items (shown for illustration with black and white labels, but unknown to the sampler) are distributed randomly. The test sample (below the arrow) is a random sample. (b) The most extreme effect of systematic sampling combined with systematic variations in the bulk sample. The selected test sample is heavily biased; this particular sampling interval on this laboratory sample will only select one item type.

groups or 'clusters', usually on the basis of location. A number of these clusters are then selected at random for further evaluation. This differs from stratified random sampling in that in stratified random sampling, all strata are sampled; in cluster sampling, only selected sub-groups are sampled. A further important difference is that cluster sampling is most effective when groups are internally as inhomogeneous as possible, while stratified sampling works best on highly homogeneous strata.

Cluster sampling is often used in geographical surveys because of the large costs of travelling between very different locations. It may also be useful in situations such as product QC sampling, where packaged products are boxed for distribution, and taking several units from a small number of distribution packages is less costly than opening a fresh distribution package for every unit taken.

'Multi-stage sampling' usually refers to a more complex form of cluster sampling in which sampling within each cluster is again formed by cluster sampling instead of simple random sampling.

13.4.4.2 Advantages and Disadvantages

Where the cost of sampling distinct locations or groups is large, cluster sampling provides a random sample at lower cost than stratified random or simple random sampling. However, because it selects a small number of clusters which may be significantly different, it will often deliver larger (that is, poorer) sampling variance than simple random sampling with the same total number of individual test samples.

Within a laboratory, it is unusual to incur disproportionate costs for sampling from different groups, so cluster sampling is unlikely to be useful in a laboratory setting.

13.4.4.3 *Statistical Treatment*

If the clusters are of equal size and all units within a cluster are tested, the cluster means can be treated as a simple random sample, so that the estimated population mean is the mean of cluster means and the standard error is based on the standard deviation of cluster means. For more complex situations, the calculations for cluster sampling are considerably more intricate, requiring weighting by cluster proportions and nested ANOVA to provide within- and between-cluster variances.

13.4.5 Quota Sampling

13.4.5.1 *Description*

The population is first divided into different groups or strata, as in the case of stratified sampling. Test samples are then collected until a predefined number (the quota) are selected from each group.

13.4.5.2 *Advantages and Disadvantages*

Quota sampling provides an approximation to stratified random sampling and is therefore expected to improve on opportunity or 'convenience' sampling (see Section 13.4.8). Unlike stratified sampling, however, the selection of samples from within each group is not random; the selection in each group is essentially a 'convenience sample' (see Section 13.4.8). The result can be unintentional bias. For example, the first five samples in a production run are unlikely to be representative of the complete run, while items produced at the end of a production run might never be sampled at all because sampling quotas have already been met. This strategy is often used in opinion polls, with varying degrees of control and cross-validation, but is rarely useful in a laboratory setting.

13.4.5.3 *Statistical Treatment*

Quota sampling is used in situations similar to stratified random sampling and the same statistical treatment is generally appropriate provided that there is no reason to believe that the procedure has introduced bias.

13.4.6 Sequential Sampling

13.4.6.1 *Description*

Sequential sampling is used when the number of samples required cannot be estimated before the start of the study. Samples are selected one at a time in sequence. After each sample has been tested, a decision is made as to whether the cumulative result is sufficient to make a decision (for example, whether to accept or reject a batch of material) or whether further sampling is required.

13.4.6.2 *Advantages and Statistical Treatment*

Sequential sampling has, provided that a suitable upper limit is set, the advantage of minimising the number of samples to be taken to reach a conclusion. However, because each decision on further sampling depends on measurement results on previous samples, it can be very slow. Further, it requires comparatively sophisticated statistical treatment (not covered in this book, but available in many quality assurance texts). Sequential sampling is accordingly rarely justified unless the costs of sampling and analysis are very high.

13.4.7 Judgement Sampling

13.4.7.1 Description

This approach involves using knowledge about the material to be sampled and about the reason for sampling, to select specific samples for testing.

13.4.7.2 Advantages and Disadvantages

Deliberate selection of particular items can be informative. However, it clearly does not constitute random sampling and the opportunity for randomisation is small. The results are therefore likely to be biased. It is consequently better to use judgement to inform a stratified random sampling strategy than to judge which individual items are tested.

This does not mean, of course, that it is never useful to supplement a random sample with additional information on apparently anomalous items. Further, anomalous samples are often important; for example, selecting apparently spoiled foodstuffs in a display to provide evidence that at least some are spoiled is perfectly acceptable provided that the average for those materials is not taken as representative of the bulk.

13.4.7.3 Statistical Treatment

Judgement sampling is typically used to identify test samples of particular individual interest. It is not usually appropriate to average the results or consider the sampling variance other than to summarise the range of results found. Instead, the individual results for particular test samples should be reported.

13.4.8 Convenience Sampling

13.4.8.1 Description

Convenience sampling involves selecting test samples on the basis of availability and/or accessibility.

13.4.8.2 Advantages and Disadvantages

Convenience sampling is the cheapest and simplest form of sampling and requires essentially no planning. However, it has no statistical advantages and is very likely to be unrepresentative. It is only defensible where the bulk to be tested is likely to be highly homogeneous.

13.4.8.3 Statistical Treatment

Averaging and estimation of dispersion of convenience samples are unwise except as summaries of the items collected.

Simple Sampling Strategies

13.4.9 Sampling in Two Dimensions

13.4.9.1 Description

Two-dimensional regions are common in environmental and geographical sampling. In the laboratory, the problem is less common, but can be important when sampling from sheet materials. This section lists some common strategies.

1. *Simple random sampling* over a two-dimensional area usually involves generating pairs of random numbers to serve as coordinates (Figure 13.5a). It generates a random sample, and so is appropriate for estimating average values over the area, but is comparatively inefficient for locating 'hot spots' because the random distances between sampling points can lead to comparatively large areas which are not sampled.
2. *Systematic sampling* involves taking samples at regular points on a (usually square) grid (Figure 13.5b). It is among the most efficient methods of locating 'hot spots' on a scale close to or larger than the grid interval. As with other systematic sampling, it can lead to bias in estimates of mean value if the sampled region shows systematic variation.
3. *Stratified random sampling in two dimensions:* The region is systematically divided into equal-sized regions, usually using a square grid, and a specified number of random points sampled within each. Figure 13.5c shows a single point per grid square. This gives 'strata' of equal size, with equal numbers of observations in each, so the sample can be treated as random and unbiased for mean values. The method is intermediate between systematic and simple random sampling for identifying 'hot spots'.
4. *W pattern:* Convenient for quick surveys, a W pattern is a form of systematic sampling which involves choosing a point at one corner of a region and taking sampling points in a W pattern across the region (Figure 13.5d). It typically generates a small number of observations and leaves comparatively large voids, making it inefficient for 'hot spot' location. It is particularly likely to be biased where gradients in analyte concentration are present; note in Figure 13.5d that the W pattern takes three observations at one edge and only two at the opposite edge, so the average will be biased towards the edge with more sampling points.
5. *Herringbone sampling:* A systematic sampling pattern which can be generated using alternating offsets from a square grid (Figure 13.5e). This is sometimes recommended for environmental sampling because, with careful planning, it can provide the 'hot spot' detection efficiency of systematic sampling with reduced likelihood of bias from the types of regularity most common in environmental studies.
6. *Area sampling and grid sampling* are phrases often used to describe cluster sampling based on prior subdivision of a region and random selection of sub-regions. Area sampling uses any convenient subdivision (such as electoral ward or county). Grid sampling is a variant of area sampling which uses a regular grid to divide the region, so 'area sampling' sometimes also refers to sampling from a grid. Both are usually associated with cluster sampling, that is, sub-sampling is performed on a random sample of sub-regions, each studied in detail.

13.4.9.2 Statistical Treatment

Statistical treatment depends on the particular strategy. Simple random sampling in two dimensions and, where equal numbers are taken from equal-sized areas, stratified random samples can be treated as random samples. Systematic strategies, including systematic sampling from a square grid, herringbone or W pattern, are usually treated as random, with the caveat that bias remains possible. Area sampling is a variant on cluster sampling and the treatment follows that for cluster sampling.

a) Simple random

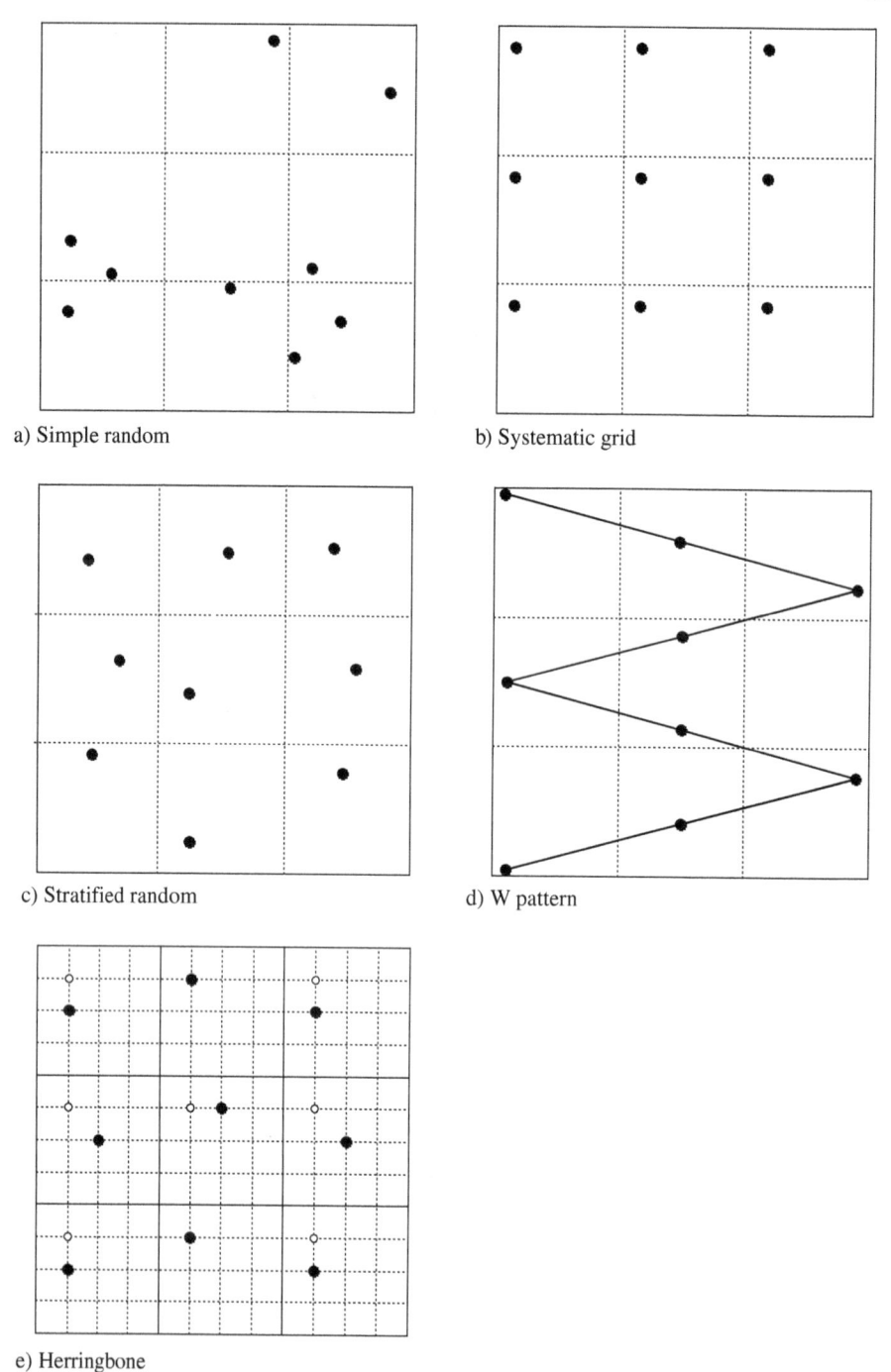

b) Systematic grid

c) Stratified random

d) W pattern

e) Herringbone

Figure 13.5 Sampling in two dimensions. All illustrations use nine sampling points. In (b), the initial sampling point is arbitrary; here it is at the top left corner of the grid square. In (d), the W pattern shown is one of four possible orientations; in practice, the orientation should be chosen at random. The solid line shows the 'W'. In (e), the sampling locations (filled circles) are generated using alternating offsets in the x and y directions from the original systematic grid points (shown as open circles where different from the sampling point); the offsets are arbitrary in either direction. Here, an offset of one-quarter of the primary grid interval is used (solid lines show the primary grid, dotted lines show quarter intervals).

13.5 Uncertainties Associated with Sampling

Sections 13.4.1 and 13.4.2 provide some basic calculations for estimating the uncertainty for a mean value derived from a simple random sample or a stratified random sample. Many of the other strategies can be treated in the same way as one or other of these. For many applications within a laboratory, these calculations will provide adequate estimates of the uncertainty arising from sampling or at least of that part of the uncertainty that is visible as sample-to-sample variation. It may, however, be useful to assess the uncertainties in the complete sampling and measurement process, from field sampling to analysis. Although beyond the scope of this book, guidance on the estimation of uncertainties for the complete sampling and analytical process has recently been developed and made available.[3] The aim of the guidance is to establish the uncertainty associated with using a particular sampling protocol on a particular sampling 'target'. The guidance provides a range of methods of assessing different contributions to the combined uncertainty in addition to providing comparatively economical approaches to estimating the associated uncertainty. The basis of the simplest approach is a balanced nested design, which can be analysed by the methods in Chapter 6, Section 6.6. This provides separate estimates of the sampling and analytical variances. These in turn can be used to estimate an uncertainty in the mean values, to allocate effort optimally between sampling and analysis or to inform sampling and method development.

13.6 Conclusion

Sampling often introduces the majority of the uncertainty in measurements on bulk materials. Proper planning and implementation of sampling are a crucial part of the overall measurement process.

Within a laboratory, analysts are often faced with the problem of sub-sampling from a larger laboratory sample. Two strategies described here, simple random sampling and stratified random sampling, are often sufficient for this purpose.

Additional detail on practical sampling methods and on the estimation of uncertainty in sampling can be found in the books and guidance cited below.

References

1. ISO/IEC 17025:2005, *General Requirements for the Competence of Testing and Calibration Laboratories*, International Organization for Standardization, Geneva, 2005.
2. See, for example, N. T. Crosby and I. Patel, *General Principles of Good Sampling Practice*, Royal Society of Chemistry, Cambridge, 1995, ISBN 0-85404-412-4, and references therein.
3. M. H. Ramsey and S. L. R. Ellison (Eds), Eurachem/CITAC/Nordtest Guide, *Measurement Uncertainty Arising from Sampling: a Guide to Methods and Approaches*, Eurachem, 2007. Available via http://www.eurachem.org (accessed 11 December 2008).
4. P. Gy (translated by A. G. Royle), *Sampling for Analytical Purposes*, Wiley, Chichester, 1998, ISBN 0-47197-956-2.
5. W. Horwitz, *Pure Appl. Chem.*, 1990, **62**, 1193.
6. AMC Technical Brief No. 19, *Terminology – the Key to Understanding Analytical Science. Part 2: Sampling and Sample Preparation*, Analytical Methods Committee, Royal Society of Chemistry, Cambridge, 2005. Available via http://www.rsc.org/amc/ (accessed 11 December 2008).
7. ISO 78-2:1999, *Chemistry – Layouts for Standards – Part 2: Methods of Chemical Analysis*, International Organization for Standardization, Geneva, 1999.
8. ISO 3534-2:2006, *Statistics – Vocabulary and Symbols. Part 2: Applied Statistics*, International Organization for Standardization, Geneva, 2006.

APPENDICES

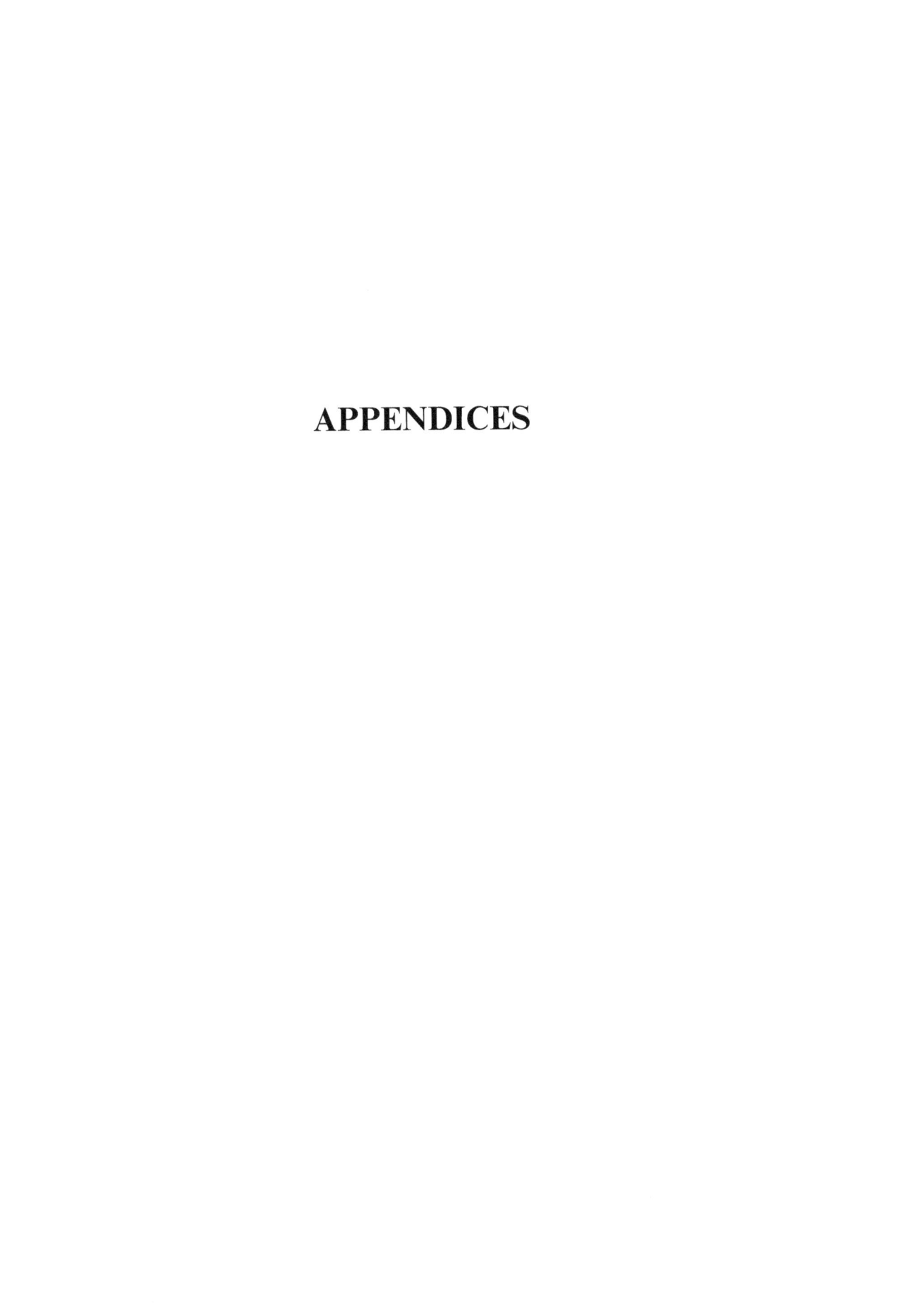

APPENDIX A
Statistical Tables

References for Appendix A

1. D. B. Rorabacher, *Anal. Chem.*, 1991, **63**, 139.
2. L. Komsta, Outliers: Tests for Outliers. Package Version 0.12, 2005. Available at http://www.r-project.org, http://www.komsta.net/.
3. E. S. Pearson and C. C. Sekar, *Biometrika*, 1936, **28**, 308.
4. H. A. David, H. O. Hartley and E. S. Pearson, *Biometrika*, 1954, **41**, 482.
5. F. E. Grubbs and G. Beck, *Technometrics*, 1972, **14**, 847.

Table A.1a Critical values for the Dixon test (95% confidence, two-tailed).[a]

n	Q (r_{10})	r_{11}	r_{12}	r_{20}	r_{21}	r_{22}
3	**0.970**	—	—	—	—	—
4	**0.829**	0.977	—	0.983	—	—
5	**0.710**	0.863	0.980	0.890	0.987	—
6	**0.625**	0.748	0.878	0.786	0.913	0.990
7	**0.568**	0.673	0.773	0.716	0.828	0.909
8	0.526	**0.615**	0.692	0.657	0.763	0.846
9	0.493	**0.570**	0.639	0.614	0.710	0.787
10	0.466	**0.534**	0.594	0.579	0.664	0.734
11	0.444	0.505	0.559	0.551	**0.625**	0.688
12	0.426	0.481	0.529	0.527	**0.592**	0.648
13	0.410	0.461	0.505	0.506	**0.565**	0.616
14	0.396	0.445	0.485	0.489	0.544	**0.590**
15	0.384	0.430	0.467	0.473	0.525	**0.568**
16	0.374	0.417	0.452	0.460	0.509	**0.548**
17	0.365	0.406	0.438	0.447	0.495	**0.531**
18	0.356	0.396	0.426	0.437	0.482	**0.516**
19	0.349	0.386	0.415	0.427	0.469	**0.503**
20	0.342	0.379	0.405	0.418	0.460	**0.491**
21	0.337	0.371	0.396	0.410	0.450	**0.480**
22	0.331	0.364	0.388	0.402	0.441	**0.470**
23	0.326	0.357	0.381	0.395	0.434	**0.461**
24	0.321	0.352	0.374	0.390	0.427	**0.452**
25	0.317	0.346	0.368	0.383	0.420	**0.445**
26	0.312	0.341	0.362	0.379	0.414	**0.438**
27	0.308	0.337	0.357	0.374	0.407	**0.432**
28	0.305	0.332	0.352	0.370	0.402	**0.426**
29	0.301	0.328	0.347	0.365	0.396	**0.419**
30	0.298	0.324	0.343	0.361	0.391	**0.414**

[a]Figures in bold show Dixon's recommended tests for different n.
[b]Test types:

$$Q = r_{10} = \frac{x_2 - x_1}{x_n - x_1} \left(\text{OR } \frac{x_n - x_{n-1}}{x_n - x_1} \right) \qquad r_{11} = \frac{x_2 - x_1}{x_{n-1} - x_1} \left(\text{OR } \frac{x_n - x_{n-1}}{x_n - x_2} \right)$$

$$r_{12} = \frac{x_2 - x_1}{x_{n-2} - x_1} \left(\text{OR } \frac{x_n - x_{n-1}}{x_n - x_3} \right) \qquad r_{20} = \frac{x_3 - x_1}{x_n - x_1} \left(\text{OR } \frac{x_n - x_{n-2}}{x_n - x_1} \right)$$

$$r_{21} = \frac{x_3 - x_1}{x_{n-1} - x_1} \left(\text{OR } \frac{x_n - x_{n-2}}{x_n - x_2} \right) \qquad r_{22} = \frac{x_3 - x_1}{x_{n-2} - x_1} \left(\text{OR } \frac{x_n - x_{n-2}}{x_n - x_3} \right)$$

Statistical Tables

Table A.1b Critical values for the Dixon test (99% confidence, two-tailed).[a]

n	Q (r_{10})	r_{11}	r_{12}	r_{20}	r_{21}	r_{22}
3	**0.994**	—	—	—	—	—
4	**0.926**	0.995	—	0.996	—	—
5	**0.821**	0.937	0.996	0.950	0.998	—
6	**0.740**	0.839	0.951	0.865	0.970	0.998
7	**0.680**	0.782	0.875	0.814	0.919	0.970
8	0.634	**0.725**	0.797	0.746	0.868	0.922
9	0.598	**0.677**	0.739	0.700	0.816	0.873
10	0.568	**0.639**	0.694	0.664	0.760	0.826
11	0.542	0.606	0.658	0.627	**0.713**	0.781
12	0.522	0.580	0.629	0.612	**0.675**	0.740
13	0.503	0.558	0.602	0.590	**0.649**	0.705
14	0.488	0.539	0.580	0.571	0.627	**0.674**
15	0.475	0.522	0.560	0.554	0.607	**0.647**
16	0.463	0.508	0.544	0.539	0.580	**0.624**
17	0.452	0.495	0.529	0.526	0.573	**0.605**
18	0.442	0.484	0.516	0.514	0.559	**0.589**
19	0.433	0.473	0.504	0.503	0.547	**0.575**
20	0.425	0.464	0.493	0.494	0.536	**0.562**
21	0.418	0.455	0.483	0.485	0.526	**0.551**
22	0.411	0.446	0.474	0.477	0.517	**0.541**
23	0.404	0.439	0.465	0.469	0.509	**0.532**
24	0.399	0.432	0.457	0.462	0.501	**0.524**
25	0.393	0.426	0.450	0.456	0.493	**0.516**
26	0.388	0.420	0.443	0.450	0.486	**0.508**
27	0.384	0.414	0.437	0.444	0.479	**0.501**
28	0.380	0.409	0.431	0.439	0.472	**0.495**
29	0.376	0.404	0.426	0.434	0.466	**0.489**
30	0.372	0.399	0.420	0.428	0.460	**0.483**

[a]Figures in bold show Dixon's recommended tests for different n.
[b]For test types see Table A.1a.

The Dixon test tables are as published by Rorabacher[1] and implemented in software by Komsta.[2] Publication is with permission from the American Chemical Society. The two-tailed tables are appropriate when testing for an outlier at either end of the data set, or for testing the more extreme value.

Table A.2 Critical values for the Grubbs tests.

n	95% G'	95% G''	95% G'''	99% G'	99% G''	99% G'''
3	1.154	1.993	–	1.155	2.000	–
4	1.481	2.429	0.0002	1.496	2.445	0.0000
5	1.715	2.755	0.0090	1.764	2.803	0.0018
6	1.887	3.012	0.0349	1.973	3.095	0.0116
7	2.020	3.222	0.0708	2.139	3.338	0.0308
8	2.127	3.399	0.1101	2.274	3.543	0.0563
9	2.215	3.552	0.1492	2.387	3.720	0.0851
10	2.290	3.685	0.1864	2.482	3.875	0.1150
11	2.355	3.803	0.2213	2.564	4.012	0.1448
12	2.412	3.909	0.2537	2.636	4.134	0.1738
13	2.462	4.005	0.2836	2.699	4.244	0.2016
14	2.507	4.093	0.3112	2.755	4.344	0.2280
15	2.548	4.173	0.3367	2.806	4.435	0.2530
16	2.586	4.247	0.3603	2.852	4.519	0.2767
17	2.620	4.316	0.3822	2.894	4.597	0.2990
18	2.652	4.380	0.4025	2.932	4.669	0.3200
19	2.681	4.440	0.4214	2.968	4.737	0.3398
20	2.708	4.496	0.4391	3.001	4.800	0.3585
21	2.734	4.549	0.4556	3.031	4.859	0.3761
22	2.758	4.599	0.4711	3.060	4.914	0.3927
23	2.780	4.646	0.4857	3.087	4.967	0.4085
24	2.802	4.691	0.4994	3.112	5.017	0.4234
25	2.822	4.734	0.5123	3.135	5.064	0.4376
26	2.841	4.775	0.5245	3.158	5.109	0.4510
27	2.859	4.814	0.5360	3.179	5.151	0.4638
28	2.876	4.851	0.5470	3.199	5.192	0.4759
29	2.893	4.886	0.5574	3.218	5.231	0.4875
30	2.908	4.921	0.5672	3.236	5.268	0.4985
40	3.036	5.201	0.6445	3.381	5.571	0.5862
50	3.128	5.407	0.6966	3.482	5.790	0.6462
60	3.200	5.568	0.7343	3.560	5.960	0.6901
70	3.258	5.700	0.7630	3.622	6.098	0.7236
80	3.306	5.811	0.7856	3.673	6.213	0.7501
90	3.348	5.906	0.8040	3.716	6.311	0.7717
100	3.384	5.990	0.8192	3.754	6.397	0.7896

For G' and G''', the table shows critical values for a two-tailed test; that is, a test applied to both the highest and lowest values (or pairs, for G''') in a single data set, or to the apparently more extreme of G_{high} and G_{low} in each case. G'', which tests for a pair of simultaneous outliers at opposite ends of the set, is not sensitive to the direction of the outliers. The test statistics are described in Chapter 5, Section 5.2.4.

Warning: The critical values for G''' are *lower* critical values; for G''', a value *less than* the tabulated critical value should be considered statistically significant. For G' and G'', a value *higher than* the tabulated critical value should be considered significant.

The critical values for G' are calculated from a formula originally given by Pearson and Sekar;[3] those for G'' use the formula for cumulative probability described by David et al.,[4] as implemented by Komsta.[2] Critical values for G''' are reprinted with permission from *Technometrics*[5] (Copyright 1972 by the American Statistical Association).

Statistical Tables

Table A.3a Critical values for the Cochran test (95% confidence).

l	n = 2	n = 3	n = 4	n = 5	n = 6	n = 7	n = 8	n = 9	n = 10	n = 11
2	0.998	0.975	0.939	0.906	0.877	0.853	0.833	0.816	0.801	0.788
3	0.967	0.871	0.798	0.746	0.707	0.677	0.653	0.633	0.617	0.603
4	0.906	0.768	0.684	0.629	0.589	0.560	0.536	0.518	0.502	0.488
5	0.841	0.684	0.598	0.544	0.506	0.478	0.456	0.439	0.424	0.412
6	0.781	0.616	0.532	0.480	0.445	0.418	0.398	0.382	0.368	0.357
7	0.727	0.561	0.480	0.431	0.397	0.373	0.354	0.338	0.326	0.315
8	0.680	0.516	0.438	0.391	0.359	0.336	0.318	0.304	0.293	0.283
9	0.638	0.477	0.403	0.358	0.328	0.307	0.290	0.277	0.266	0.257
10	0.602	0.445	0.373	0.331	0.303	0.282	0.267	0.254	0.244	0.235
11	0.570	0.417	0.348	0.308	0.281	0.262	0.247	0.235	0.225	0.217
12	0.541	0.392	0.326	0.288	0.262	0.244	0.230	0.219	0.210	0.202
13	0.515	0.371	0.307	0.271	0.246	0.229	0.215	0.205	0.196	0.189
14	0.492	0.352	0.291	0.255	0.232	0.215	0.202	0.192	0.184	0.177
15	0.471	0.335	0.276	0.242	0.220	0.203	0.191	0.182	0.174	0.167
16	0.452	0.319	0.262	0.230	0.208	0.193	0.181	0.172	0.164	0.158
17	0.434	0.305	0.250	0.219	0.198	0.183	0.172	0.163	0.156	0.150
18	0.418	0.293	0.240	0.209	0.189	0.175	0.164	0.156	0.149	0.143
19	0.403	0.281	0.230	0.200	0.181	0.167	0.157	0.149	0.142	0.136
20	0.389	0.270	0.221	0.192	0.174	0.160	0.150	0.142	0.136	0.130
21	0.377	0.261	0.212	0.185	0.167	0.154	0.144	0.136	0.130	0.125
22	0.365	0.252	0.204	0.178	0.160	0.148	0.138	0.131	0.125	0.120
23	0.354	0.243	0.197	0.171	0.155	0.142	0.133	0.126	0.120	0.116
24	0.343	0.235	0.191	0.166	0.149	0.137	0.129	0.122	0.116	0.111
25	0.334	0.228	0.185	0.160	0.144	0.133	0.124	0.117	0.112	0.107
26	0.325	0.221	0.179	0.155	0.139	0.128	0.120	0.114	0.108	0.104
27	0.316	0.215	0.174	0.150	0.135	0.124	0.116	0.110	0.105	0.100
28	0.308	0.209	0.168	0.146	0.131	0.121	0.113	0.106	0.101	0.097
29	0.300	0.203	0.164	0.142	0.127	0.117	0.109	0.103	0.098	0.094
30	0.293	0.198	0.159	0.138	0.124	0.114	0.106	0.100	0.095	0.092
31	0.286	0.193	0.155	0.134	0.120	0.111	0.103	0.097	0.093	0.089
32	0.279	0.188	0.151	0.130	0.117	0.108	0.100	0.095	0.090	0.086
33	0.273	0.184	0.147	0.127	0.114	0.105	0.098	0.092	0.088	0.084
34	0.267	0.179	0.144	0.124	0.111	0.102	0.095	0.090	0.086	0.082
35	0.262	0.175	0.140	0.121	0.108	0.100	0.093	0.088	0.083	0.080
36	0.256	0.171	0.137	0.118	0.106	0.097	0.091	0.085	0.081	0.078
37	0.251	0.168	0.134	0.116	0.103	0.095	0.088	0.083	0.079	0.076
38	0.246	0.164	0.131	0.113	0.101	0.093	0.086	0.082	0.078	0.074
39	0.241	0.161	0.128	0.111	0.099	0.091	0.085	0.080	0.076	0.073
40	0.237	0.158	0.126	0.108	0.097	0.089	0.083	0.078	0.074	0.071
41	0.233	0.154	0.123	0.106	0.095	0.087	0.081	0.076	0.073	0.069
42	0.228	0.151	0.121	0.104	0.093	0.085	0.079	0.075	0.071	0.068
43	0.224	0.149	0.118	0.102	0.091	0.083	0.078	0.073	0.070	0.067
44	0.221	0.146	0.116	0.100	0.089	0.082	0.076	0.072	0.068	0.065
45	0.217	0.143	0.114	0.098	0.087	0.080	0.075	0.070	0.067	0.064
46	0.213	0.141	0.112	0.096	0.086	0.079	0.073	0.069	0.065	0.063
47	0.210	0.138	0.110	0.094	0.084	0.077	0.072	0.068	0.064	0.061
48	0.206	0.136	0.108	0.093	0.083	0.076	0.070	0.066	0.063	0.060
49	0.203	0.134	0.106	0.091	0.081	0.074	0.069	0.065	0.062	0.059
50	0.200	0.131	0.104	0.089	0.080	0.073	0.068	0.064	0.061	0.058

Values in Table A.3a are calculated using $C_{crit}(p,l,n) = 1/[1 + (l-1) \times f]$, where p is the desired level of confidence (0.95 in this table), l is the number of groups and n is the number of observations in each group; f is the critical value of the F distribution for probability $(1-p)/l$, with degrees of freedom $(l-1)(n-1)$ and $(n-1)$ for numerator and denominator respectively.

Table A.3b Critical values for the Cochran test (99% confidence).

l	n = 2	n = 3	n = 4	n = 5	n = 6	n = 7	n = 8	n = 9	n = 10	n = 11
2	1.000	0.995	0.979	0.959	0.937	0.917	0.899	0.882	0.867	0.854
3	0.993	0.942	0.883	0.833	0.793	0.761	0.734	0.711	0.691	0.674
4	0.968	0.864	0.781	0.721	0.676	0.641	0.613	0.590	0.570	0.554
5	0.928	0.789	0.696	0.633	0.588	0.553	0.526	0.504	0.485	0.470
6	0.883	0.722	0.626	0.563	0.520	0.487	0.461	0.440	0.423	0.408
7	0.838	0.664	0.568	0.508	0.466	0.435	0.411	0.391	0.375	0.362
8	0.794	0.615	0.521	0.463	0.423	0.393	0.370	0.352	0.337	0.325
9	0.754	0.573	0.481	0.425	0.387	0.359	0.338	0.321	0.307	0.295
10	0.717	0.536	0.447	0.393	0.357	0.331	0.311	0.295	0.281	0.270
11	0.684	0.504	0.418	0.366	0.332	0.307	0.288	0.272	0.260	0.250
12	0.653	0.475	0.392	0.343	0.310	0.286	0.268	0.254	0.242	0.232
13	0.624	0.450	0.369	0.322	0.291	0.268	0.251	0.237	0.226	0.217
14	0.599	0.427	0.350	0.304	0.274	0.252	0.236	0.223	0.212	0.204
15	0.575	0.407	0.332	0.288	0.259	0.239	0.223	0.210	0.200	0.192
16	0.553	0.389	0.316	0.274	0.246	0.226	0.211	0.199	0.190	0.182
17	0.532	0.372	0.301	0.261	0.234	0.215	0.201	0.189	0.180	0.172
18	0.514	0.357	0.288	0.249	0.223	0.205	0.191	0.180	0.171	0.164
19	0.496	0.343	0.276	0.239	0.214	0.196	0.183	0.172	0.164	0.156
20	0.480	0.330	0.265	0.229	0.205	0.188	0.175	0.165	0.156	0.150
21	0.465	0.318	0.255	0.220	0.197	0.180	0.168	0.158	0.150	0.143
22	0.451	0.307	0.246	0.212	0.189	0.173	0.161	0.152	0.144	0.138
23	0.437	0.297	0.237	0.204	0.182	0.167	0.155	0.146	0.138	0.132
24	0.425	0.287	0.229	0.197	0.176	0.161	0.149	0.141	0.133	0.127
25	0.413	0.278	0.222	0.190	0.170	0.155	0.144	0.136	0.129	0.123
26	0.402	0.270	0.215	0.184	0.164	0.150	0.139	0.131	0.124	0.119
27	0.391	0.262	0.209	0.179	0.159	0.145	0.135	0.127	0.120	0.115
28	0.382	0.255	0.202	0.173	0.154	0.141	0.131	0.123	0.116	0.111
29	0.372	0.248	0.197	0.168	0.150	0.137	0.127	0.119	0.113	0.108
30	0.363	0.241	0.191	0.164	0.145	0.133	0.123	0.116	0.110	0.105
31	0.355	0.235	0.186	0.159	0.141	0.129	0.120	0.112	0.106	0.102
32	0.347	0.229	0.181	0.155	0.138	0.126	0.116	0.109	0.104	0.099
33	0.339	0.224	0.177	0.151	0.134	0.122	0.113	0.106	0.101	0.096
34	0.332	0.218	0.173	0.147	0.131	0.119	0.110	0.104	0.098	0.094
35	0.325	0.213	0.168	0.144	0.127	0.116	0.108	0.101	0.096	0.091
36	0.318	0.209	0.165	0.140	0.124	0.113	0.105	0.098	0.093	0.089
37	0.312	0.204	0.161	0.137	0.121	0.111	0.102	0.096	0.091	0.087
38	0.306	0.200	0.157	0.134	0.119	0.108	0.100	0.094	0.089	0.085
39	0.300	0.196	0.154	0.131	0.116	0.106	0.098	0.092	0.087	0.083
40	0.294	0.192	0.151	0.128	0.114	0.103	0.096	0.090	0.085	0.081
41	0.289	0.188	0.148	0.125	0.111	0.101	0.094	0.088	0.083	0.079
42	0.283	0.184	0.145	0.123	0.109	0.099	0.092	0.086	0.081	0.077
43	0.278	0.181	0.142	0.120	0.107	0.097	0.090	0.084	0.080	0.076
44	0.274	0.177	0.139	0.118	0.105	0.095	0.088	0.082	0.078	0.074
45	0.269	0.174	0.136	0.116	0.102	0.093	0.086	0.081	0.076	0.073
46	0.264	0.171	0.134	0.114	0.101	0.091	0.084	0.079	0.075	0.071
47	0.260	0.168	0.132	0.112	0.099	0.090	0.083	0.078	0.073	0.070
48	0.256	0.165	0.129	0.109	0.097	0.088	0.081	0.076	0.072	0.069
49	0.252	0.162	0.127	0.108	0.095	0.086	0.080	0.075	0.071	0.067
50	0.248	0.160	0.125	0.106	0.093	0.085	0.078	0.073	0.069	0.066

Values in Table A.3b are calculated using $C_{crit}(p,l,n)=1/[1+(l-1)\times f]$, where p is the desired level of confidence (0.99 in this table), l is the number of groups and n is the number of observations in each group; f is the critical value of the F distribution for probability $(1-p)/l$, with degrees of freedom $(l-1)(n-1)$ and $(n-1)$ for numerator and denominator respectively.

Statistical Tables

Table A.4 Critical values for Student's *t*-test: critical values are given for both one-tailed tests (1T) and two-tailed tests (2T).

ν	1T: 2T:	Confidence level (1 − α) (%)							
	1T:	85	90	95	97.5	99	99.5	99.9	99.95
	2T:	70	80	90	95	98	99	99.8	99.9
1		1.963	3.078	6.314	12.706	31.821	63.657	318.309	636.619
2		1.386	1.886	2.920	4.303	6.965	9.925	22.327	31.599
3		1.250	1.638	2.353	3.182	4.541	5.841	10.215	12.924
4		1.190	1.533	2.132	2.776	3.747	4.604	7.173	8.610
5		1.156	1.476	2.015	2.571	3.365	4.032	5.893	6.869
6		1.134	1.440	1.943	2.447	3.143	3.707	5.208	5.959
7		1.119	1.415	1.895	2.365	2.998	3.499	4.785	5.408
8		1.108	1.397	1.860	2.306	2.896	3.355	4.501	5.041
9		1.100	1.383	1.833	2.262	2.821	3.250	4.297	4.781
10		1.093	1.372	1.812	2.228	2.764	3.169	4.144	4.587
11		1.088	1.363	1.796	2.201	2.718	3.106	4.025	4.437
12		1.083	1.356	1.782	2.179	2.681	3.055	3.930	4.318
13		1.079	1.350	1.771	2.160	2.650	3.012	3.852	4.221
14		1.076	1.345	1.761	2.145	2.624	2.977	3.787	4.140
15		1.074	1.341	1.753	2.131	2.602	2.947	3.733	4.073
16		1.071	1.337	1.746	2.120	2.583	2.921	3.686	4.015
17		1.069	1.333	1.740	2.110	2.567	2.898	3.646	3.965
18		1.067	1.330	1.734	2.101	2.552	2.878	3.610	3.922
19		1.066	1.328	1.729	2.093	2.539	2.861	3.579	3.883
20		1.064	1.325	1.725	2.086	2.528	2.845	3.552	3.850
21		1.063	1.323	1.721	2.080	2.518	2.831	3.527	3.819
22		1.061	1.321	1.717	2.074	2.508	2.819	3.505	3.792
23		1.060	1.319	1.714	2.069	2.500	2.807	3.485	3.768
24		1.059	1.318	1.711	2.064	2.492	2.797	3.467	3.745
25		1.058	1.316	1.708	2.060	2.485	2.787	3.450	3.725
26		1.058	1.315	1.706	2.056	2.479	2.779	3.435	3.707
27		1.057	1.314	1.703	2.052	2.473	2.771	3.421	3.690
28		1.056	1.313	1.701	2.048	2.467	2.763	3.408	3.674
29		1.055	1.311	1.699	2.045	2.462	2.756	3.396	3.659
30		1.055	1.310	1.697	2.042	2.457	2.750	3.385	3.646
31		1.054	1.309	1.696	2.040	2.453	2.744	3.375	3.633
32		1.054	1.309	1.694	2.037	2.449	2.738	3.365	3.622
33		1.053	1.308	1.692	2.035	2.445	2.733	3.356	3.611
34		1.052	1.307	1.691	2.032	2.441	2.728	3.348	3.601
35		1.052	1.306	1.690	2.030	2.438	2.724	3.340	3.591
36		1.052	1.306	1.688	2.028	2.434	2.719	3.333	3.582
37		1.051	1.305	1.687	2.026	2.431	2.715	3.326	3.574
38		1.051	1.304	1.686	2.024	2.429	2.712	3.319	3.566
39		1.050	1.304	1.685	2.023	2.426	2.708	3.313	3.558
40		1.050	1.303	1.684	2.021	2.423	2.704	3.307	3.551
41		1.050	1.303	1.683	2.020	2.421	2.701	3.301	3.544
42		1.049	1.302	1.682	2.018	2.418	2.698	3.296	3.538
43		1.049	1.302	1.681	2.017	2.416	2.695	3.291	3.532
44		1.049	1.301	1.680	2.015	2.414	2.692	3.286	3.526
45		1.049	1.301	1.679	2.014	2.412	2.690	3.281	3.520
46		1.048	1.300	1.679	2.013	2.410	2.687	3.277	3.515
47		1.048	1.300	1.678	2.012	2.408	2.685	3.273	3.510
48		1.048	1.299	1.677	2.011	2.407	2.682	3.269	3.505
49		1.048	1.299	1.677	2.010	2.405	2.680	3.265	3.500
50		1.047	1.299	1.676	2.009	2.403	2.678	3.261	3.496
60		1.045	1.296	1.671	2.000	2.390	2.660	3.232	3.460
70		1.044	1.294	1.667	1.994	2.381	2.648	3.211	3.435
80		1.043	1.292	1.664	1.990	2.374	2.639	3.195	3.416
90		1.042	1.291	1.662	1.987	2.368	2.632	3.183	3.402
100		1.042	1.290	1.660	1.984	2.364	2.626	3.174	3.390
∞		1.036	1.282	1.645	1.960	2.326	2.576	3.090	3.291

Table A.5a Critical values for the F-test (95% confidence) [this table is suitable for use for a *one-tailed F*-test at 95% confidence ($\alpha = 0.05$)].

v_2 \ v_1	1	2	3	4	5	6	7	8	10	12	24	∞
1	161.4	199.5	215.7	224.6	230.2	234.0	236.8	238.9	241.9	243.9	249.1	254.3
2	18.51	19.00	19.16	19.25	19.30	19.33	19.35	19.37	19.40	19.41	19.45	19.50
3	10.13	9.552	9.277	9.117	9.013	8.941	8.887	8.845	8.786	8.745	8.639	8.526
4	7.709	6.944	6.591	6.388	6.256	6.163	6.094	6.041	5.964	5.912	5.774	5.628
5	6.608	5.786	5.409	5.192	5.050	4.950	4.876	4.818	4.735	4.678	4.527	4.365
6	5.987	5.143	4.757	4.534	4.387	4.284	4.207	4.147	4.060	4.000	3.841	3.669
7	5.591	4.737	4.347	4.120	3.972	3.866	3.787	3.726	3.637	3.575	3.410	3.230
8	5.318	4.459	4.066	3.838	3.687	3.581	3.500	3.438	3.347	3.284	3.115	2.928
9	5.117	4.256	3.863	3.633	3.482	3.374	3.293	3.230	3.137	3.073	2.900	2.707
10	4.965	4.103	3.708	3.478	3.326	3.217	3.135	3.072	2.978	2.913	2.737	2.538
11	4.844	3.982	3.587	3.357	3.204	3.095	3.012	2.948	2.854	2.788	2.609	2.404
12	4.747	3.885	3.490	3.259	3.106	2.996	2.913	2.849	2.753	2.687	2.505	2.296
13	4.667	3.806	3.411	3.179	3.025	2.915	2.832	2.767	2.671	2.604	2.420	2.206
14	4.600	3.739	3.344	3.112	2.958	2.848	2.764	2.699	2.602	2.534	2.349	2.131
15	4.543	3.682	3.287	3.056	2.901	2.790	2.707	2.641	2.544	2.475	2.288	2.066
16	4.494	3.634	3.239	3.007	2.852	2.741	2.657	2.591	2.494	2.425	2.235	2.010
17	4.451	3.592	3.197	2.965	2.810	2.699	2.614	2.548	2.450	2.381	2.190	1.960
18	4.414	3.555	3.160	2.928	2.773	2.661	2.577	2.510	2.412	2.342	2.150	1.917
19	4.381	3.522	3.127	2.895	2.740	2.628	2.544	2.477	2.378	2.308	2.114	1.878
20	4.351	3.493	3.098	2.866	2.711	2.599	2.514	2.447	2.348	2.278	2.082	1.843
21	4.325	3.467	3.072	2.840	2.685	2.573	2.488	2.420	2.321	2.250	2.054	1.812
22	4.301	3.443	3.049	2.817	2.661	2.549	2.464	2.397	2.297	2.226	2.028	1.783
23	4.279	3.422	3.028	2.796	2.640	2.528	2.442	2.375	2.275	2.204	2.005	1.757
24	4.260	3.403	3.009	2.776	2.621	2.508	2.423	2.355	2.255	2.183	1.984	1.733
25	4.242	3.385	2.991	2.759	2.603	2.490	2.405	2.337	2.236	2.165	1.964	1.711
26	4.225	3.369	2.975	2.743	2.587	2.474	2.388	2.321	2.220	2.148	1.946	1.691
27	4.210	3.354	2.960	2.728	2.572	2.459	2.373	2.305	2.204	2.132	1.930	1.672
28	4.196	3.340	2.947	2.714	2.558	2.445	2.359	2.291	2.190	2.118	1.915	1.654
29	4.183	3.328	2.934	2.701	2.545	2.432	2.346	2.278	2.177	2.104	1.901	1.638
30	4.171	3.316	2.922	2.690	2.534	2.421	2.334	2.266	2.165	2.092	1.887	1.622
32	4.149	3.295	2.901	2.668	2.512	2.399	2.313	2.244	2.142	2.070	1.864	1.594
34	4.130	3.276	2.883	2.650	2.494	2.380	2.294	2.225	2.123	2.050	1.843	1.569
36	4.113	3.259	2.866	2.634	2.477	2.364	2.277	2.209	2.106	2.033	1.824	1.547
38	4.098	3.245	2.852	2.619	2.463	2.349	2.262	2.194	2.091	2.017	1.808	1.527
40	4.085	3.232	2.839	2.606	2.449	2.336	2.249	2.180	2.077	2.003	1.793	1.509
60	4.001	3.150	2.758	2.525	2.368	2.254	2.167	2.097	1.993	1.917	1.700	1.389
120	3.920	3.072	2.680	2.447	2.290	2.175	2.087	2.016	1.910	1.834	1.608	1.254
∞	3.841	2.996	2.605	2.372	2.214	2.099	2.010	1.938	1.831	1.752	1.517	1.000

Statistical Tables

Table A.5b Critical values for the F-test (97.5% confidence) [this table is suitable for use for a *two-tailed F*-test at 95% confidence ($\alpha = 0.05$)].

v_2	v_1											
	1	*2*	*3*	*4*	*5*	*6*	*7*	*8*	*10*	*12*	*24*	*∞*
1	647.8	799.5	864.2	899.6	921.8	937.1	948.2	956.7	968.6	976.7	997.2	1018.0
2	38.51	39.00	39.17	39.25	39.30	39.33	39.36	39.37	39.40	39.41	39.46	39.50
3	17.44	16.04	15.44	15.10	14.88	14.73	14.62	14.54	14.42	14.34	14.12	13.90
4	12.22	10.65	9.979	9.605	9.364	9.197	9.074	8.980	8.844	8.751	8.511	8.257
5	10.01	8.434	7.764	7.388	7.146	6.978	6.853	6.757	6.619	6.525	6.278	6.015
6	8.813	7.260	6.599	6.227	5.988	5.820	5.695	5.600	5.461	5.366	5.117	4.849
7	8.073	6.542	5.890	5.523	5.285	5.119	4.995	4.899	4.761	4.666	4.415	4.142
8	7.571	6.059	5.416	5.053	4.817	4.652	4.529	4.433	4.295	4.200	3.947	3.670
9	7.209	5.715	5.078	4.718	4.484	4.320	4.197	4.102	3.964	3.868	3.614	3.333
10	6.937	5.456	4.826	4.468	4.236	4.072	3.950	3.855	3.717	3.621	3.365	3.080
11	6.724	5.256	4.630	4.275	4.044	3.881	3.759	3.664	3.526	3.430	3.173	2.883
12	6.554	5.096	4.474	4.121	3.891	3.728	3.607	3.512	3.374	3.277	3.019	2.725
13	6.414	4.965	4.347	3.996	3.767	3.604	3.483	3.388	3.250	3.153	2.893	2.595
14	6.298	4.857	4.242	3.892	3.663	3.501	3.380	3.285	3.147	3.050	2.789	2.487
15	6.200	4.765	4.153	3.804	3.576	3.415	3.293	3.199	3.060	2.963	2.701	2.395
16	6.115	4.687	4.077	3.729	3.502	3.341	3.219	3.125	2.986	2.889	2.625	2.316
17	6.042	4.619	4.011	3.665	3.438	3.277	3.156	3.061	2.922	2.825	2.560	2.247
18	5.978	4.560	3.954	3.608	3.382	3.221	3.100	3.005	2.866	2.769	2.503	2.187
19	5.922	4.508	3.903	3.559	3.333	3.172	3.051	2.956	2.817	2.720	2.452	2.133
20	5.871	4.461	3.859	3.515	3.289	3.128	3.007	2.913	2.774	2.676	2.408	2.085
21	5.827	4.420	3.819	3.475	3.250	3.090	2.969	2.874	2.735	2.637	2.368	2.042
22	5.786	4.383	3.783	3.440	3.215	3.055	2.934	2.839	2.700	2.602	2.331	2.003
23	5.750	4.349	3.750	3.408	3.183	3.023	2.902	2.808	2.668	2.570	2.299	1.968
24	5.717	4.319	3.721	3.379	3.155	2.995	2.874	2.779	2.640	2.541	2.269	1.935
25	5.686	4.291	3.694	3.353	3.129	2.969	2.848	2.753	2.613	2.515	2.242	1.906
26	5.659	4.265	3.670	3.329	3.105	2.945	2.824	2.729	2.590	2.491	2.217	1.878
27	5.633	4.242	3.647	3.307	3.083	2.923	2.802	2.707	2.568	2.469	2.195	1.853
28	5.610	4.221	3.626	3.286	3.063	2.903	2.782	2.687	2.547	2.448	2.174	1.829
29	5.588	4.201	3.607	3.267	3.044	2.884	2.763	2.669	2.529	2.430	2.154	1.807
30	5.568	4.182	3.589	3.250	3.026	2.867	2.746	2.651	2.511	2.412	2.136	1.787
32	5.531	4.149	3.557	3.218	2.995	2.836	2.715	2.620	2.480	2.381	2.103	1.750
34	5.499	4.120	3.529	3.191	2.968	2.808	2.688	2.593	2.453	2.353	2.075	1.717
36	5.471	4.094	3.505	3.167	2.944	2.785	2.664	2.569	2.429	2.329	2.049	1.687
38	5.446	4.071	3.483	3.145	2.923	2.763	2.643	2.548	2.407	2.307	2.027	1.661
40	5.424	4.051	3.463	3.126	2.904	2.744	2.624	2.529	2.388	2.288	2.007	1.637
60	5.286	3.925	3.343	3.008	2.786	2.627	2.507	2.412	2.270	2.169	1.882	1.482
120	5.152	3.805	3.227	2.894	2.674	2.515	2.395	2.299	2.157	2.055	1.760	1.310
∞	5.024	3.689	3.116	2.786	2.567	2.408	2.288	2.192	2.048	1.945	1.640	1.000

Table A.5c Critical values for the *F*-test (99% confidence) [this table is suitable for use for a *one-tailed F*-test at 99% confidence ($\alpha = 0.01$)].

v_2	v_1 1	2	3	4	5	6	7	8	10	12	24	∞
1	4052	4999	5403	5625	5764	5859	5928	5981	6056	6106	6235	6366
2	98.50	99.00	99.17	99.25	99.30	99.33	99.36	99.37	99.40	99.42	99.46	99.50
3	34.12	30.82	29.46	28.71	28.24	27.91	27.67	27.49	27.23	27.05	26.60	26.13
4	21.20	18.00	16.69	15.98	15.52	15.21	14.98	14.80	14.55	14.37	13.93	13.46
5	16.26	13.27	12.06	11.39	10.97	10.67	10.46	10.29	10.05	9.888	9.466	9.020
6	13.75	10.92	9.780	9.148	8.746	8.466	8.260	8.102	7.874	7.718	7.313	6.880
7	12.25	9.547	8.451	7.847	7.460	7.191	6.993	6.840	6.620	6.469	6.074	5.650
8	11.26	8.649	7.591	7.006	6.632	6.371	6.178	6.029	5.814	5.667	5.279	4.859
9	10.56	8.022	6.992	6.422	6.057	5.802	5.613	5.467	5.257	5.111	4.729	4.311
10	10.04	7.559	6.552	5.994	5.636	5.386	5.200	5.057	4.849	4.706	4.327	3.909
11	9.646	7.206	6.217	5.668	5.316	5.069	4.886	4.744	4.539	4.397	4.021	3.602
12	9.330	6.927	5.953	5.412	5.064	4.821	4.640	4.499	4.296	4.155	3.780	3.361
13	9.074	6.701	5.739	5.205	4.862	4.620	4.441	4.302	4.100	3.960	3.587	3.165
14	8.862	6.515	5.564	5.035	4.695	4.456	4.278	4.140	3.939	3.800	3.427	3.004
15	8.683	6.359	5.417	4.893	4.556	4.318	4.142	4.004	3.805	3.666	3.294	2.868
16	8.531	6.226	5.292	4.773	4.437	4.202	4.026	3.890	3.691	3.553	3.181	2.753
17	8.400	6.112	5.185	4.669	4.336	4.102	3.927	3.791	3.593	3.455	3.084	2.653
18	8.285	6.013	5.092	4.579	4.248	4.015	3.841	3.705	3.508	3.371	2.999	2.566
19	8.185	5.926	5.010	4.500	4.171	3.939	3.765	3.631	3.434	3.297	2.925	2.489
20	8.096	5.849	4.938	4.431	4.103	3.871	3.699	3.564	3.368	3.231	2.859	2.421
21	8.017	5.780	4.874	4.369	4.042	3.812	3.640	3.506	3.310	3.173	2.801	2.360
22	7.945	5.719	4.817	4.313	3.988	3.758	3.587	3.453	3.258	3.121	2.749	2.305
23	7.881	5.664	4.765	4.264	3.939	3.710	3.539	3.406	3.211	3.074	2.702	2.256
24	7.823	5.614	4.718	4.218	3.895	3.667	3.496	3.363	3.168	3.032	2.659	2.211
25	7.770	5.568	4.675	4.177	3.855	3.627	3.457	3.324	3.129	2.993	2.620	2.169
26	7.721	5.526	4.637	4.140	3.818	3.591	3.421	3.288	3.094	2.958	2.585	2.131
27	7.677	5.488	4.601	4.106	3.785	3.558	3.388	3.256	3.062	2.926	2.552	2.097
28	7.636	5.453	4.568	4.074	3.754	3.528	3.358	3.226	3.032	2.896	2.522	2.064
29	7.598	5.420	4.538	4.045	3.725	3.499	3.330	3.198	3.005	2.868	2.495	2.034
30	7.562	5.390	4.510	4.018	3.699	3.473	3.304	3.173	2.979	2.843	2.469	2.006
32	7.499	5.336	4.459	3.969	3.652	3.427	3.258	3.127	2.934	2.798	2.423	1.956
34	7.444	5.289	4.416	3.927	3.611	3.386	3.218	3.087	2.894	2.758	2.383	1.911
36	7.396	5.248	4.377	3.890	3.574	3.351	3.183	3.052	2.859	2.723	2.347	1.872
38	7.353	5.211	4.343	3.858	3.542	3.319	3.152	3.021	2.828	2.692	2.316	1.837
40	7.314	5.179	4.313	3.828	3.514	3.291	3.124	2.993	2.801	2.665	2.288	1.805
60	7.077	4.977	4.126	3.649	3.339	3.119	2.953	2.823	2.632	2.496	2.115	1.601
120	6.851	4.787	3.949	3.480	3.174	2.956	2.792	2.663	2.472	2.336	1.950	1.381
∞	6.635	4.605	3.782	3.319	3.017	2.802	2.639	2.511	2.321	2.185	1.791	1.000

Statistical Tables

Table A.5d Critical values for the *F*-test (99.5% confidence) [this table is suitable for use for a *two-tailed F*-test at 99% confidence ($\alpha = 0.01$)].

v_2	v_1											
	1	2	3	4	5	6	7	8	10	12	24	∞
1	16211	19999	21615	22500	23056	23437	23715	23925	24224	24426	24940	25464
2	198.5	199.0	199.2	199.2	199.3	199.3	199.4	199.4	199.4	199.4	199.5	199.5
3	55.55	49.80	47.47	46.19	45.39	44.84	44.43	44.13	43.69	43.39	42.62	41.83
4	31.33	26.28	24.26	23.15	22.46	21.97	21.62	21.35	20.97	20.70	20.03	19.32
5	22.78	18.31	16.53	15.56	14.94	14.51	14.20	13.96	13.62	13.38	12.78	12.14
6	18.63	14.54	12.92	12.03	11.46	11.07	10.79	10.57	10.25	10.03	9.474	8.879
7	16.24	12.40	10.88	10.05	9.522	9.155	8.885	8.678	8.380	8.176	7.645	7.076
8	14.69	11.04	9.596	8.805	8.302	7.952	7.694	7.496	7.211	7.015	6.503	5.951
9	13.61	10.11	8.717	7.956	7.471	7.134	6.885	6.693	6.417	6.227	5.729	5.188
10	12.83	9.427	8.081	7.343	6.872	6.545	6.302	6.116	5.847	5.661	5.173	4.639
11	12.23	8.912	7.600	6.881	6.422	6.102	5.865	5.682	5.418	5.236	4.756	4.226
12	11.75	8.510	7.226	6.521	6.071	5.757	5.525	5.345	5.085	4.906	4.431	3.904
13	11.37	8.186	6.926	6.233	5.791	5.482	5.253	5.076	4.820	4.643	4.173	3.647
14	11.06	7.922	6.680	5.998	5.562	5.257	5.031	4.857	4.603	4.428	3.961	3.436
15	10.80	7.701	6.476	5.803	5.372	5.071	4.847	4.674	4.424	4.250	3.786	3.260
16	10.58	7.514	6.303	5.638	5.212	4.913	4.692	4.521	4.272	4.099	3.638	3.112
17	10.38	7.354	6.156	5.497	5.075	4.779	4.559	4.389	4.142	3.971	3.511	2.984
18	10.22	7.215	6.028	5.375	4.956	4.663	4.445	4.276	4.030	3.860	3.402	2.873
19	10.07	7.093	5.916	5.268	4.853	4.561	4.345	4.177	3.933	3.763	3.306	2.776
20	9.944	6.986	5.818	5.174	4.762	4.472	4.257	4.090	3.847	3.678	3.222	2.690
21	9.830	6.891	5.730	5.091	4.681	4.393	4.179	4.013	3.771	3.602	3.147	2.614
22	9.727	6.806	5.652	5.017	4.609	4.322	4.109	3.944	3.703	3.535	3.081	2.545
23	9.635	6.730	5.582	4.950	4.544	4.259	4.047	3.882	3.642	3.475	3.021	2.484
24	9.551	6.661	5.519	4.890	4.486	4.202	3.991	3.826	3.587	3.420	2.967	2.428
25	9.475	6.598	5.462	4.835	4.433	4.150	3.939	3.776	3.537	3.370	2.918	2.377
26	9.406	6.541	5.409	4.785	4.384	4.103	3.893	3.730	3.492	3.325	2.873	2.330
27	9.342	6.489	5.361	4.740	4.340	4.059	3.850	3.687	3.450	3.284	2.832	2.287
28	9.284	6.440	5.317	4.698	4.300	4.020	3.811	3.649	3.412	3.246	2.794	2.247
29	9.230	6.396	5.276	4.659	4.262	3.983	3.775	3.613	3.377	3.211	2.759	2.210
30	9.180	6.355	5.239	4.623	4.228	3.949	3.742	3.580	3.344	3.179	2.727	2.176
32	9.090	6.281	5.171	4.559	4.166	3.889	3.682	3.521	3.286	3.121	2.670	2.114
34	9.012	6.217	5.113	4.504	4.112	3.836	3.630	3.470	3.235	3.071	2.620	2.060
36	8.943	6.161	5.062	4.455	4.065	3.790	3.585	3.425	3.191	3.027	2.576	2.013
38	8.882	6.111	5.016	4.412	4.023	3.749	3.545	3.385	3.152	2.988	2.537	1.970
40	8.828	6.066	4.976	4.374	3.986	3.713	3.509	3.350	3.117	2.953	2.502	1.932
60	8.495	5.795	4.729	4.140	3.760	3.492	3.291	3.134	2.904	2.742	2.290	1.689
120	8.179	5.539	4.497	3.921	3.548	3.285	3.087	2.933	2.705	2.544	2.089	1.431
∞	7.879	5.298	4.279	3.715	3.350	3.091	2.897	2.744	2.519	2.358	1.898	1.000

APPENDIX B
Symbols, Abbreviations and Notation

Symbols are often conventionally used for several different purposes. In this book, conventional symbols are used, so the same symbol may be used for more than one purpose.

Symbols are listed here if they occur at several locations in the text; otherwise, they are defined as they occur.

Frequently Used Symbols

α	significance level or probability for a Type I error
β	probability for a Type II error
δ	deviation or bias from a true or reference value
Δ	factor effect in a ruggedness test
θ	an angle; in Chapter 11, the angle between lines in a V-mask
λ	mean and variance of a Poisson distribution
μ	population mean
μ_0	target/reference value
ν	number of degrees of freedom
ν_b, ν_w	respectively, the degrees of freedom for the between-group and within-group mean squares in one-way analysis of variance
σ	population standard deviation
σ^2	population variance
σ_b^2	between-group component of variance in one-way analysis of variance
σ_L^2	between-laboratory component of variance
σ_P	standard deviation for proficiency assessment
σ_w^2	within-group component of variance in one-way analysis of variance
χ^2	chi-squared value (also used in the name of the associated distribution)
a	(i) intercept with the y-axis for the fitted straight line $y = a + bx$
	(ii) half-width of a rectangular or triangular distribution
b	(i) gradient of the fitted straight line $y = a + bx$
	(ii) (with subscripts) coefficient in a general polynomial expression
C	(i) (Chapter 5) the test statistic for Cochran's test for homogeneity of variances
	(ii) half-width of a confidence interval, as in $\bar{x} \pm C$
	(iii) (Chapter 12) analyte mass fraction in the Horwitz function

---

Practical Statistics for the Analytical Scientist: A Bench Guide, 2nd Edition
Stephen L R Ellison, Vicki J Barwick, Trevor J Duguid Farrant
© LGC Limited 2009
Published by the Royal Society of Chemistry, www.rsc.org

Symbols, Abbreviations and Notation

c_i	(Chapter 10) sensitivity coefficient in a calculation of combined uncertainty
C_i	(Chapter 11) cumulative sum of the first i values in a sequence of observations
d	difference between a pair of data points
$\bar{d}$	mean of differences between pairs of data points
e	random error
F	ratio of variances in statistical tests involving comparisons of variances
F_{crit}	critical value of the F distribution
G', G'', G'''	test statistics for the various Grubbs tests
H_0	null hypothesis in a significance test
H_1	alternative hypothesis in a significance test
i, j	indices for members in a data set (for example, x_i is the ith member of the set of values of x)
k	(i) (Chapter 10) conventionally, in measurement uncertainty estimation, the *coverage factor* in calculating expanded uncertainty $U=ku$ from standard uncertainty u
	(ii) further index for members of a grouped data set
l	number of groups in a Cochran test for outlying variance
m	mass
m_i	(Chapter 13) mass of a stratum
m_{tot}	(Chapter 13) total mass of a bulk sample
M	mean square (in analysis of variance)
$M_0, M_1, \ldots$	different mean squares in two-factor analysis of variance
M_w, M_b	within- and between-group mean squares in one-way analysis of variance
n	number of values in a data set or in one group within a data set
N	total number of values in a data set consisting of a number of groups
N_i	(Chapter 13) total number of items in a particular stratum
p	(i) a probability
	(ii) (Chapter 6) number of levels of a factor in analysis of variance
	(iii) purity (as mass fraction)
p, q, r	(Chapter 10) quantities in an uncertainty evaluation
P_i	(Chapter 13) proportion of each stratum in a bulk sample
q	(Chapter 6) number of levels of a second factor in analysis of variance
Q	(i) (Chapter 5) Dixon's test statistic
	(ii) (Chapter 12) a Q-score
Q_i	the ith quartile of a data set
r	correlation coefficient
r	repeatability limit
R	reproducibility limit
s	sample standard deviation
S	sum of squares (in analysis of variance)
S_w, S_b, S_{tot}	respectively, the within-group, between-group and total sum of squares in one-way analysis of variance
s^2	sample variance
s_a	standard deviation of the intercept a
s_b	standard deviation of the gradient b
s_d	standard deviation of the differences between pairs of data points
s_I	standard deviation under intermediate conditions of measurement
s_r	repeatability standard deviation
s_R	reproducibility standard deviation
$s(\bar{x})$	standard deviation of the mean

$s_{y/x}$	residual standard deviation for a regression of y on x
t	Student's t value (used in statistical tests involving comparisons of mean values)
t_{crit}	critical value of the t-distribution
u	standard uncertainty
U	expanded uncertainty
$u(x_i)$	uncertainty associated with a value x_i
$u(y)$	uncertainty in measurement result y
$u_i(y)$	uncertainty in measurement result y arising from uncertainty in value x_i
x_a	target value in a proficiency testing scheme
$\bar{x}$	sample mean
$\tilde{x}$	median
$\hat{x}$	predicted value of x obtained from a measured value of y
$\bar{X}$	(Chapter 13) mean concentration in a sampling target
$\hat{y}$	predicted value of y for a known value of x
z	(i) z-score
	(ii) quantile of the normal distribution

Abbreviations

ANOVA	analysis of variance
CRM	certified reference material
CuSum	cumulative sum (in CuSum charts)
CV	coefficient of variation
ISO	International Organization for Standardization
IQR	inter-quartile range
IUPAC	International Union of Pure and Applied Chemistry
LOD	limit of detection
LOQ	limit of quantitation
MAD	median absolute deviation
MAD_E	robust estimate of the standard deviation
PDF	probability density function
PT	proficiency testing
QA	quality assurance
QC	quality control
Q-score	performance score used in proficiency testing schemes
RSD	relative standard deviation
RSZ	rescaled sum of z-scores
SSZ	sum of squared z-scores
z-score	performance score used in proficiency testing schemes

Mathematical Notation

$\sum_{i=1}^{n} x_i$ or $\sum_{i=1,n} x_i$ sum of the values $x_1, x_2, \ldots, x_n$

$\sum_{i=1}^{n} x_i^2$ sum of the squares of the values $x_1, x_2, \ldots, x_n$

Symbols, Abbreviations and Notation

$\left(\sum_{i=1}^{n} x_i\right)^2$ square of the sum of the values $x_1, x_2, \ldots, x_n$

$f(x_1, x_2, \ldots)$ denotes a function of (that is, an equation or calculation involving) $x_1, x_2,$ etc.
$\partial y/\partial x_i$ partial differential of y with respect to x_i
$|x|$ 'mod x'; the absolute value of x
p^f notation for factorial experimental design with f factors each with p levels

APPENDIX C
Questions and Solutions

Questions

Question 1

Ten samples of flour taken at random from a large batch of material were analysed to determine the amount of moisture present. The results are shown in Table C.1.

1. Calculate the sample mean, variance and standard deviation of the data in Table C.1.
2. What is the median value of the data set?
3. Calculate the 95% confidence interval for the mean.

Question 2

As part of a training exercise, two analysts were asked to carry out replicate analyses of a test solution. Their results are shown in Table C.2.

1. For each set of results produce a dot plot, a histogram and a box-and-whisker plot.
2. Based on the plots, are there any suspect values in either data set?
3. Carry out Dixon's test to determine whether there are any outliers in either data set.
4. Carry out the appropriate Grubbs test(s) to check for outliers in both sets of data.

Table C.1 Results from the determination of the moisture content of samples of flour.

Moisture content (% m/m)									
14.06	13.76	13.99	13.94	13.94	13.95	13.96	14.17	14.20	13.86

Table C.2 Results from the analysis of a test solution.

Analyst	Concentration (mg L^{-1})						
Analyst 1	63.86	64.48	65.66	60.12	63.93	65.45	62.96
Analyst 2	63.91	65.92	68.98	64.24	65.45	65.82	65.11

Practical Statistics for the Analytical Scientist: A Bench Guide, 2nd Edition
Stephen L R Ellison, Vicki J Barwick, Trevor J Duguid Farrant
© LGC Limited 2009
Published by the Royal Society of Chemistry, www.rsc.org

Question 3

An experiment was undertaken to determine whether filtering a detergent to remove insoluble components would make any difference to the results of the spectrophotometric determination of phosphorus.

Six different samples were analysed. Each sample was divided into two portions. One portion was filtered, the other portion was not. The rest of the analysis was identical for both portions of the sample. The results are shown in Table C.3.

Does filtering the samples have a significant effect on the results? (Assume a significance level of $\alpha = 0.05$.)

Question 4

Lablink Limited requires all new analysts to carry out a trial using a standard operating procedure to analyse a quality control solution of known concentration. Two recruits, Adams and Baker, were asked to analyse batches of test samples which also contained a quality control solution with a concentration of 65.8 mg L^{-1}. The results that each analyst obtained for the quality control solution are shown in Table C.4.

1. Calculate the mean and standard deviation for each set of results.
2. Plot the results produced by each analyst. Are there any suspect points?
3. Carry out the Dixon Q test on each set of results and take appropriate action with any outliers identified.
4. How well do the analysts' results compare with the known concentration of the quality control solution?

Question 5

Three methods for the determination of blood alcohol levels are being evaluated. To check whether there is any significant bias in the results produced by the methods, each method is

Table C.3 Results from a study to determine the effect of filtering the sample on the determination of the phosphorus content of detergent samples (results expressed as mg L^{-1}).

	Sample					
	A	B	C	D	E	F
Filtered	37.1	26.4	26.2	33.2	24.3	34.7
Unfiltered	35.2	26.0	25.7	32.8	24.7	33.1

Table C.4 Results from the analysis of a quality control solution with a concentration of 65.8 mg L^{-1}.

Analyst	Results (mg L^{-1})						
Adams	63.9	65.9	62.8	64.1	64.6	65.2	62.9
Baker	64.8	65.9	65.2	63.6	65.7	66.2	65.1

used to analyse 10 portions of a certified reference material. The results are summarised in Table C.5.

The certified concentration of alcohol in the reference material is 80 mg per 100 mL of solution (mg/100 mL).

Carry out an appropriate statistical test to determine whether the results produced by any of the methods are significantly greater than the certified value at the 95% confidence level.

Question 6

An automated densitometer is used for the determination of the alcohol content (% v/v) of samples. Ten replicate analyses of a certified reference material (CRM) were carried out using the densitometer. The certified alcohol content was 40.1% v/v. The results are shown in Table C.6.

Are the results produced by the densitometer biased at the 95% confidence level?

Question 7

An area of land is being investigated to determine the concentration of lead present in the soil. Ten samples of soil are submitted to the laboratory for analysis. The results are shown in Table C.7. The concentration of lead in the soil must not exceed 500 mg kg^{-1}. Based on the results obtained, is there any evidence to suggest that the concentration of lead in the soil exceeds this limit? (Assume a significance level of $\alpha = 0.05$.)

Table C.5 Results from the analysis of a reference material certified for blood alcohol content [results expressed as mg alcohol per 100 mL of blood (mg/100 mL)].

Method	Mean	Number of observations	Standard deviation
Method 1	80.055	10	0.17
Method 2	80.651	10	0.16
Method 3	80.099	10	0.14

Table C.6 Results from the determination of the alcohol content of a CRM using an automated densitometer.

Alcohol content (% v/v)									
40.3	40.2	40.1	40.2	40.0	40.1	40.3	40.1	40.0	40.2

Table C.7 Results from the determination of lead in soil samples.

Sample No.	1	2	3	4	5	6	7	8	9	10
Lead (mg kg^{-1})	521	531	508	503	512	528	516	503	526	490

Questions and Solutions

Question 8

A method validation exercise is comparing the precision of two analytical methods. A quality control material is analysed 11 times, under repeatability conditions, using each of the methods. The results are summarised in Table C.8.

Is there a significant difference in the repeatability of the two methods at the 95% confidence level?

Question 9

A study was designed to investigate two approaches to preparing dilute solutions. The aim of the study was to determine whether there was a significant difference in the absorbance readings or the precision of the data obtained using the two approaches. A stock solution of Methyl Orange in propan-2-ol/water was prepared and the following two dilution schemes applied:

Dilution Scheme 1. Pipette 1 mL of stock solution into a 100 mL volumetric flask and make up to volume with demineralised water.

Dilution Scheme 2. Pipette 50 µL of stock solution into a 5 mL volumetric flask and make up to volume with demineralised water.

Each procedure was carried out 13 times and the absorbance of each of the solutions at 473 nm was recorded. The measurements were made by the same analyst, using the same spectrophotometer, over a short period of time and the order of analysis was randomised. The results are presented in Table C.9.

Table C.8 Results from a study to compare the repeatability of two test methods.

Method	Mean ($mg\,kg^{-1}$)	Standard deviation ($mg\,kg^{-1}$)
Method 1	31.83	0.424
Method 2	29.32	0.648

Table C.9 Results from the analysis of dilute solutions of Methyl Orange.

Solution No.	Absorbance at 473 nm	
	Scheme 1	Scheme 2
1	0.299	0.304
2	0.297	0.305
3	0.298	0.305
4	0.297	0.304
5	0.296	0.305
6	0.297	0.305
7	0.297	0.304
8	0.297	0.302
9	0.297	0.304
10	0.298	0.304
11	0.297	0.302
12	0.297	0.304
13	0.297	0.302
Mean	0.2972	0.3038
Standard deviation	0.000725	0.00114

Use the data in Table C.9 to determine whether there is a significant difference (at the 95% confidence level) between the means and between the standard deviations of the absorbances obtained using the two dilution schemes.

Question 10

Table C.10 shows some dissolution data from a study of dissolution testing using a well-established pharmacopoeia method. Laboratories were supplied with reference tablets and ran replicate determinations under repeatability conditions. Six replicate observations of the mass of active ingredient extracted per tablet are shown for each laboratory, rounded to two decimal places. A dot plot of the rounded data is shown in Figure C.1.

1. Calculate the mean of each laboratory's results.
2. From inspection of the figure, is there any reason to believe laboratories differ significantly in precision?
3. Using software or the equations given in Chapter 6, carry out an analysis of variance to determine whether there is a significant difference between the results produced by the different laboratories. (Assume a significance level of $\alpha = 0.05$.)
4. Calculate the repeatability standard deviation and the reproducibility standard deviation of the results.

Table C.10 Results from an interlaboratory study of dissolution testing (results expressed in mg).

Laboratory			
A	B	C	D
6.77	6.46	6.41	5.79
6.79	6.30	7.26	6.09
6.84	6.00	6.46	5.96
6.58	6.76	7.15	6.32
6.89	5.96	7.04	6.66
6.43	6.28	6.63	5.79

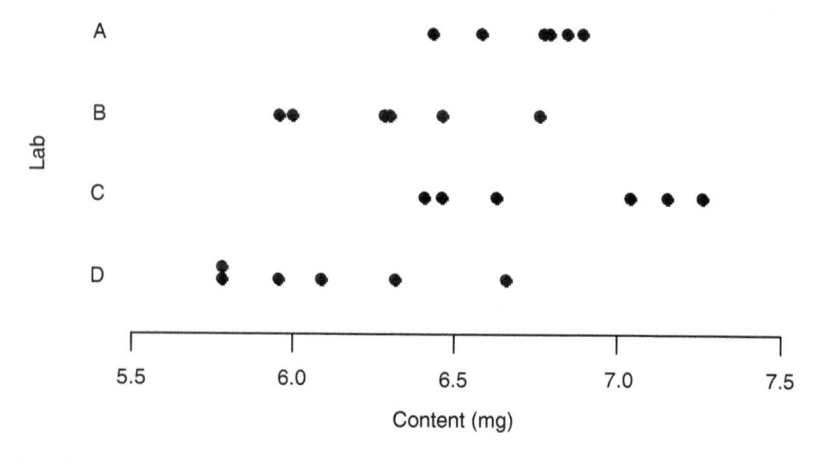

Figure C.1 Dissolution data from Table C.10.

Question 11

The quality manager of Fast Fertilisers has returned from a statistics course intent on evaluating the sources of variation in the manufacture of fertilisers containing phosphorus oxide. Fast Fertilisers have four shifts of workers who produce fertilisers 24 h per day. The quality of the fertilisers is monitored on a routine basis during each of the shifts. Each batch of fertiliser is processed into three forms (block, powder and granule) and each is then tested in the laboratory. From prior observations the quality manager decides that two possible sources of variation are differences between shifts and the different processed forms of the material. A trial is designed in which two samples from each of the three solid forms are taken in each of the four shifts during a single day's run, using a single raw material stock. The results from the study are given in Table C.11.

Using software or the equations given in the Appendix to Chapter 6, carry out an analysis of variance to determine whether there are any significant differences between the results produced by the different shifts or the results obtained for the different forms of the material. Use the interpretation in Chapter 6, Section 6.5.2.4, and assume a significance level of $\alpha = 0.05$.

Question 12

Four analysts take part in a study to evaluate the repeatability and intermediate precision of a method for the determination of arsenic in a sample of PVC. The analysts each analysed five portions of the same, essentially homogeneous, PVC sample under repeatability conditions, each analyst working on a different day. The results of the study are presented in Table C.12.

1. Using software or the equations given in Chapter 6, complete the one-way ANOVA table for the data in Table C.12.
2. Is there a significant difference between the analysts' results at the 95% level of confidence? If so, is it also significant at the 99% level?

Table C.11 Results from the determination of phosphorus oxide in fertilisers (results expressed as %m/m).

Form of material	Shift A	Shift B	Shift C	Shift D
Block	7.2	7.8	7.2	7.3
	7.5	7.8	7.8	7.5
Powder	7.5	7.5	7.5	7.1
	7.1	7.6	7.5	7.8
Granule	7.5	7.4	7.6	7.1
	7.0	7.5	7.1	7.5

Table C.12 Results obtained by different analysts on different days for the determination of arsenic in PVC (results expressed in mg kg^{-1}).

Analyst 1	Analyst 2	Analyst 3	Analyst 4
8.35	8.36	8.30	8.57
8.41	8.35	8.37	8.55
8.22	8.32	8.23	8.60
8.42	8.31	8.40	8.61
8.43	8.37	8.42	8.59

3. Calculate the repeatability standard deviation and intermediate precision standard deviation for the method for the determination of arsenic in PVC.
4. Calculate the repeatability limit for the method at the 95% confidence level.

Question 13

The data in Table C.13 are taken from a study to test whether the results of elemental analysis by two different inductively coupled plasma (ICP) methods are independent of the digestion methods used to prepare the test solution. If the results are independent of method, the laboratory would be able to simplify its method documentation and validation procedures.

The data are for copper in a bovine liver reference material. The digestion methods were microwave extraction and wet oxidation; the instrumental 'determination' methods were ICP-mass spectrometry and ICP-optical emission spectrometry.

Are the results independent of digestion method and determination method? Are the effects of digestion and determination interdependent?

Question 14

QuickLab regularly analyses soil samples to determine the selenium content and has developed a new method of quantitation. It has already been established that instrument response is linearly related to the selenium concentration but a calibration experiment is required to establish a calibration curve and associated 95% confidence interval for a single sample. The results of the experiment are given in Table C.14.

1. Construct a scatter plot of instrument response against concentration.
2. Using software or the equations given in Chapter 7, obtain the values of the coefficients a and b of the least-squares regression equation, $y = a + bx$, and plot the residuals.

Table C.13 Copper (mg kg^{-1}) in bovine liver determined by different methods.

Digestion method	Determination method	
	ICP-MS	ICP-OES
Microwave	151.7	161.6
	155.1	166.2
	162.0	166.0
	157.3	164.6
	155.9	168.8
Wet oxidation	176.6	145.7
	178.4	151.1
	179.1	153.4
	173.2	154.8
	175.4	149.6

Table C.14 Results from a calibration experiment for the determination of selenium (results expressed as instrument response in arbitrary units).

Concentration (mg L^{-1})					
0	20	40	80	120	160
−0.0011	17.6	30.1	61.4	91.4	121.3
0.00044	16.6	30.0	61.9	92.5	119.2
−0.0015	17.1	29.9	62.2	92.5	119.4

Questions and Solutions

3. Using software or the equation given in Chapter 7, obtain the correlation coefficient, r.
4. Calculate the residual standard deviation.
5. Calculate the 95% confidence intervals for (a) the gradient, (b) the intercept and (c) the predicted values of x when $y_0 = \bar{y}$ and when $y_0 = 120$.

Question 15

The data shown in Table C.15 were produced from a linearity study carried out during method validation. Seven standard solutions were each analysed in triplicate.

Least-squares linear regression has been carried out and the results are presented in Table C.16. A plot of the data is shown in Figure C.2. A plot of the residual values is shown in Figure C.3.

1. Comment on the design of the experiment. What if anything would you do differently if you had to cover the same concentration range?
2. Comment on any features of note in the plots.
3. Is the intercept significantly different from zero?

Table C.15 Data from a linearity study.

Solution concentration ($mg\ L^{-1}$)	Instrument response (arbitrary units)		
0.01	1.00	0.98	0.99
0.05	4.90	4.99	5.01
0.1	10.10	9.95	9.95
0.2	20.11	19.90	19.99
0.4	40.00	39.90	39.85
0.8	79.79	80.21	80.30
1.6	140.00	140.00	140.00

Table C.16 Results from least-squares linear regression of the data shown in Table C.15.

Regression statistics:

r	0.997
Observations	21

ANOVA:

	Degrees of freedom v	Sum of squares	Mean square	F	p-Value
Regression	1	46627.0	46627.0	3765.9	3×10^{-23}
Residual	19	235.2	12.4		
Total	20	46862.2			

Coefficients:

	Value	t_{calc}	p-Value	Lower 95%	Upper 95%
Intercept (a)	2.35	2.33	0.031	0.24	4.45
Gradient (b)	88.46	61.37	3×10^{-23}	85.45	91.48

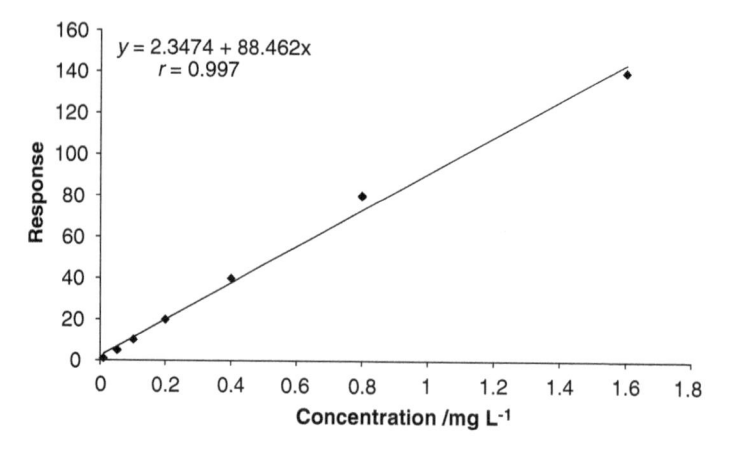

Figure C.2 Scatter plot of data presented in Table C.15.

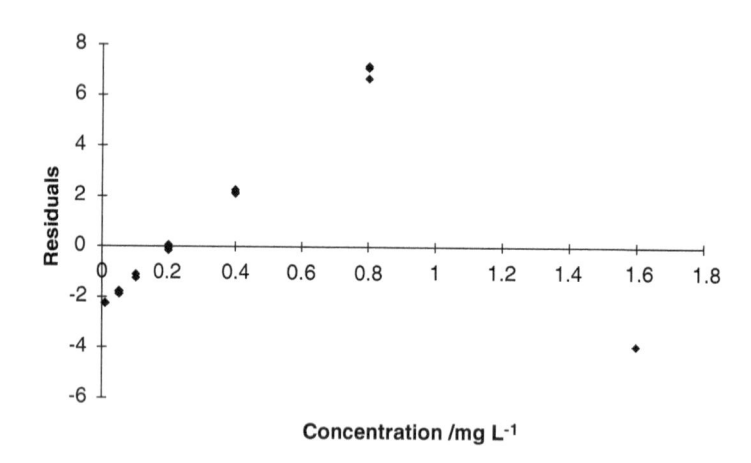

The fitted line is $y = 2.3474 + 88.462x$ with $r = 0.997$.

Figure C.3 Plot of residual values.

Question 16

In a method for the determination of an additive in diesel oil, the additive is extracted using silica cartridges and then quantified by HPLC. For the extraction, the sample is loaded onto a silica extraction cartridge and the diesel oil eluted with hexane; the additive is then eluted with butanol in hexane (5% v/v). A ruggedness test is carried out to evaluate the extraction stage of the method (the HPLC stage having already been validated). The ruggedness test was applied to a sample of diesel oil spiked with the additive at a concentration of approximately $1.2\,\text{mg}\,\text{L}^{-1}$ (a typical concentration). The parameters investigated in the ruggedness study and the values used are shown in Table C.17.

The experimental design used and the results obtained are shown in Table C.18.

The standard deviation of results for samples containing approximately $1.2\,\text{mg}\,\text{L}^{-1}$ of the additive has been estimated previously as $0.02\,\text{mg}\,\text{L}^{-1}$, based on 10 determinations.

Do any of the parameters studied have a significant effect on the extraction of the additive from diesel oil at the 95% confidence level?

Questions and Solutions

Table C.17 Parameters studied in a ruggedness test of the extraction of an additive from diesel oil.

Parameter		Value		
Brand of cartridge	A	Brand 1	a	Brand 2
Rate of elution of additive	B	1.5 mL min^{-1}	b	2.5 mL min^{-1}
Concentration of butanol in hexane used to elute additive	C	4.5% v/v	c	5.5% v/v
Volume of butanol–hexane used to elute additive	D	8 mL	d	12 mL
Mass of silica in extraction cartridge	E	5 g	e	2 g
Rate of elution of oil	F	2 mL min^{-1}	f	4 mL min^{-1}
Volume of hexane used to elute oil from cartridge	G	15 mL	g	25 mL

Table C.18 Results from a ruggedness study of the extraction of an additive from diesel oil.

Experimental parameter	Experiment number							
	1	2	3	4	5	6	7	8
A or a	A	A	A	A	a	a	a	a
B or b	B	B	b	b	B	B	b	b
C or c	C	c	C	c	C	c	C	c
D or d	D	D	d	d	d	d	D	D
E or e	E	e	E	e	e	E	e	E
F or f	F	f	f	F	F	f	f	F
G or g	G	g	g	G	g	G	G	g
Observed result (mg L^{-1})	1.10	1.06	0.96	0.84	0.96	0.86	1.06	1.05

Table C.19 Results from the replicate determination of ammonia in a water sample (results expressed in mg L^{-1}).

0.084		0.073	0.070	0.070	0.070	0.070
0.077		0.076	0.080	0.071	0.077	0.069
Mean				0.0739		
Standard deviation				0.0048		

Question 17

The performance of a method for the determination of low concentrations of ammonia in water samples was checked by carrying out replicate analyses of a low-level sample. The results, all corrected against a single sample blank reading, are shown in Table C.19. During routine use of the method, results for test samples are corrected by a single blank reading determined with each batch of samples.

Calculate the limit of detection for ammonia assuming a false positive probability of $\alpha = 0.05$ and a false negative probability of $\beta = 0.05$.

Question 18

Convert the following information to standard uncertainties:

1. The manufacturer's specification for a 100 mL Class A volumetric flask is quoted as ±0.08 mL.
2. The calibration certificate for a four-figure balance states that the measurement uncertainty is ±0.0004 g with a level of confidence of not less than 95%.
3. The purity of a compound is quoted by the supplier as $(99.9 \pm 0.1)\%$.
4. The standard deviation of repeat weighings of a 0.3 g check weight is 0.00021 g.
5. The calibration certificate for a 25 mL Class A pipette quotes an uncertainty of 0.03 mL. The reported uncertainty is based on a standard uncertainty multiplied by a coverage factor $k = 2$, providing a level of confidence of approximately 95%.

Question 19

A method requires 2 mL of an aqueous solution to be dispensed using a Class A glass pipette. The following sources of uncertainty have been identified as contributing to the uncertainty in the volume of the liquid delivered by the pipette:

1. Manufacturing tolerance for a 2 mL Class A pipette: quoted by manufacturer as ±0.01 mL.
2. Random variation (precision) in filling the pipette to the calibration line and dispensing the liquid from the pipette: estimated as 0.0016 mL, expressed as a standard deviation (estimate obtained from 10 repeat dispensings of liquid from the pipette).
3. Difference between the laboratory temperature and the calibration temperature of the pipette: estimated as 0.00061 mL, expressed as a standard uncertainty in measured volume (based on an estimate of the possible difference between laboratory temperature and the calibration temperature and on knowledge of the coefficient of volume expansion of the liquid).

If necessary, convert the data given for the 2 mL pipette to standard uncertainties. Combine the standard uncertainties to obtain an estimate of the combined standard uncertainty in the volume of liquid delivered by the pipette.

Question 20

A standard solution is prepared by dissolving approximately 100 mg of material (weighed on a four-figure balance) in water and making up to 100 mL in a volumetric flask. An estimate of the uncertainty in the concentration of the solution is required. The relevant data are given in Table C.20.

1. Where necessary, convert the data in Table C.20 to standard uncertainties.
2. Combine the relevant sources of uncertainty to obtain an estimate of the standard uncertainty in the mass of the material and the volume of the solution.
3. Combine the standard uncertainties for the mass, purity and volume to obtain an estimate of the uncertainty in the concentration of the solution.

Question 21

A solution of sodium hydroxide is standardised against the titrimetric standard potassium hydrogen phthalate (KHP). A sample of KHP is dissolved and then titrated using the sodium

Questions and Solutions

Table C.20 Data for the preparation of a standard solution.

Parameter	Value	Source of uncertainty	Data for evaluating uncertainty
Mass of material (m)	100.5 mg	Balance calibration	0.0004 g (expressed as an expanded uncertainty, with $k=2$ for approximately 95% confidence)
		Precision of balance	0.000041 g (expressed as a standard deviation)
Purity of material (p)	0.999	Purity	±0.001 (supplier's specification)
Volume of solution (V)	100 mL	Tolerance of flask	±0.08 mL
		Precision of filling the flask to the calibration line	0.017 mL (data from replicate 'fill-and-weigh' experiments, expressed as a standard deviation)
		Effect of the laboratory temperature differing from the flask calibration temperature	0.031 mL (expressed as a standard uncertainty)

Table C.21 Data for the standardisation of a sodium hydroxide solution.

	Parameter	Value	Standard uncertainty
m_{KHP}	Mass of KHP	0.38880 g	0.00013 g
p_{KHP}	Purity of KHP	1.0	0.00029
M_{KHP}	Molar mass of KHP	204.2212 g mol^{-1}	0.0038 g mol^{-1}
V_T	Titration volume	18.64 mL	0.013 mL
P_{method}	Method precision	—	0.0005 (expressed as a relative standard deviation)

hydroxide solution. The concentration of the sodium hydroxide solution is calculated from

$$c_{NaOH} = \frac{1000 \times m_{KHP} \times p_{KHP}}{M_{KHP} \times V_T}$$

where:

c_{NaOH} is the concentration of the sodium hydroxide solution (mol L^{-1})
m_{KHP} is the mass of the titrimetric standard KHP (g)
p_{KHP} is the purity of the KHP given as a mass fraction
M_{KHP} is the molar mass of KHP (g mol^{-1})
V_T is the titration volume of the sodium hydroxide solution (mL)

The values for each of the parameters, and their associated standard uncertainties, are given in Table C.21. The precision of the entire method has been estimated by carrying out the whole standardisation procedure a number of times. This is given in the table as P_{method} (expressed as a relative standard deviation).

1. The equation shown above is the 'model' for calculating the concentration of the sodium hydroxide solution. How would you include the precision term in this model? (Hint: the uncertainty estimate for method precision is expressed as a relative standard deviation.)

2. Construct a spreadsheet, following the approach described in Chapter 10, Section 10.3.1, to calculate the standard uncertainty in the concentration of the sodium hydroxide solution.
3. Calculate the expanded uncertainty in the concentration of the sodium hydroxide solution, assuming a level of confidence of approximately 95%.

Question 22

The performance of a test method for the determination of nickel in edible fats by atomic absorption spectrometry was monitored by analysing a quality control material with each batch of test samples. Table C.22 shows the results obtained for the quality control material over a 6 month period. Each result represents a single analysis of the quality control material. The target value for the quality control material has been established previously as 3.8 mg kg^{-1}. The method validation studies established that the standard deviation of results when the method is operating correctly is 0.08 mg kg^{-1}.

1. Construct a Shewhart chart with warning and action limits equivalent to approximately the 95% and 99.7% confidence limits.
2. Would you consider the method to be under statistical control?
3. Construct a CuSum chart using a target value of $\mu = 3.8$ mg kg^{-1}.
4. Construct a V-mask using $d = 2$ units along the x-axis and $\theta = 22°$. Are there any points where you would consider the method not to be under statistical control?

Question 23

The data shown in Table C.23 are from one round of a proficiency testing scheme for the determination of copper in soil using aqua regia extraction.

1. Calculate robust estimates of the mean and standard deviation using the median and adjusted median absolute deviation (MAD$_E$) of the data.
2. The scheme organisers, in consultation with the scheme participants, have set a standard deviation for proficiency assessment (σ_p) of 12 mg kg^{-1}. Choose an appropriate assigned value and calculate a z-score for each of the laboratories.
3. How many of the laboratories would have their performance judged as satisfactory, questionable and unsatisfactory, respectively?

Table C.22 Results from the determination of nickel in an edible fat quality control material (results expressed in mg kg^{-1}).

Measurement No.	Ni	Measurement No.	Ni	Measurement No.	Ni	Measurement No.	Ni
1	3.83	11	3.74	21	3.90	31	3.70
2	3.73	12	3.92	22	3.65	32	3.65
3	3.74	13	3.63	23	3.74	33	3.63
4	3.80	14	3.71	24	3.82	34	3.74
5	3.70	15	3.65	25	3.82	35	3.84
6	3.92	16	3.95	26	3.77	36	3.71
7	3.86	17	3.74	27	3.80	37	3.71
8	3.71	18	3.70	28	3.85	38	3.83
9	3.77	19	3.92	29	3.77	39	3.80
10	3.82	20	3.63	30	3.83	40	3.77

Questions and Solutions 233

Table C.23 Results from a proficiency testing round for the determination of copper in soil by aqua regia extraction. Results are shown to the number of digits reported by participants.

Laboratory ID No.	Copper concentration ($mg\,kg^{-1}$)	Laboratory ID No.	Copper concentration ($mg\,kg^{-1}$)
1	90	15	124.16
2	127	16	116
3	121	17	112
4	126.2	18	130
5	134	19	146
6	123	20	130
7	109.92	21	117
8	122	22	78.5
9	143	23	121.3
10	137	24	127.4
11	145	25	116.2
12	118	26	123
13	141	27	140.23
14	135		

Table C.24 Data from a proficiency testing round for the determination of the alcoholic strength of a spirit.

Laboratory ID No.	%abv	Laboratory ID No.	%abv
1	42.98	18	42.96
2	42.96	19	42.97
3	42.99	20	42.97
4	43.02	21	42.99
5	42.99	22	42.98
6	42.99	23	43.01
7	42.97	24	42.96
8	43.00	25	43.00
9	42.96	26	42.74
10	43.00	27	42.98
11	42.90	28	42.88
12	42.96	29	42.96
13	43.00	30	42.98
14	43.01	31	42.99
15	42.97	32	42.98
16	42.95	33	42.98
17	42.99		

Question 24

The data in Table C.24 are from one round of a proficiency testing scheme for the determination of the alcoholic strength (%abv) of a spirit.

1. Obtain the assigned value (x_a) for the round by calculating the median of the participants' results.
2. The scheme organisers have set a standard deviation for proficiency assessment (σ_p) of 0.03 %abv. Use this standard deviation and the assigned value calculated in (1) to calculate a z-score for each of the participants.
3. Identify any laboratories whose performance would be judged unsatisfactory or questionable.

Solutions

Calculations and solutions for the questions are shown below. Note that to aid calculation checking and prevent excessive rounding error, intermediate and final values are given to several significant figures beyond those normally necessary for reporting analytical data.

Question 1

1. Mean

$$\bar{x} = \frac{\sum_{i=1}^{n} x_i}{n} = \frac{139.83}{10} = 13.983\% \text{ m/m}$$

   Variance

$$s^2 = \frac{\sum_{i=1}^{n}(x_i - \bar{x})^2}{n-1} = \frac{0.1582}{9} = 0.0176$$

   Standard deviation

$$s = \sqrt{s^2} = \sqrt{0.0176} = 0.133\% \text{ m/m}$$

2. Median
   Arrange the data in order of magnitude:

13.76	13.86	13.94	13.94	13.95	13.96	13.99	14.06	14.17	14.20

$$\tilde{x} = \begin{cases} n \text{ odd}: & x_{(n+1)/2} \\ n \text{ even}: & \dfrac{x_{n/2} + x_{(n+2)/2}}{2} \end{cases}$$

$$\tilde{x} = \frac{x_5 + x_6}{2} = \frac{13.95 + 13.96}{2} = 13.955\% \text{ m/m}$$

3. Confidence interval

$$\bar{x} \pm \frac{t \times s}{\sqrt{n}}$$

   The two-tailed critical value of t at the 95% confidence level with $v = 9$ is 2.262 (see Appendix A, Table A.4). The confidence interval is therefore:

$$13.983 \pm \frac{2.262 \times 0.133}{\sqrt{10}} = 13.983 \pm 0.095\% \text{ m/m}$$

   With appropriate rounding, this could reasonably be expressed as $14.0 \pm 0.1\%$ m/m

Question 2

1. Figure C.4 shows the dot plots, Figure C.5 shows example histograms and Figure C.6 shows the box-and-whisker plots. Note that histograms may look very different if the bin widths and locations are chosen differently.

Questions and Solutions

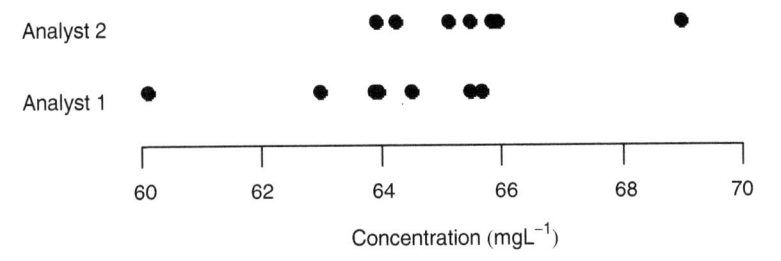

Figure C.4 Dot plots of data presented in Table C.2.

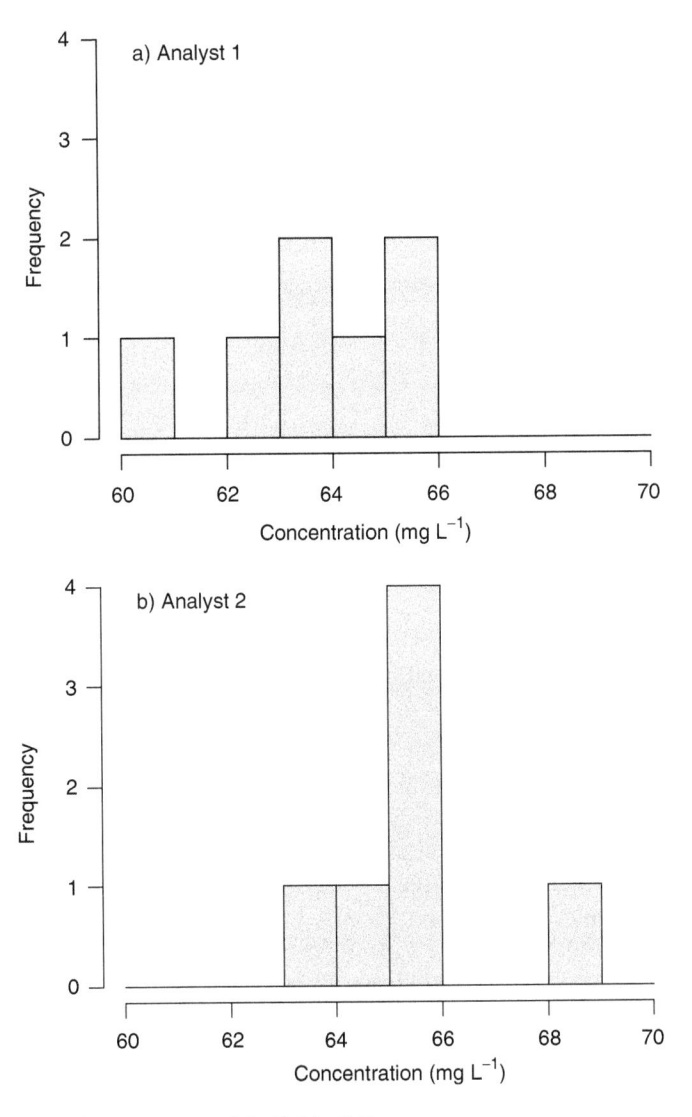

Figure C.5 Histograms of data presented in Table C.2.

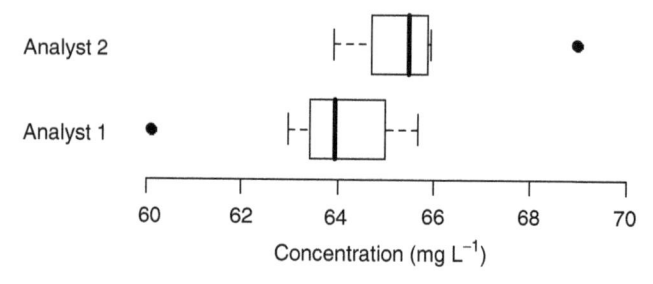

```
                    Analyst 2         |--[ | ]      •


                    Analyst 1   •     |-[ |  ]-|


                         ┌─────┬─────┬─────┬─────┬─────┐
                        60    62    64    66    68    70
                              Concentration (mg L$^{-1}$)
```

Figure C.6 Box-and-whisker plots of data presented in Table C.2. The interpretation of the box, whiskers, median line and separate points is described in Chapter 2, Section 2.8.

2. In each case, one extreme value, the lowest for Analyst 1 and the highest for Analyst 2, appears to be inconsistent with the other data in all three plots.

3. The appropriate equations for the Dixon Q test are:

$$Q = r_{10} = \frac{x_2 - x_1}{x_n - x_1} \text{ (Lowest observation)}$$

$$Q = r_{10} = \frac{x_n - x_{n-1}}{x_n - x_1} \text{ (Highest observation)}$$

Analyst 1:

$$\text{Lowest observation: } Q = \frac{62.96 - 60.12}{65.66 - 60.12} = 0.513$$

$$\text{Highest observation: } Q = \frac{65.66 - 65.45}{65.66 - 60.12} = 0.0379$$

Analyst 2:

$$\text{Lowest observation: } Q = \frac{64.24 - 63.91}{68.98 - 63.91} = 0.0651$$

$$\text{Highest observation: } Q = \frac{68.98 - 65.92}{68.98 - 63.91} = 0.604$$

The critical value for the Dixon Q test at the 95% confidence level for $n = 7$ is 0.568 (Appendix A, Table A.1a). The calculated Q value for the highest result for Analyst 2 exceeds this value. The critical value at the 99% confidence level is 0.680 (Appendix A, Table A.1b). Dixon's test therefore recognises the highest result produced by Analyst 2 as an outlier at the 95% confidence level but not at the 99% confidence level.

4. In each data set there is a single suspect value so G' is the appropriate test:

$$G'_{lowest} = \frac{\bar{x} - x_1}{s} \text{ or } G'_{highest} = \frac{x_n - \bar{x}}{s}$$

Analyst 1:

$$\text{Suspect low value}: G'_{\text{lowest}} = \frac{63.78 - 60.12}{1.867} = 1.961$$

Analyst 2:

$$\text{Suspect high value}: G'_{\text{highest}} = \frac{68.98 - 65.63}{1.660} = 2.016$$

The critical value at the 95% confidence level for $n = 7$ is 2.020 (Appendix A, Table A.2). The critical value at the 99% confidence level is 2.139. The Grubbs test therefore identifies no outliers at the 95% confidence level, although the decision is marginal in the case of the highest result for Analyst 2.

Question 3

The results are from matched pairs so a paired t-test is appropriate. The calculated differences are shown in Table C.25.

Mean difference

$$\bar{d} = \frac{\sum_{i=1}^{n} d_i}{n} = \frac{4.4}{6} = 0.733 \text{ mg L}^{-1}$$

Standard deviation of the differences

$$s_d = \sqrt{\frac{\sum_{i=1}^{n}(d_i - \bar{d})^2}{n-1}} = \sqrt{\frac{3.673}{5}} = 0.857 \text{ mg L}^{-1}$$

The null hypothesis is that filtering the samples has no effect on the results obtained. Mathematically, this can be restated as

$$H_0: \mu_d = 0$$

that is, the null hypothesis H_0 is that the population mean difference μ_d is equal to zero. The question being asked is whether filtering the samples makes any difference to the results obtained. The alternative hypothesis is therefore stated as

$$H_1: \mu_d \neq 0$$

that is, the population mean difference μ_d differs from zero. This requires a two-tailed test as we are interested only in whether there is a change in the results, not in the direction of any change.

Table C.25 Calculated differences for use in paired t-test.

Filtered	37.1	26.4	26.2	33.2	24.3	34.7
Unfiltered	35.2	26.0	25.7	32.8	24.7	33.1
Difference	1.9	0.4	0.5	0.4	−0.4	1.6

The t statistic is calculated as follows:

$$t = \frac{0.733}{0.857/\sqrt{6}} = 2.095$$

The two-tailed critical value for t at the 95% confidence level with $v = 5$ is 2.571 (see Appendix A, Table A.4).

Since the calculated t value is less than the critical value, we can conclude that filtering the samples does not have a statistically significant effect on the results obtained.

Question 4

1. Means and standard deviations
   Adams:

   $$\text{Mean} \quad \bar{x}_A = \frac{\sum_{i=1}^{n} x_i}{n} = \frac{449.4}{7} = 64.20 \text{ mg L}^{-1}$$

   $$\text{Sample standard deviation} \quad s_A = \sqrt{\frac{\sum_{i=1}^{n}(x_i - \bar{x})^2}{n-1}} = \sqrt{\frac{7.8}{6}} = 1.14 \text{ mg L}^{-1}$$

   Baker:

   $$\text{Mean} \quad \bar{x}_B = \frac{\sum_{i=1}^{n} x_i}{n} = \frac{456.5}{7} = 65.21 \text{ mg L}^{-1}$$

   $$\text{Sample standard deviation} \quad s_B = \sqrt{\frac{\sum_{i=1}^{n}(x_i - \bar{x})^2}{n-1}} = \sqrt{\frac{4.46857}{6}} = 0.863 \text{ mg L}^{-1}$$

2. Plots
   There are a number of ways of plotting the data. Two appropriate examples are shown in Figure C.7. The spread of the results in the dot plots is similar for the two analysts, although the box plot suggests that one point for Baker might be an outlier among an otherwise more precise set of data. This is not unusual in box plots of small data sets, but it does suggest an outlier test to see if the anomalous point is a serious outlier or simply a chance observation in a small data set.

3. Dixon tests
   The appropriate equations for the Dixon Q test are:

   $$Q = r_{10} = \frac{x_2 - x_1}{x_n - x_1} \text{ (Lowest observation)} \quad \text{and} \quad Q = r_{10} = \frac{x_n - x_{n-1}}{x_n - x_1} \text{ (Highest observation)}$$

   Calculating both for each analyst:

	Adams	Baker
Lowest observation:	$Q = \dfrac{62.9 - 62.8}{65.9 - 62.8} = 0.0323$	$Q = \dfrac{64.8 - 63.6}{66.2 - 63.6} = 0.462$
Highest observation:	$Q = \dfrac{65.9 - 65.2}{65.9 - 62.8} = 0.226$	$Q = \dfrac{66.2 - 65.9}{66.2 - 63.6} = 0.115$

Questions and Solutions

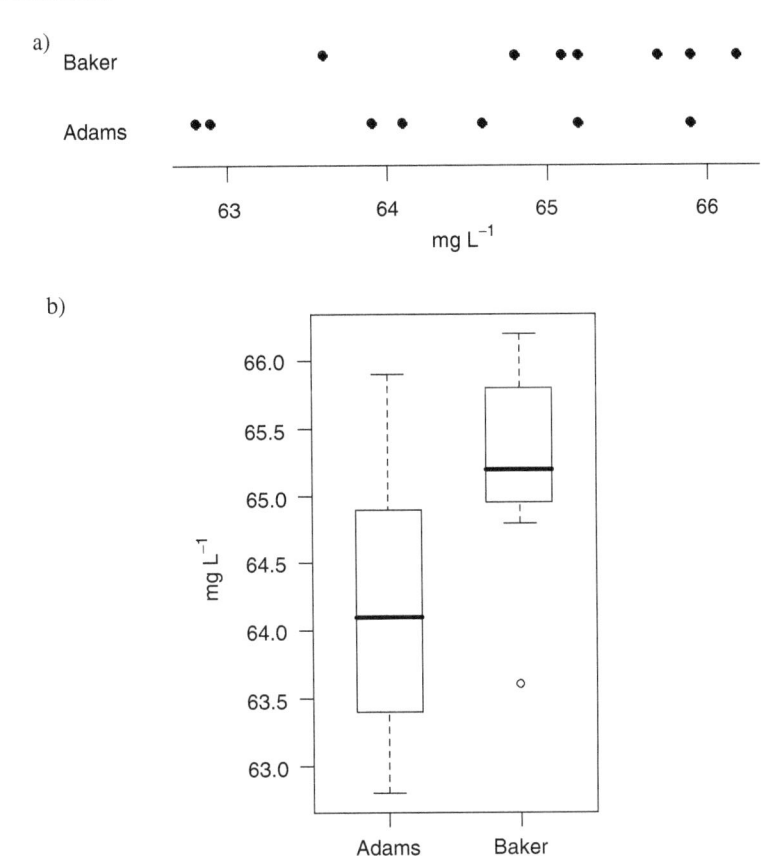

Figure C.7 Plots of data produced by Adams and Baker. a) Dot plot. b) Box plot.

The critical value for the Dixon Q test at the 95% confidence level with $n = 7$ is 0.568 (Appendix A, Table A.1a). In all cases, the calculated value is less than the critical value, so there are no significant outliers in either data set at the 95% level of confidence. Since the critical value at 99% is greater than the 95% critical value, there is no need to retest at the 99% level.

4. How well do the values compare with the target of 65.8 mg L^{-1}?
Since there are no marked outliers and the distributions appear otherwise normal, a *t*-test can be used. The question asks about how well the results agree with the target value without regard to direction, so a two-tailed test is appropriate.

The null hypothesis H$_0$ is $\mu = 65.8$.
The alternative hypothesis H$_1$ is $\mu \neq 65.8$.

$$\text{Adams:} \quad t = \frac{|\bar{x}_A - \mu_0|}{s_A/\sqrt{n_A}} = \frac{|64.20 - 65.8|}{1.14/\sqrt{7}} = 3.713$$

$$\text{Baker:} \quad t = \frac{|\bar{x}_B - \mu_0|}{s_B/\sqrt{n_B}} = \frac{|65.21 - 65.8|}{0.863/\sqrt{7}} = 1.809$$

The two-tailed critical value of *t* at the 95% confidence level with $\nu = 6$ is 2.447 (see Appendix A, Table A.4).

The t value calculated from Adams' data is greater than the critical value. We can therefore conclude that the mean of Adams' results is significantly different from the target concentration of the quality control solution.

Question 5

The appropriate hypotheses for the test are:

$$H_0: \mu = 80\,\text{mg}/100\,\text{mL}$$
$$H_1: \mu > 80\,\text{mg}/100\,\text{mL}$$

Since the hypotheses involve the comparison of a mean value with a reference value a single sample t-test is appropriate. The alternative hypothesis states that the mean is greater than the reference value; the test is therefore a one-tailed test. The equation for calculating the t value is:

$$t = \frac{\bar{x} - \mu_0}{s/\sqrt{n}}$$

$$\text{Method 1}: t = \frac{80.055 - 80}{0.17/\sqrt{10}} = 1.023$$

$$\text{Method 2}: t = \frac{80.651 - 80}{0.16/\sqrt{10}} = 12.867$$

$$\text{Method 3}: t = \frac{80.099 - 80}{0.14/\sqrt{10}} = 2.236$$

The critical value for t (one-tailed, 95% confidence, $v = 9$) is 1.833 (from Appendix A, Table A.4). The means of the results produced by methods 2 and 3 are therefore significantly greater than the reference value of 80 mg/100 mL.

Question 6

Mean of the 10 results

$$\bar{x} = \frac{\sum_{i=1}^{n} x_i}{n} = \frac{401.5}{10} = 40.15\%\ \text{v/v}$$

Sample standard deviation

$$s = \sqrt{\frac{\sum_{i=1}^{n}(x_i - \bar{x})^2}{n-1}} = \sqrt{\frac{0.105}{9}} = 0.108\%\ \text{v/v}$$

The certified concentration of the CRM is 40.1% v/v.
The relevant hypotheses are:

$$H_0: \mu = 40.1$$
$$H_1: \mu \neq 40.1$$

$$t = \frac{|\bar{x} - \mu_0|}{s/\sqrt{n}} = \frac{|40.15 - 40.1|}{0.108/\sqrt{10}} = 1.464$$

The critical value for t (two-tailed, 95% confidence, $v = 9$) is 2.262 (from Appendix A, Table A.4). The calculated t value is less than the critical value. There is therefore insufficient evidence to

Questions and Solutions

reject the null hypothesis and we conclude that the results produced by the densitometer are not significantly biased at the 95% level of confidence.

Question 7

Mean of the 10 results

$$\bar{x} = \frac{\sum_{i=1}^{n} x_i}{n} = \frac{5138}{10} = 513.8 \text{ mg kg}^{-1}$$

Sample standard deviation

$$s = \sqrt{\frac{\sum_{i=1}^{n}(x_i - \bar{x})^2}{n-1}} = \sqrt{\frac{1539.6}{9}} = 13.1 \text{ mg kg}^{-1}$$

The null hypothesis, H_0, is that the mean is equal to 500 mg kg^{-1}, that is, $\mu = 500$ mg kg^{-1}. The alternative hypothesis, H_1 is that the mean is greater than 500 mg kg^{-1}, that is, $\mu > 500$ mg kg^{-1}. This requires a one-sided test, using the calculation

$$t = \frac{\bar{x} - \mu_0}{s/\sqrt{n}} = \frac{513.8 - 500}{13.1/\sqrt{10}} = 3.33$$

A significance level of $\alpha = 0.05$ corresponds to 95% confidence. The critical value for t (one-tailed, 95% confidence, $v = 9$) is 1.833, from Appendix A, Table A.4. The calculated t value exceeds the critical value. The null hypothesis is therefore rejected so there is evidence to suggest that the concentration of lead in the soil exceeds the limit of 500 mg kg^{-1}.

Question 8

The experiment involves the comparison of standard deviations; the correct test is therefore the F-test. The relevant hypotheses are:

$$H_0: \sigma_1^2 = \sigma_2^2$$
$$H_1: \sigma_1^2 \neq \sigma_2^2$$

(Note that the hypotheses are set out in terms of variances.) As the alternative hypothesis states that the variances are not equal, a two-tailed F-test is appropriate. The F value is calculated from

$$F = \frac{s_{max}^2}{s_{min}^2}$$

Remember that for a two-tailed F-test the larger variance must be the numerator in the equation for F.

$$F = \frac{0.648^2}{0.424^2} = \frac{0.420}{0.180} = 2.333$$

For a two-tailed F-test at a significance level $\alpha = 0.05$, the critical value is obtained from the table for a two-tailed test at 95% confidence (Appendix A, Table A.5b). The critical value for $v_{max} = v_{min} = 10$ is 3.717. The calculated F ratio is less than the critical value, so we can conclude that the difference in repeatability of the methods is not significant at the 95% level of confidence.

Question 9

The means are compared using a two-sample t-test. To ensure that the correct test is used, an F-test should be carried out first to determine whether there is a significant difference between the variances of the data.

The hypotheses for the F-test are:

$$H_0: \sigma_1^2 = \sigma_2^2$$
$$H_1: \sigma_1^2 \neq \sigma_2^2$$

The form of the alternative hypothesis indicates that a two-tailed test is required.

$$F = \frac{s_{max}^2}{s_{min}^2} = \frac{0.00114^2}{0.000725^2} = 2.472$$

Remember that the larger variance is the numerator in a two-tailed F-test. Since the test is two-tailed and the significance level is $\alpha = 0.05$ (95% confidence), the critical value is found from the F table for a two-tailed test at 95% confidence (Appendix A, Table A.5b) with degrees of freedom $v_{max} = v_{min} = 12$. The critical value is therefore 3.277. The calculated F value is less than the critical value, so we can conclude that the difference in standard deviations of the results obtained from the two dilution schemes is not significant at the 95% level of confidence.

As there is no significant difference between the standard deviations, the equal-variance form of the two-sample t-test is used to compare the mean values. In this case, $n_1 = n_2$, so the simplified calculation for s_{diff} in equation (4.14) in Chapter 4 can be used, giving

$$t = \frac{|\bar{x}_1 - \bar{x}_2|}{\sqrt{\frac{s_1^2 + s_2^2}{n}}}$$

The relevant hypotheses are:

$$H_0: \mu_1 = \mu_2$$
$$H_1: \mu_1 \neq \mu_2$$

The test is therefore two-tailed.

$$t = \frac{|0.2972 - 0.3038|}{\sqrt{\frac{0.000725^2 + 0.00114^2}{13}}} = 17.61$$

Note: the means have been rounded to four places; carrying out the t test on the raw data from Table C.9 may show a slightly different value for t.

The critical value for t (two-tailed, 95% confidence, $v = n_1 + n_2 - 2 = 24$) is 2.064 (see Appendix A, Table A.4). The calculated value for t substantially exceeds the critical value, so the null hypothesis is rejected and we can conclude that there is a strongly significant difference between the means of the results obtained using the two dilution schemes.

Question 10

1.

$$\text{Laboratory A} \quad \bar{x}_A = \frac{\sum_{i=1}^{n} x_{A_i}}{n} = \frac{40.30}{6} = 6.717$$

$$\text{Laboratory B} \quad \bar{x}_B = \frac{\sum_{i=1}^{n} x_{B_i}}{n} = \frac{37.76}{6} = 6.293$$

$$\text{Laboratory C} \quad \bar{x}_C = \frac{\sum_{i=1}^{n} x_{C_i}}{n} = \frac{40.95}{6} = 6.825$$

$$\text{Laboratory D} \quad \bar{x}_D = \frac{\sum_{i=1}^{n} x_{D_i}}{n} = \frac{36.61}{6} = 6.102$$

2. The data plotted in Figure C.1 indicate that the within-laboratory precision is broadly similar across the four laboratories.

3. The data are grouped by a single factor (laboratory), so one-way ANOVA is appropriate. The ANOVA table is shown as Table C.26. Detailed manual calculations are given below the answer to this question.
The critical value for F (one-tailed test, 95% confidence, $v_1 = 3$, $v_2 = 20$) is 3.098 (see Appendix A, Table A.5a). As the calculated value of F exceeds the critical value, we conclude that there is a significant difference between the results produced by the different laboratories.

4. The repeatability standard deviation s_r is obtained from the within-group mean square term:

$$s_r = \sqrt{M_w} = \sqrt{0.0928} = 0.305 \text{ mg L}^{-1}$$

The reproducibility standard deviation is a combination of the within- and between-group standard deviations. The between-group standard deviation is calculated from

$$s_b = \sqrt{\frac{M_b - M_w}{n}} = \sqrt{\frac{0.706 - 0.0928}{6}} = 0.320 \text{ mg L}^{-1}$$

The reproducibility standard deviation s_R is calculated from

$$s_R = \sqrt{s_r^2 + s_b^2} = \sqrt{0.305^2 + 0.320^2} = 0.442 \text{ mg L}^{-1}$$

Table C.26 Completed ANOVA table for data in Table C.10.

Source of variation	Sum of squares	Degrees of freedom v	Mean square	F
Between groups	2.118	3	0.706	7.608
Within groups	1.856	20	0.0928	
Total	3.973	23		

Question 10: Detailed Calculations

The calculations below follow the manual calculation given in the Appendix to Chapter 6. The number of laboratories (p) is equal to 4 and the number of results per laboratory (n) is equal to 6. There are $N = np = 24$ observations in total. The summations required for the manual calculation are shown in Table C.27.

To complete the one-way ANOVA table manually, the following calculations are required:

$$S_1 = \frac{\sum_{i=1}^{p}\left(\sum_{k=1}^{n} x_{ik}\right)^2}{n} = \frac{(40.30^2 + 37.76^2 + 40.95^2 + 36.61^2)}{6} = 1011.184$$

$$S_2 = \sum_{i=1}^{p}\sum_{k=1}^{n} x_{ik}^2 = 6.77^2 + 6.79^2 + 6.84^2 + 6.58^2 + 6.89^2 + 6.43^2 + 6.46^2 + 6.30^2$$
$$+ 6.00^2 + 6.76^2 + 5.96^2 + 6.28^2 + 6.41^2 + 7.26^2 + 6.46^2 + 7.15^2$$
$$+ 7.04^2 + 6.63^2 + 5.79^2 + 6.09^2 + 5.96^2 + 6.32^2 + 6.66^2 + 5.79^2$$
$$= 1013.039$$

$$S_3 = \frac{\left(\sum_{i=1}^{p}\sum_{k=1}^{n} x_{ik}\right)^2}{N} = \frac{155.62^2}{24} = 1009.066$$

The ANOVA table is calculated as shown in Table C.28.

Table C.27 Summations of data in Table C.10.

	Laboratory (p)			
	A	B	C	D
Replicates (n)	6.77	6.46	6.41	5.79
	6.79	6.30	7.26	6.09
	6.84	6.00	6.46	5.96
	6.58	6.76	7.15	6.32
	6.89	5.96	7.04	6.66
	6.43	6.28	6.63	5.79
$\sum_{k=1}^{n} x_{ik}$	40.30	37.76	40.95	36.61	$\sum_{i=1}^{p}\sum_{k=1}^{n} x_{ik} = 155.62$

Table C.28 Calculations for the one-way ANOVA table for data in Table C.10.

Source of variation	Sum of squares	Degrees of freedom v	Mean square	F
Between groups	$S_b = S_1 - S_3$ $= 1011.184 - 1009.066$ $= 2.118$	$p - 1 = 3$	$M_b = S_b/(p-1)$ $= 2.118/3$ $= 0.706$	M_b/M_w $= 0.706/0.0928$ $= 7.608$
Within groups	$S_w = S_2 - S_1$ $= 1013.039 - 1011.184$ $= 1.856$	$N - p = 20$	$M_w = S_w/(N-p)$ $= 1.856/20$ $= 0.0928$	
Total	$S_{tot} = S_b + S_w$ $= S_2 - S_3$ $= 1013.039 - 1009.066$ $= 3.973$	$N - 1 = 23$		

Question 11

This study is an example of a cross-classified ('factorial') design with replication. There are two factors – shift and form of material – so the data can be analysed using a two-factor ANOVA. The completed ANOVA table is shown as Table C.29; detailed manual calculations can be found below the answer to this question.

Interpretation follows the method outlined in Chapter 6, Section 6.5.2. First, compare the interaction mean square term with the residual mean square term:

$$F = \frac{M_1}{M_0} = \frac{0.0113}{0.0758} = 0.149$$

The critical value for F (95% confidence, one-tailed, $v_1 = 6$, $v_2 = 12$) is 2.996. The interaction term is therefore not significant; the effects of shifts and product forms are not strongly interdependent. The between-column and between-row mean square terms can therefore be tested individually.

Comparing each with the residual mean square:

Between shifts:

$$F = \frac{M_3}{M_0} = \frac{0.0967}{0.0758} = 1.276$$

The critical value for F (95% confidence, one-tailed, $v_1 = 3$, $v_2 = 12$) is 3.490. There is therefore no significant difference between the results produced by these four different shifts.

Between forms of material:

$$F = \frac{M_2}{M_0} = \frac{0.0629}{0.0758} = 0.830$$

The critical value for F (95% confidence, one-tailed, $v_1 = 2$, $v_2 = 12$) is 3.885 (Appendix A, Table A.5a). There is therefore no significant difference between the results obtained for the different forms of the material.

Table C.29 Completed ANOVA table for data in Table C.11.

Source of variation	Sum of squares	Degrees of freedom v	Mean square[a]
Between columns (shifts)	0.2900	3	0.0967
Between rows (form of material)	0.1258	2	0.0629
Interaction	0.0675	6	0.0113
Residual	0.9100	12	0.0758
Total	1.3933	23	

[a]To three significant figures.

Question 11: Detailed calculations

The number of shifts (p) is 4, the number of forms of material (q) is 3 and the number of replicates (n) is 2. The sums of the pairs of results and the grand totals for shift and fertiliser form are shown in Table C.30. These are required to carry out the calculations for the two-factor ANOVA.

The calculations required to evaluate the sum of squares terms manually are shown below, using the raw data:

1. $$\frac{\sum_{i=1}^{p}\left(\sum_{j=1}^{q}\sum_{k=1}^{n} x_{ijk}\right)^2}{qn} = \frac{(43.8^2 + 45.6^2 + 44.7^2 + 44.3^2)}{3 \times 2} = \frac{7958.38}{6} = 1326.39667$$

2. $$\frac{\sum_{j=1}^{q}\left(\sum_{i=1}^{p}\sum_{k=1}^{n} x_{ijk}\right)^2}{pn} = \frac{(60.1^2 + 59.6^2 + 58.7^2)}{4 \times 2} = \frac{10609.86}{8} = 1326.2325$$

3. $$\frac{\sum_{i=1}^{p}\sum_{j=1}^{q}\left(\sum_{k=1}^{n} x_{ijk}\right)^2}{n} = \frac{\left(\begin{array}{c}14.7^2 + 14.6^2 + 14.5^2 + 15.6^2 + 15.1^2 + 14.9^2 \\ +15.0^2 + 15.0^2 + 14.7^2 + 14.8^2 + 14.9^2 + 14.6^2\end{array}\right)}{2}$$
$$= \frac{2653.18}{2} = 1326.59$$

4. $$\sum_{i=1}^{p}\sum_{j=1}^{q}\sum_{k=1}^{n} x_{ijk}^2 = 7.2^2 + 7.5^2 + 7.5^2 + 7.1^2 + 7.5^2 + 7.0^2 + 7.8^2 + 7.8^2$$
$$+ 7.5^2 + 7.6^2 + 7.4^2 + 7.5^2 + 7.2^2 + 7.8^2 + 7.5^2 + 7.5^2$$
$$+ 7.6^2 + 7.1^2 + 7.3^2 + 7.5^2 + 7.1^2 + 7.8^2 + 7.1^2 + 7.5^2$$
$$= 1327.5$$

5. $$\frac{\left(\sum_{i=1}^{p}\sum_{j=1}^{q}\sum_{k=1}^{n} x_{ijk}\right)^2}{N} = \frac{178.4^2}{24} = \frac{31826.56}{24} = 1326.10667$$

Table C.30 Summations of data in Table C.11.

	Sums				Total
	A	B	C	D	
Block	14.7	15.6	15.0	14.8	60.1
Powder	14.6	15.1	15.0	14.9	59.6
Granules	14.5	14.9	14.7	14.6	58.7
Total	43.8	45.6	44.7	44.3	178.4

Questions and Solutions

Table C.31 Calculations for the two-factor ANOVA table for data in Table C.11.

Source of variation	Sum of squares	Degrees of freedom v	Mean square[a]
Between columns (shifts)	$S_3 = (1)-(5) = 1326.39667-1326.10667$ = **0.2900**	$p-1 = 3$	$M_3 = S_3/(p-1)$ $= 0.29/3 =$ **0.0967**
Between rows (form of material)	$S_2 = (2)-(5) = 1326.2325-1326.10667$ = **0.1258**	$q-1 = 2$	$M_2 = S_2/(q-1)$ $= 0.1258/2 =$ **0.0629**
Interaction	$S_1 = [(3)+(5)]-[(1)+(2)]$ $= (1326.59 + 1326.10667)$ $-(1326.39667 + 1326.2325)$ $=$ **0.0675**	$(p-1)(q-1) = 6$	$M_1 = S_1/(p-1)(q-1)$ $= 0.0675/6 =$ **0.0113**
Residual	$S_0 = (4)-(3) = 1327.5-1326.59 =$ **0.9100**	$N-pq = 12$	$M_0 = S_0/(N-pq)$ $= 0.91/12 =$ **0.0758**
Total	$S_3 + S_2 + S_1 + S_0 = (4)-(5)$ $= 1327.5-1326.10667 =$ **1.3933**	$N-1 = 23$	

[a]To three significant figures.

The ANOVA table (based on the raw data) is shown in Table C.31.

Note: the raw data range is about 0.8 and the maximum is 7.8, so coding as suggested in the Appendix to Chapter 6 could be useful. In this case, coding by subtracting 7 (that is, removing the leading digit from the raw data) reduces sums 1–5 to 4.7967, 4.6325, 4.9900, 5.9000 and 4.5067 respectively, reducing the required numerical precision by three digits.

Question 12

1. The completed ANOVA table is shown as Table C.32. Detailed manual calculations can be found below the answer to this question.
2. The 95% one-tailed critical value for F with 3 and 16 degrees of freedom for numerator (v_1) and denominator (v_2) is 3.239 (Appendix A, Table A.5a). The calculated value of F in Table C.32 is 18.3. The calculated value of F exceeds the critical value substantially, so there is a significant difference between the analysts at the 95% level of confidence. The critical value at 99% confidence is 5.292 (Appendix A, Table A.5c), so the difference is also significant at the more stringent 99% confidence level.
3. The repeatability standard deviation is obtained from the within-group mean square term:

$$s_r = \sqrt{M_w} = \sqrt{0.00375} = 0.0612 \text{ mg kg}^{-1}$$

The intermediate precision (different analysts working on different days) is a combination of the within- and between-group standard deviations. The between-group standard deviation is

Table C.32 Completed ANOVA table for data in Table C.12.

Source of variation	Sum of squares	Degrees of freedom v	Mean square	F
Between groups	0.20954	3	0.0686	18.3
Within groups	0.06004	16	0.00375	
Total	0.26598	19		

calculated from:

$$s_b = \sqrt{\frac{M_b - M_w}{n}} = \sqrt{\frac{0.0686 - 0.00375}{5}} = 0.114 \text{ mg kg}^{-1}$$

The intermediate precision is calculated from:

$$s_I = \sqrt{s_r^2 + s_b^2} = \sqrt{0.0612 + 0.114} = 0.129 \text{ mg kg}^{-1}$$

4. The repeatability limit is calculated from:

$$r = t \times \sqrt{2} \times s_r$$

The repeatability standard deviation has 16 degrees of freedom, and looking up the two-tailed value of Student's t for $\alpha = 0.05$ and $v = 16$ gives $t = 2.120$. Therefore,

$$r = 2.120 \times \sqrt{2} \times 0.0612 = 0.18 \text{ mg kg}^{-1}$$

Note that in most practical circumstances, the approximation $r = 2.8s_r$ (0.172, here) would be considered sufficient.

Question 12: Detailed calculations

The sums of observations for each analyst are:

	Analyst 1	Analyst 2	Analyst 3	Analyst 4	Total
$\sum_{k=1}^{n} x_{ik}$	41.83	41.71	41.72	42.92	$\sum_{i=1}^{p}\sum_{k=1}^{n} x_{ik} = 168.18$

To complete the ANOVA table, the following calculations (shown using raw data) are required. Intermediate values after coding by subtracting 8 from the raw data are given in the Note at the end of this answer.

$$S_1 = \frac{\sum_{i=1}^{p}\left(\sum_{k=1}^{n} x_{ik}\right)^2}{n} = \frac{(41.83^2 + 41.71^2 + 41.72^2 + 42.92^2)}{5} = 1414.432$$

$$S_2 = \sum_{i=1}^{p}\sum_{k=1}^{n} x_{ik}^2 = 8.35^2 + 8.41^2 + 8.22^2 + 8.42^2 + 8.43^2 + 8.36^2 + 8.35^2$$
$$+ 8.32^2 + 8.31^2 + 8.37^2 + 8.30^2 + 8.37^2 + 8.23^2 + 8.40^2$$
$$+ 8.42^2 + 8.57^2 + 8.55^2 + 8.60^2 + 8.61^2 + 8.59^2$$
$$= 1414.492$$

$$S_3 = \frac{\left(\sum_{i=1}^{p}\sum_{k=1}^{n} x_{ik}\right)^2}{N} = \frac{168.18^2}{20} = 1414.226$$

The calculations for the ANOVA table are shown in Table C.33.
Note: coding by removing 8 from the raw data would reduce the intermediate sums substantially; with coding, S_1, S_2 and S_3 are 3.552, 3.612 and 3.346, respectively.

Questions and Solutions

Table C.33 One-way ANOVA table for data in Table C.12.

Source of variation	Sum of squares	Degrees of freedom v	Mean square	F
Between groups	$S_b = S_1 - S_3$ $= 1414.432 - 1414.226$ $= \mathbf{0.20954}$	$p - 1 = \mathbf{3}$	$M_b = S_b/(p-1)$ $= 0.20954/3$ $= \mathbf{0.0686}$	M_b/M_w $= 0.0686/0.00375$ $= \mathbf{18.3}$
Within groups	$S_w = S_2 - S_1$ $= 1414.492 - 1414.432$ $= \mathbf{0.06004}$	$N - p = \mathbf{16}$	$M_w = S_w/(N-p)$ $= 0.06004/16$ $= \mathbf{0.00375}$	
Total	$S_{tot} = S_b + S_w$ $= S_2 - S_3 = 1414.492$ $- 1414.226 = \mathbf{0.26598}$	$N - 1 = \mathbf{19}$		

Table C.34 ANOVA table for data in Table C.13.

Source of variation	Sum of squares	Degrees of freedom v	Mean square	F	p-Value	F_{crit}
Sample (digestion)	39.5	1	39.5	4.0	0.06	4.494
Columns (determination)	343.6	1	343.6	35.1	2.1×10^{-5}	4.494
Interaction	1501.6	1	1501.6	153.5	1.3×10^{-9}	4.494
Within	156.5	16	9.8			
Total	2041.3	19				

Question 13

Table C.13 shows a two-factor experiment with replicated observations in each cell, so a two-factor ANOVA with replication is appropriate. The completed ANOVA table, including the critical values for F and the associated p-values, is shown as Table C.34.

Following the interpretation in Section 6.5.2.4, the interaction term is examined first. The interaction is very strongly significant (the calculated F ratio is 153.5 against a critical value of 4.494 from Appendix A, Table A.5a, and the p-value is very small – far less than the 0.05 required for significance at the 95% level of confidence). It should therefore be concluded that the digestion and instrumental determination methods are both important and that they are strongly interdependent.

The relative size of the effects can be judged by examining the mean squares. The interaction mean square is clearly larger than the other two effects, suggesting that this is the dominant effect. This is supported by inspection of an interaction plot, as shown in Figure C.8. The lines connecting means for the same digestion method are clearly far from parallel and cross over near the centre; this is typical of a dominant interaction.

Note also that the two digestion methods show very different mean values for each instru;mental method; the differences are almost equal and opposite in sign. The mean difference between digestion methods is therefore small. This is why the apparent significance of the digestion method is marginal in the ANOVA table, nicely illustrating the need to assess the interaction first.

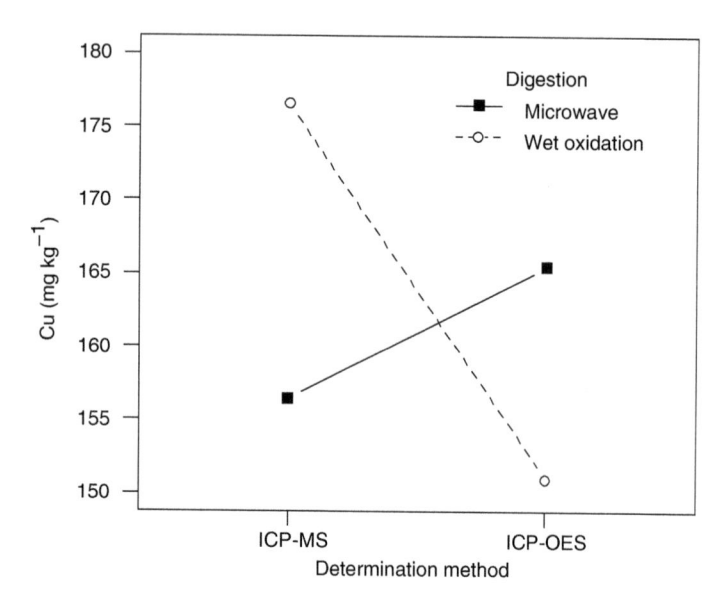

Figure C.8 Interaction plot for copper data.

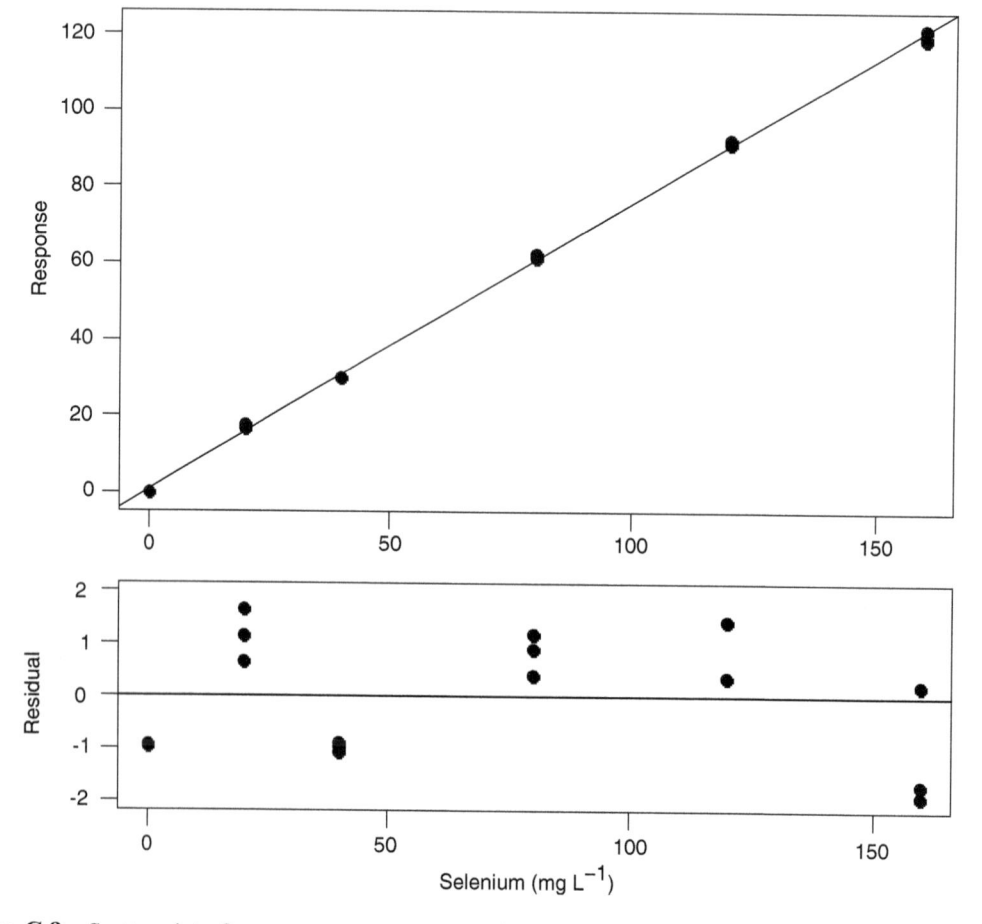

Figure C.9 Scatter plot of response versus concentration for the data shown in Table C.14, with fitted line residuals.

Questions and Solutions

Question 14

1. The scatter plot, with fitted line and residuals plot, is shown in Figure C.9.
2. The calculations required for the estimation of the gradient and intercept are shown in Table C.35.
   The gradient is calculated as follows:

$$b = \frac{\sum_{i=1}^{n}[(x_i - \bar{x})(y_i - \bar{y})]}{\sum_{i=1}^{n}(x_i - \bar{x})^2} = \frac{42801.15}{57000} = 0.750897$$

   The intercept is calculated as follows:

$$a = \bar{y} - b\bar{x} = 53.50544 - (0.750897 \times 70) = 0.9426$$

   The equation of the best fit straight line is therefore $y = 0.9426 + 0.7509x$.

3. The correlation coefficient is calculated as follows:

$$r = \frac{\sum_{i=1}^{n}[(x_i - \bar{x})(y_i - \bar{y})]}{\sqrt{\left[\sum_{i=1}^{n}(x_i - \bar{x})^2\right]\left[\sum_{i=1}^{n}(y_i - \bar{y})^2\right]}} = \frac{42801.15}{\sqrt{57000 \times 32162.34}} = 0.9996$$

4. The residual standard deviation is calculated from

$$s_{y/x} = \sqrt{\frac{\sum_{i=1}^{n}(y_i - \hat{y}_i)^2}{n-2}}$$

   The residual values are shown in Table C.36, giving

$$s_{y/x} = \sqrt{\frac{23.0679}{18-2}} = 1.201$$

5. The standard deviation for the estimate of the gradient is calculated as follows:

$$s_b = \frac{s_{y/x}}{\sqrt{\sum_{i=1}^{n}(x_i - \bar{x})^2}} = \frac{1.201}{\sqrt{57000}} = 0.00503$$

   The standard deviation for the estimate of the intercept is calculated as follows:

$$s_a = s_{y/x}\sqrt{\frac{\sum_{i=1}^{n}x_i^2}{n\sum_{i=1}^{n}(x_i - \bar{x})^2}} = 1.201\sqrt{\frac{145200}{18 \times 57000}} = 0.452$$

Table C.35 Calculations required for the estimation of the gradient and intercept for the data shown in Table C.14.

x_i	y_i	$(x_i-\bar{x})$	$(x_i-\bar{x})^2$	$(y_i-\bar{y})$	$(y_i-\bar{y})^2$	$(x_i-\bar{x})(y_i-\bar{y})$
0	−0.0011	−70	4900	−53.5065	2862.949	3745.457
0	0.00044	−70	4900	−53.5050	2862.785	3745.350
0	−0.0015	−70	4900	−53.5069	2862.992	3745.485
20	17.6	−50	2500	−35.9054	1289.200	1795.272
20	16.6	−50	2500	−36.9054	1362.011	1845.272
20	17.1	−50	2500	−36.4054	1325.356	1820.272
40	30.1	−30	900	−23.4054	547.814	702.163
40	30.0	−30	900	−23.5054	552.506	705.163
40	29.9	−30	900	−23.6054	557.217	708.163
80	61.4	10	100	7.8946	62.324	78.946
80	61.9	10	100	8.3946	70.469	83.946
80	62.2	10	100	8.6946	75.595	86.946
120	91.4	50	2500	37.8946	1435.998	1894.728
120	92.5	50	2500	38.9946	1520.576	1949.728
120	92.5	50	2500	38.9946	1520.576	1949.728
160	121.3	90	8100	67.7946	4596.103	6101.511
160	119.2	90	8100	65.6946	4315.776	5912.511
160	119.4	90	8100	65.8946	4342.094	5930.511
$\sum_{i=1}^{n}$			57000		32162.34	42801.15

Table C.36 Residual values calculated for the data in Table C.14.

y_i	$\hat{y}_i$ ($\hat{y}_i = a+bx_i$)	$(y_i-\hat{y}_i)$	$(y_i-\hat{y}_i)^2$
−0.0011	0.9426	−0.9437	0.8906
0.00044	0.9426	−0.9422	0.8877
−0.0015	0.9426	−0.9441	0.8914
17.6	15.9606	1.6394	2.6877
16.6	15.9606	0.6394	0.4089
17.1	15.9606	1.1394	1.2983
30.1	30.9785	−0.8785	0.7718
30.0	30.9785	−0.9785	0.9575
29.9	30.9785	−1.0785	1.1632
61.4	61.0144	0.3856	0.1487
61.9	61.0144	0.8856	0.7843
62.2	61.0144	1.1856	1.4056
91.4	91.0503	0.3497	0.1223
92.5	91.0503	1.4497	2.1016
92.5	91.0503	1.4497	2.1016
121.3	121.0862	0.2138	0.0457
119.2	121.0862	−1.8862	3.5578
119.4	121.0862	−1.6862	2.8433
$\sum_{i=1}^{n}$			23.0679

The 95% confidence intervals for the gradient and the intercept are:

$$b \pm t \times s_b = 0.7509 \pm 2.120 \times 0.00503 = 0.7509 \pm 0.01066$$

$$a \pm t \times s_a = 0.9426 \pm 2.120 \times 0.452 = 0.9426 \pm 0.958$$

Questions and Solutions

where t is the two-tailed Student's t value for 95% confidence and $v = 16$, obtained from Appendix A, Table A.4.

The prediction interval for x values is calculated from

$$s_{\hat{x}} = \frac{s_{y/x}}{b}\sqrt{\frac{1}{N} + \frac{1}{n} + \frac{(\bar{y}_0 - \bar{y})^2}{b^2 \sum_{i=1}^{n}(x_i - \bar{x})^2}}$$

$\bar{y} = 53.505$. If $y_0 = \bar{y}$, the predicted value is

$$\hat{x} = \frac{53.505 - 0.9426}{0.7509} = 70\,\text{mg L}^{-1}.$$

The prediction interval is

$$s_{\hat{x}} = \frac{1.201}{0.7509}\sqrt{\frac{1}{1} + \frac{1}{18} + \frac{(53.505 - 53.505)^2}{0.7509^2 \times 57000}} = 1.64\,\text{mg L}^{-1}$$

If $y_0 = 120$, the predicted value is

$$\hat{x} = \frac{120 - 0.9426}{0.7509} = 158.6\,\text{mg L}^{-1}$$

The prediction interval is

$$s_{\hat{x}} = \frac{1.201}{0.7509}\sqrt{\frac{1}{1} + \frac{1}{18} + \frac{(120 - 53.505)^2}{0.7509^2 \times 57000}} = 1.75\,\text{mg L}^{-1}$$

The 95% confidence interval is calculated by multiplying the prediction interval by the appropriate Student's t value (two-tailed, 95% confidence, $v = 16$, obtained from Appendix A, Table A.4).

If $y_0 = \bar{y} = 53.505$, the confidence interval for $\hat{x}$ is $70.0 \pm 2.120 \times 1.64 = 70.0 \pm 3.48\,\text{mg L}^{-1}$.
If $y_0 = 120$, the confidence interval for $\hat{x}$ is $158.6 \pm 2.120 \times 1.75 = 158.6 \pm 3.71\,\text{mg L}^{-1}$.

Question 15

1. There are sufficient concentration levels and sufficient replication at each level. However, the spacing of the concentrations of the standard solutions it not ideal. The concentrations should be approximately evenly spaced across the range of interest.
2. The regression statistics are consistent with a linear relationship. However, a visual examination indicates that data up to $x = 0.8\,\text{mg L}^{-1}$ would fit a steeper line. The fitted line is pulled down by the excessive leverage of the data at $x = 1.6\,\text{mg L}^{-1}$. This effect is also clear from the plot of the residual values. The leverage is exaggerated by the uneven spacing of

the concentration levels. The reason for the leverage is found by examination of the data itself, which shows that while data at levels 0.01–0.8 mg L^{-1} are generally reported to one or two decimal places, all three replicates at 1.6 mg L^{-1} give a response of 140. This suggests detector saturation or that the maximum output reading of the detector has been reached.

3. The intercept, a, is estimated as 2.347. The upper and lower 95% confidence limits for the intercept do not cross zero and the p-value for the intercept is less than 0.05. We can therefore conclude that the intercept is significantly different from zero. This bias arises from the leverage caused by the responses recorded for the most concentrated standard solution.

Question 16

The observed results were as follows:

	Experiment							
	1	2	3	4	5	6	7	8
Result identifier (Table C.37)	s	t	u	v	w	x	y	z
Observed result (mg L^{-1})	1.10	1.06	0.96	0.84	0.96	0.86	1.06	1.05

The calculated effects (Δ_i) on the results due to changing each parameter are shown in Table C.37. The critical difference is calculated from

$$|\Delta_{crit}| > \frac{ts}{\sqrt{2}} = \frac{2.262 \times 0.02}{\sqrt{2}} = 0.032$$

where t is the two-tailed-Student's t value at the 95% confidence level for nine degrees of freedom taken from Appendix A, Table A.4 (remember that the degrees of freedom relate to the estimate of s).

The differences calculated for parameters D, C and G are greater than $|\Delta_{crit}|$. We can therefore conclude that changing these parameters by the amount specified in the experimental plan has a significant effect on the amount of the additive extracted from diesel oil.

Question 17

The general form of the equation for calculating the limit of detection is

$$LOD = x_0 + k_l st_{(\nu,\alpha)} + k_l st_{(\nu,\beta)}$$

In this case results are corrected for a blank reading so $x_0 = 0$. However, results were not corrected by independent blank readings – each result in Table C.19 was corrected by the same blank reading. Therefore $k_l = \sqrt{1 + 1/n_B} = \sqrt{2}$ ($n_B = 1$).

An approximate estimate of the LOD can be obtained assuming $\nu = \infty$. The one-tailed Student's t value for $\alpha = \beta = 0.05$ and $\nu = \infty$ is 1.645 (Appendix A, Table A.4). The LOD is given by

$$LOD = (1.41 \times 0.0048 \times 1.645) + (1.41 \times 0.0048 \times 1.645) = 0.022 \text{ mg L}^{-1}$$

A more rigorous estimate can be obtained using the t value for the number of degrees of freedom associated with the estimate of s. The t value for $\nu = 11$ is 1.796. The LOD becomes

$$LOD = (1.41 \times 0.0048 \times 1.796) + (1.41 \times 0.0048 \times 1.796) = 0.024 \text{ mg L}^{-1}$$

Questions and Solutions

Table C.37 Evaluation of results from a ruggedness study of the extraction of an additive from diesel oil.

Parameter	Mean of results at normal value	Mean of results at alternative value	Difference
A	$\dfrac{(s+t+u+v)}{4} = 0.99$	$\dfrac{(w+x+y+z)}{4} = 0.9825$	$\Delta_A = 0.0075$
B	$\dfrac{(s+t+w+x)}{4} = 0.995$	$\dfrac{(u+v+y+z)}{4} = 0.9775$	$\Delta_B = 0.0175$
C	$\dfrac{(s+u+w+y)}{4} = 1.02$	$\dfrac{(t+v+x+z)}{4} = 0.9525$	$\Delta_C = 0.0675$
D	$\dfrac{(s+t+y+z)}{4} = 1.0675$	$\dfrac{(u+v+w+x)}{4} = 0.905$	$\Delta_D = 0.1625$
E	$\dfrac{(s+u+x+z)}{4} = 0.9925$	$\dfrac{(t+v+w+y)}{4} = 0.98$	$\Delta_E = 0.0125$
F	$\dfrac{(s+v+w+z)}{4} = 0.9875$	$\dfrac{(t+u+x+y)}{4} = 0.985$	$\Delta_F = 0.0025$
G	$\dfrac{(s+v+x+y)}{4} = 0.965$	$\dfrac{(t+u+w+z)}{4} = 1.0075$	$\Delta_G = -0.0425$

Question 18

A standard uncertainty is an uncertainty expressed as a standard deviation.

1. This information is a tolerance in the form '$x \pm a$'. If a rectangular distribution is assumed, the standard uncertainty is calculated by dividing a by $\sqrt{3}$:

$$u = 0.08/\sqrt{3} = 0.046 \text{ mL}$$

If a triangular distribution is assumed (which would be appropriate if there is evidence to suggest that values closer to 100 mL are more likely than values at the extremes of the tolerance range), the standard uncertainty is calculated by dividing a by $\sqrt{6}$:

$$u = 0.08/\sqrt{6} = 0.033 \text{ mL}$$

2. This information is a confidence interval of the form $x \pm d$ at 95% confidence. Since the number of degrees of freedom is unknown, the standard uncertainty is calculated assuming large degrees of freedom so d is divided by 1.96:

$$u = 0.0004/1.96 = 0.0002 \text{ g}$$

3. This information is a tolerance in the form '$x \pm a$'. Assuming a rectangular distribution:

$$u = 0.1/\sqrt{3} = 0.058\%$$

4. This information is expressed as a standard deviation. Since a standard uncertainty is defined as an uncertainty estimate expressed as a standard deviation, no conversion is necessary:

$$u = 0.00021 \text{ g}$$

5. This information is in the form of an expanded uncertainty. The standard uncertainty is obtained by dividing the expanded uncertainty by the stated coverage factor:

$$u = 0.03/2 = 0.015 \text{ mL}$$

Question 19

The manufacturing tolerance is expressed in the form $x \pm a$. Assuming a rectangular distribution, the standard uncertainty is $u = 0.01/\sqrt{3} = 0.0058$ mL.

The precision of filling and emptying the pipette is expressed as a standard deviation, so no conversion of the data is required.

The estimate of the effect of the laboratory temperature differing from the flask calibration temperature is already expressed as a standard uncertainty.

All three uncertainty estimates are expressed in terms of their effect on the measurement result. They can therefore be combined using the basic rule for the combination of variances to give the uncertainty in the volume of liquid delivered by the pipette, $u(V)$:

$$u(V) = \sqrt{0.0058^2 + 0.0016^2 + 0.00061^2} = 0.0060 \text{ mL}$$

Question 20

1. The standard uncertainties are shown in Table C.38.
2. The uncertainty in the mass of the material is obtained by combining the standard uncertainties associated with the balance calibration and the balance precision:

$$u(m) = \sqrt{0.0002^2 + 0.000041^2} = 0.0002 \text{ g} = 0.2 \text{ mg}$$

Table C.38 Standard uncertainties associated with estimating the uncertainty in the concentration of a standard solution.

Source of uncertainty	Conversion required	Standard uncertainty
Balance calibration	Data expressed as expanded uncertainty. Divide by stated coverage factor, $k=2$	0.0002 g
Precision of balance	None – expressed as a standard deviation	0.000041 g
Purity	Data expressed as a tolerance. Assume a rectangular distribution so divide by $\sqrt{3}$	0.00058
Tolerance of flask	Assume a rectangular distribution so divide by $\sqrt{3}$	0.046 mL
Precision of filling the flask to the calibration line	None – expressed as a standard deviation	0.017 mL
Effect of the laboratory temperature differing from the flask calibration temperature	None – expressed as a standard uncertainty	0.031 mL

Questions and Solutions 257

The uncertainty in the volume of the solution is obtained by combining the standard uncertainties associated with the flask tolerance, precision of filling the flask and temperature effects:

$$u(V) = \sqrt{0.046^2 + 0.017^2 + 0.031^2} = 0.058 \text{ mL}$$

3. The concentration of the solution C, expressed in mg L^{-1}, is calculated from

$$C = \frac{m \times p}{V} \times 1000 = \frac{100.5 \times 0.999}{100} \times 1000 = 1004.0 \text{ mg L}^{-1}$$

The equation used to calculate the concentration involves multiplication and division. The standard uncertainties in m, p and V are therefore combined as relative uncertainties:

$$\frac{u(C)}{C} = \sqrt{\left[\frac{u(m)}{m}\right]^2 + \left[\frac{u(p)}{p}\right]^2 + \left[\frac{u(V)}{V}\right]^2}$$

Therefore:

$$u(C) = 1004 \times \sqrt{\left(\frac{0.2}{100.5}\right)^2 + \left(\frac{0.00058}{0.999}\right)^2 + \left(\frac{0.058}{100}\right)^2} = 2.16 \text{ mg L}^{-1}$$

Question 21

1. The equation for calculating the concentration of the sodium hydroxide solution involves multiplication and division. The uncertainty associated with precision is expressed as a relative standard deviation. The most straightforward way to incorporate precision into the model is therefore to include a multiplicative term which is equal to 1 with a relative uncertainty equal to 0.0005.
The extended model including the precision term is therefore

$$c_{NaOH} = \frac{1000 \times m_{KHP} \times p_{KHP}}{M_{KHP} \times V_T} \times P_{method}$$

2. The spreadsheet is shown in Figures C.10a and C.10b. The former shows the values and the latter shows the cell references.
3. The expanded uncertainty is calculated by multiplying the standard uncertainty by an appropriate coverage factor. For approximately 95% confidence a coverage factor of $k=2$ is used:

$$U = 2 \times 0.0001 = 0.0002 \text{ mol L}^{-1}$$

Question 22

1. The target value μ is 3.8 mg kg^{-1} and the standard deviation is 0.08 mg kg^{-1}.
Warning limits (~95% confidence) are calculated as

$$\mu \pm 2s = 3.8 \pm 2 \times 0.08 = 3.8 \pm 0.16$$

The warning limits are therefore drawn at 3.96 and 3.64 mg kg^{-1}.

	A	B	C	D	E	F	G	H
1	Parameter	Value	u					
2	m_{KHP}	0.3888	0.00013	0.38893	0.3888	0.3888	0.3888	0.3888
3	p_{KHP}	1.0	0.00029	1	1.00029	1	1	1
4	M_{KHP}	204.2212	0.0038	204.2212	204.2212	204.225	204.2212	204.2212
5	V_T	18.64	0.013	18.64	18.64	18.64	18.653	18.64
6	P_{method} (relative)	1	0.0005	1	1	1	1	1.0005
7								
8	c_{NaOH}	**0.10214**	**0.00010**	0.10217	0.10217	0.10213	0.10206	0.10219
9				0.0000342	0.0000296	-0.0000019	-0.00007118	0.00005107

Figure C.10a Spreadsheet for the evaluation of the uncertainty in the concentration of a sodium hydroxide solution – values

	A	B	C	D	E	F	G	H
1	Parameter	Value	u					
2	m_{KHP}	0.3888	0.00013	=$B2+$C2	=$B2	=$B2	=$B2	=$B2
3	p_{KHP}	1.0	0.00029	=$B3	=$B3+$C3	=$B3	=$B3	=$B3
4	M_{KHP}	204.2212	0.0038	=$B4	=$B4	=$B4+$C4	=$B4	=$B4
5	V_T	18.64	0.013	=$B5	=$B5	=$B5	=$B5+$C5	=$B5
6	P_{method} (relative)	1	0.0005	=$B6	=$B6	=$B6	=$B6	=$B6+$C6
7								
8	c_{NaOH}	=(1000*B2* B3*B6) /(B4*B5)	=SQRT (SUMSQ (D9:H9))	=(1000*D2* D3*D6) /(D4*D5)	=(1000*E2* E3*E6) /(E4*E5)	=(1000*F2*F3 *F6) /(F4*F5)	=(1000*G2* G3*G6) /(G4*G5)	=(1000*H2* H3*H6) /(H4*H5)
9				=D8-B8	=E8-B8	=F8-B8	=G8-B8	=H8-B8

Figure C.10b Spreadsheet for the evaluation of the uncertainty in the concentration of a sodium hydroxide solution – cell references

Action limits ($\sim 99.7\%$ confidence) are calculated as

$$\mu \pm 3s = 3.8 \pm 3 \times 0.08 = 3.8 \pm 0.24$$

The action limits are therefore drawn at 4.04 and 3.56 mg kg^{-1}.

The Shewhart chart is shown in Figure C.11.

2. On three occasions the results breached the lower warning limit but stayed within the action limit. However, on all three occasions, the next result obtained was well within the warning limits. There are no trends in the data (that is, no drift or step changes), so the system can be considered to be under control.

3. The calculation of the cumulative sum is shown in Table C.39 and the CuSum chart is shown in Figure C.12. Remember that for interpretation using a V-mask, the axes should be scaled so that the divisions on both axes are the same length. A division on the x-axis should represent a

Questions and Solutions 259

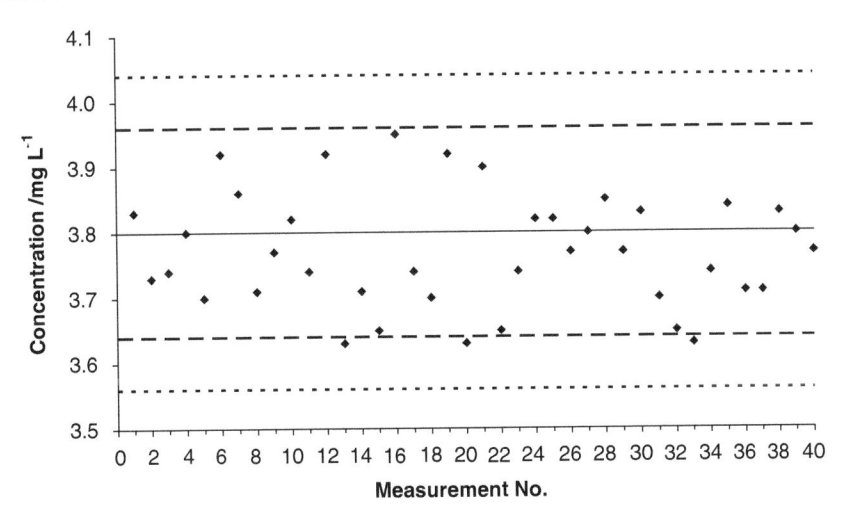

Figure C.11 Shewhart chart constructed from data in Table C.22.

Table C.39 Calculation of the cumulative sum for data shown in Table C.22.

Measure-ment No.	QC result ($\bar{x}_i$)	$\bar{x}_i-\mu$	C_i	Measure-ment No.	QC result ($\bar{x}_i$)	$\bar{x}_i-\mu$	C_i
1	3.83	0.03	0.03	21	3.90	0.10	−0.43
2	3.73	−0.07	−0.04	22	3.65	−0.15	−0.58
3	3.74	−0.06	−0.10	23	3.74	−0.06	−0.64
4	3.80	0.00	−0.10	24	3.82	0.02	−0.62
5	3.70	−0.10	−0.20	25	3.82	0.02	−0.60
6	3.92	0.12	−0.08	26	3.77	−0.03	−0.63
7	3.86	0.06	−0.02	27	3.80	0.00	−0.63
8	3.71	−0.09	−0.11	28	3.85	0.05	−0.58
9	3.77	−0.03	−0.14	29	3.77	−0.03	−0.61
10	3.82	0.02	−0.12	30	3.83	0.03	−0.58
11	3.74	−0.06	−0.18	31	3.70	−0.10	−0.68
12	3.92	0.12	−0.06	32	3.65	−0.15	−0.83
13	3.63	−0.17	−0.23	33	3.63	−0.17	−1.00
14	3.71	−0.09	−0.32	34	3.74	−0.06	−1.06
15	3.65	−0.15	−0.47	35	3.84	0.04	−1.02
16	3.95	0.15	−0.32	36	3.71	−0.09	−1.11
17	3.74	−0.06	−0.38	37	3.71	−0.09	−1.20
18	3.70	−0.10	−0.48	38	3.83	0.03	−1.17
19	3.92	0.12	−0.36	39	3.80	0.00	−1.17
20	3.63	−0.17	−0.53	40	3.77	−0.03	−1.20

single unit (that is, measurement number), whereas a division on the y-axis should be equivalent to $2\sigma/\sqrt{n} = 2 \times 0.08/\sqrt{1} = 0.16$

4. In Figure C.12, the V-mask is shown overlaying the point at measurement number 33. Some of the preceding points fall outside the arms of the mask, indicating that the mean of the measurement results has shifted significantly. If the mask is overlaid on the chart at point 15, one of the preceding points falls outside of the mask. After this point, the data continues on a downward trend.

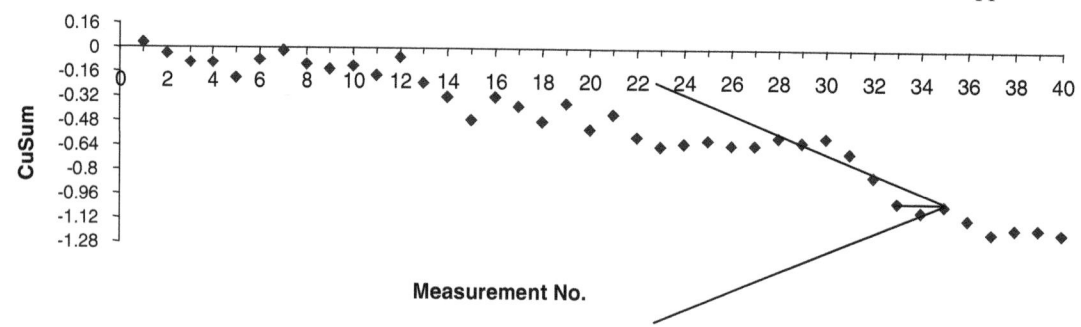

Figure C.12 CuSum chart for data shown in Table C.39.

Table C.40 Data from a proficiency testing round for the determination of copper in soil using aqua regia extraction.

Laboratory ID No.	Result x_i (mg kg^{-1})	$\|x_i - median\|$	$z = \dfrac{x_i - x_a}{\sigma_p}$
1	90	34.16	−2.85
2	127	2.84	0.24
3	121	3.16	−0.26
4	126.2	2.04	0.17
5	134	9.84	0.82
6	123	1.16	−0.10
7	109.92	14.24	−1.19
8	122	2.16	−0.18
9	143	18.84	1.57
10	137	12.84	1.07
11	145	20.84	1.74
12	118	6.16	−0.51
13	141	16.84	1.40
14	135	10.84	0.90
15	124.16	0.00	0.00
16	116	8.16	−0.68
17	112	12.16	−1.01
18	130	5.84	0.49
19	146	21.84	1.82
20	130	5.84	0.49
21	117	7.16	−0.60
22	78.5	45.66	−3.81
23	121.3	2.86	−0.24
24	127.4	3.24	0.27
25	116.2	7.96	−0.66
26	123	1.16	−0.10
27	140.23	16.07	1.34
Mean	124.22		
Standard deviation	15.35		
Median	124.16		
MAD		7.96	
MAD$_E$		11.8	

Questions and Solutions

Question 23

1. The mean, standard deviation, median and MAD_E are shown in Table C.40 together with the absolute deviations used to calculate MAD_E and the resulting z-scores.

2. To calculate MAD_E for this data set, first calculate the absolute difference (the deviation) between each result and the median and then find the median of these values. The median of the absolute deviations (MAD) is $7.96\,\text{mg kg}^{-1}$. This is converted to an estimate of standard deviation (MAD_E) by multiplying by 1.483:

$$MAD_E = 1.483 \times 7.96 = 11.8\,\text{mg kg}^{-1}$$

3. In this case, the mean and the median are very similar so either could be used as the target value. Since there appear to be some low outliers, however, the z-scores shown in Table C.40 were calculated using an assigned value $x_a = 124.16\,\text{mg kg}^{-1}$ based on the median. The standard deviation for proficiency assessment is set independently by the organiser as $\sigma_p = 12\,\text{mg kg}^{-1}$. The z-scores are therefore calculated as

$$z_i = \frac{x_i - 124.16}{12}$$

4. The result reported by laboratory number 1 would be considered questionable as the z-score is $2 < |z| \leq 3$. The result reported by laboratory number 22 would be considered unsatisfactory as $|z| > 3$. The performance of all the other laboratories is considered satisfactory as $|z| \leq 2$.

Question 24

1. The median of the results is 42.98 %abv.
2. The z-score is calculated from $z = (x - x_a)/\sigma_p$. In this case, the assigned value $x_a = 42.98$ %abv and the target standard deviation $\sigma_p = 0.03$ %abv. The z-score for each participant is shown in Table C.41.

Table C.41 z-Scores for participants from a proficiency testing scheme for the determination of the alcoholic strength of spirits.

Laboratory ID No.	z-Score	Laboratory ID No.	z-Score
1	0.00	18	−0.67
2	−0.67	19	−0.33
3	0.33	20	−0.33
4	1.33	21	0.33
5	0.33	22	0.00
6	0.33	23	1.00
7	−0.33	24	−0.67
8	0.67	25	0.67
9	−0.67	26	−8.00
10	0.67	27	0.00
11	−2.67	28	−3.33
12	−0.67	29	−0.67
13	0.67	30	0.00
14	1.00	31	0.33
15	−0.33	32	0.00
16	−1.00	33	0.00
17	0.33		

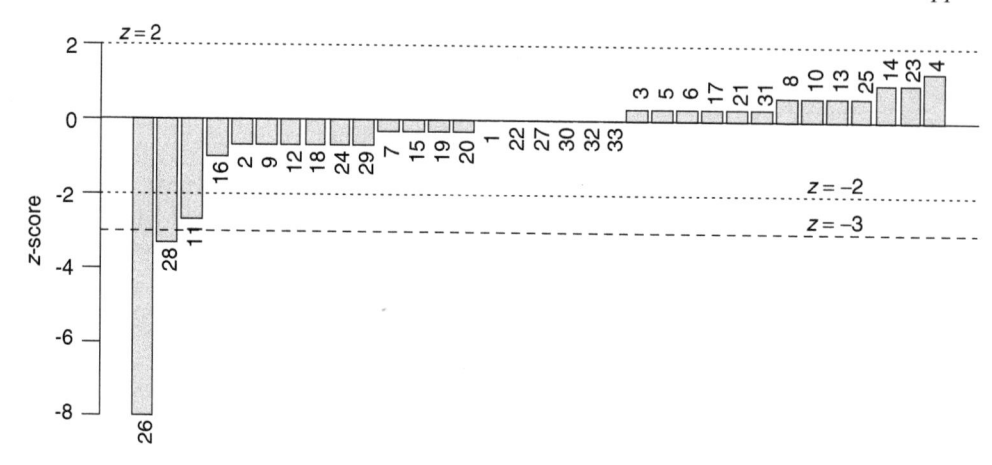

Figure C.13 Plot of z-scores from a proficiency testing scheme for the determination of the alcoholic strength of spirits. Laboratory identifiers are shown on each bar. Horizontal dashed and dotted lines are at $z=-3$ and $z=\pm 2$, respectievly.

3. A plot of the ordered z-scores is shown in Figure C.13. The z-scores show that the performance of laboratories 26 and 28 would be considered unsatisfactory ($|z|>3$). The performance of laboratory 11 is questionable ($2<|z|\leq 3$). The performance of all the other participants is considered satisfactory ($|z|\leq 2$).

Subject Index

Note: page numbers in *italic* refer to figures and tables.

A15 estimate 54–5
abbreviations, list of 216
accuracy defined 152
analysis of variance *see* ANOVA
Anderson–Darling test 23
ANOVA (analysis of variance) 22, 59–91
  assumptions 79–81
  background and basics
    factors and levels 59–60
    interpretation of results 62–3
    understanding and using tables 60–2
  homogeneity of variance (Levene's test) 80–1
  manual calculations and examples 82–91
  missing data 81
  normality checks 79–80
  one-way ANOVA 59, 63–5, 82–4
  two-factor ANOVA 59, 65–6
    with cross-classification 66–76, 84–8
    nested designs 76–9, 88–91
  uses
    in blocked experiments 130–1, 140–2
    in precision estimates 147–9
    regression data 98, 101–2, 103, 112–13
  variance components 61, 63, 64–5, 67, 77, 78, 79
area sampling 199
arithmetic mean *see* mean
assigned values (proficiency tests) 181–2

balanced incomplete block 140
balanced nested design 118
best fit line 92, 99
bias 37, 150–2
binomial distribution *18*, *19*, 20
blank samples 153
blocking and blocked designs 129–31, 139–42
  analysing a blocked experiment 130–1
    manual calculations 140–2

  confounding with higher-order interaction 139
  incomplete block designs 140
  multi-factor experiments 139
Box-Behnken design 136–7
box plots 11–12
  notched 12
bulk material 189

calibration of instruments 4, 103–7
  linear calibration design 116, *117*
  linearity checks 107, 157
  prediction uncertainty 99–100, 104–6
  and *p*-values 101
  replication 106–7
categorical data 24
central composite designs 135, *136*
central limit theorem 20
certified reference material 36, 150, 181
Chemical and Biological Metrology Programme v
chi-squared distribution 21–2
cluster sampling 195–7
Cochran test 53, 148
  tables of critical values 209–10
coefficient of variation 28
collaborative study/trial 53, 148, 183
composite samples 190
confidence intervals
  linear regression 95, 100
  for mean values 46–7
  and number of samples 120–2
confidence level 32
  and uncertainty reporting 168
control charts *see* quality control
convenience sampling 198
correlation coefficient *96*, 98–9, 109
counts 25
coverage factor 168

critical values 31, 33
   F-test 44, 102, 103
   t-test 35–6, 38–9, 42
   tables of 205–15
cross-classification (ANOVA) 66–76
cumulative distribution 10–11, 17, 19
cumulative frequency 9
CuSum Charts 175–8

D-optimal designs 138
data
   quick guide to reviewing and checking 2–3
   quick guide to summarising and describing 3
   types of 24
decision rules
   CuSum Charts 175–8
   Shewhart Charts 175
degrees of freedom 25
   ANOVA table 61, *66*, *70*, *77*
   for uncertainty estimate 168
   and t-distribution 20–1
density function 16
dependent variable 12, 92
design of experiments
   see experimental design
detection, limit of 153–5
discrete distributions 20
dispersion, measures of 25, 27–9
distribution function 16
distributions
   background and terminology 16–17
   different types of 17–22
Dixon tests 49–51
   tables of critical values 206–7
dot plots 7–8

effect defined 114
effective degree of freedom 168
equal variance t-test 38
expanded uncertainty 161, 168
experimental design 2, 115–42
   blocked designs 129–31, 133, 139–42
   D-optimal designs 138
   factorial designs *117*, 118–19
      see also ANOVA, two-factor
   fractional factorial designs 133–5, *136*
      design resolution 134–5
   Latin square designs 131–3
   linear calibration 116, *117*
      see also calibration of instruments
   measuring the right effect 115
   mixed level design 118

mixture designs 137–8
nested designs *117*, 118
   see also ANOVA, nested designs
nuisance effects, controlling 124–33
number of samples/observations required
   for desired standard deviation of the mean 119
   for desired t-test power 122–4
   for given confidence interval width 120–2
optimisation designs 135–7
paired designs 128
planning, basic 115
principles of good design 115–16
randomisation 125–8
replication, simple116, *117*
rotatable designs 135, 137
simplex designs *137*, 138
single vs multi-factor experiments 115
strategies (basic) 116–19
strategies (advanced) 133–40
validating designs 133

F-distribution *21*, 22, 45–6
F-test 42–6
   ANOVA *61*, 62, 67, 70–2, 77–9, 102
   example calculation 44–5
   tables of critical values 212–15
   and test for non-linearity 103
factorial designs *117*, 118–19
   advanced fractional designs 133–5, *136*
factors
   in ANOVA 59, 69–70
   experimental design 114–15
fractional factorial designs 133–5, *136*
frequency *9*, 25
frequency polygon 10
full factorial designs *117*, 118–19

gradient (linear regression)
   calculations 94–7
   regression statistics from software 100–1
Graeco-Latin square 131–33
graphical methods 7–15
   box-plots 11–12
   cumulative distribution 10–11
   dot plots 7–8
   frequency polygon 10
   histograms 9–10
   normal probability plots 13–15
   scatter plots 12–13
   stem-and-leaf plots 8–9
   tally charts 9
grid sampling 199

Subject Index

grouping factor (ANOVA) 60
Grubbs tests 51–3, 148
   table of critical values 208
Guide to the Expression of Uncertainty in Measurement
   (GUM) 161–2

H15 estimate 55
herringbone sampling 199
hierarchical ANOVA classification 76–9, 88–91
hierarchical experimental designs 118
histograms 9, 10, 16, *17*, 25
Horwitz function 183–4
Huber's proposal 55, 56–7
hypotheses
   in ANOVA 59, 62
   F-test 43, *44*, 62
   null and alternative described 30, 31
   one-sample t-tests 35, 36
   paired t-tests 41, 42
   two-sample t-tests 37, 39
   and Type I/II errors 114

increment of material 189
independent variable 12, 92
influence quantity 161, 165–6
instruments
   calibration *see* calibration of instruments
   drift and randomisation 125
   interaction (ANOVA) 69–70
   creating and interpreting plots 72, *73–4*
intercept (linear regression)
   calculations 94–7
   regression statistics from software 100–1
intermediate precision 146
interquartile range 28–9
   and box plot 11
interval data 24

judgement sampling 198

Kolmogorov–Smirnov test 23
Kragten spreadsheet 168–70
kurtosis 29, *30*

laboratory sample defined 190
Latin square designs 131–3
least-squares regression *see* linear regression
Levene's test 80–1
leverage 93–4
linear calibration *see* calibration of instruments
linear regression 92–108
   assumptions 92–3
   centroid 99, 104, 105

common mistakes 107–8
correlation coefficient *96*, 98–9
experimental designs 103–7
gradient and intercept calculations 94–7
interpreting statistics from software 100–2
non-linearity tests 102–3
residuals 92
   inspecting/plotting 97
   standard deviation of 94
and scatter plots 12–13
uncertainty in predicated values of x 99–100
visual inspection of data 12–13, 93–4
see also polynomial regression
linearity
   and instrument calibration 157
   and method validation 156–7
   non-linearity tests 102–3, 107, 109
location, measures of 25, 26
lognormal distribution *18*, *19*, 20

MAD and MAD$_E$ 55–6
mathematical notation 218–19
mean 26
   confidence intervals for 46–7
   robust estimators for 54–5
   standard deviation of 27–8, 119
   t-test comparisons 34–40
mean absolute deviation 55–6
mean squares (ANOVA table) 61–2, 66–7, *70*, 77, 85–91
measurement uncertainty 6, 161–72
   basic steps 162–8, 170–1
   bottom-up approach 170
   combining standard uncertainties 166–8
   definitions and terminology 161–2
   influence quantities 161, 165–6
   reporting measurement uncertainty 168
   and sampling strategies 201
   sources of uncertainty 161
      and measurement equation 162–4
      standard uncertainties for 164–5
   specifying the measurand 162
   spreadsheets 168–70
   top-down approach 170
   type A/type B estimates 161, 164
   using reproducibility data 170
median 26
   and robust statistics 53–4
median absolute deviation 55–6
method validation and performance 4–5, 144–60
   accuracy defined 152
   approach to 4–5, 144–6
   bias assessment 150–2
   limit of detection 153–5

limit of quantitation 155–6
linearity and working range 156–7
precision assessment 145–9
ruggedness testing 157–9
mixture designs 137–8
mode 26
multi-factor experiments 115
simple blocking extended to 139
multi-stage sampling 195–7

nested designs
ANOVA 77–9, 88–91
for experiments *117*, 118
precision estimates 147–9
non-linearity tests 102–3, 107, 109
non-parametric tests 32
normal distribution 16, 17, *18*, 19–20
checking normality 23–4, 79–80
derived distributions 20–2
normal probability plots 13–15
normal scores 13, 14

ogive curve 11
one-sample t-test 34–7
one-tailed test 32–3, 35, 37, 41, 43, 44, *45*
optimisation designs 135–7
ordinal data 24
outliers 2–3, 48–58
action on detecting 49
and box plot 11–12
defined and discussed 48–9
and normal probability plot 14
and normality checks 23
robust estimators 53–8
straggler 49, 50, 53
tests
Cochran 53
Dixon 49–51
Grubbs 51–3

p-values 34
in calibration 101
paired comparisons 40–2
parent distribution 16, 31–2
performance of laboratories *see* method validation and performance
Plackett–Burman design 158–9
point of influence (linear regression) 93–4
Poisson distribution *18*, *19*, 20
polynomial regression 108–13
ANOVA tables 109, 112–13
calculations 109–12
linearity and non-linearity 108–9

pooling data 27
population mean 26, 54–5
population parameters 23
population standard deviation 27
population variance 27
power (of a test)
calculations 123–4
defined 114
and number of samples 122–4
precision 145–9, *152*
experimental studies
nested designs 147–9
replication studies 146–7
intermediate precision 146, 148
precision limits 149
repeatability/reproducibility 146, 148, 149
statistical evaluation 149
target value 149
prediction uncertainty 99–100, 104–7
predictor variable 12
probability density function 16–17, *18*
probability distribution function 16
probability values 34
proficiency testing 6, 180–8
calculation of scores
assigned values 181–2
standard deviation for proficiency assessment 182–4
z- and Q-scores 184–5, *186*, 187–8
cumulative scores 187–8
rescaled sum of z-scores 187
sum of squared z-scores 187
guidelines and standards 180
interpreting/acting on results 185–7
ranking laboratories 188
propagation of uncertainty, law of 166

Q-score 185
quadratic equation 108, 109
quality control 5–6, 173
action/warning limits 173, 174, 175
CuSum Charts 175–8
Shewhart Charts 173–5
quantiles
and cumulative distribution 11
quantile–quantile (Q–Q) plot 13
quantitation, limit of 155–6
quartiles 28–9
and box plots 11
quota sampling 197

random effects (ANOVA) 59, 72, 75
random sample 23, 191–4, 199

Subject Index

randomisation 115–16, 125–8
   advantages of 125–6
   limitations of 128
   random number tables 126 7
   spreadsheet and experimental design software 127–8
range 28
ratio scale data 24, 28
rectangular distribution 22
reference material 150, 151, 181
regression *see* linear regression; polynomial regression
relative standard deviation 28
repeatability 5, 118, 146–9
   and proficiency testing 183
replication 116, *117*
   and bias assessment 151
   and precision evaluation 146–7
reporting results, quick guide to 3
representative samples 190
reproducibility 5, 118, 146, 148–9
   and proficiency testing 183
rescaled sum of z-scores (RSZ) 187
residuals 92, 94, 96
   and ANOVA 79, *80*
   inspecting 97–8
   and non-linearity testing 102–3
   standard deviation of 94
response defined 114
response surface designs 135–7
robust statistics 53–8
   for population means 54–5
   of standard deviation 55–7
   when to use 57–8
ruggedness/robustness 157–9

sample standard deviation 27
sample statistics 23
sample variance 27
sampling strategies 2, 23, 189–201
   composite samples 190
   convenience sampling 198
   definitions and nomenclature 189–90
   judgement sampling 198
   multi-stage and cluster sampling 195–7
   quota sampling 197
   representative samples 190
   sequential sampling 197–8
   simple random sampling 191–2, 199
   stratified random sampling 192–4, 199
   systematic sampling 194–5, 199
   two-dimensional sampling 199–201
   uncertainties in sampling 201
scale data 24, 28
scatter plots 12–13, 93

second-order polynomial *see* polynomial regression
sequential sampling 197–8
Shapiro–Wilk test 23
Shewhart Charts 173–5
significance level 32
significance testing
   basic procedures 29–34
   confidence intervals for mean values 46–7
   F-test 42–6
   t-tests 34–42
simple random sampling 191–2, 199
simple replication 116, *117*
skewness 29
sMAD 55, 56
spiked samples 150–1
standard deviation 27–8
   F-test comparisons 42–6
   robust estimates of 55–7
standard error 27–8
   of prediction 99–100, 104–6
standard normal distribution *19*, 20
standard uncertainty 161, 164–5, 166–8
star designs 135, *136*
statistical significance 32
statistical tables 205–15
statistical procedures
   basic techniques 25–47
   choice of 1–6
stem-and-leaf plots 8–9
straggler 49, 50, 53
stratified random sampling 192–4, 199
strip charts 7
Student's t-test 20, 34–42
   example calculations 36, 39, 42
   one-sample test procedure 34–7
   paired comparisons 40–2
   and power 122–4
   t-distribution 20–1
   table of critical values 211
   two-sample test procedure 37–40
sub-sample 190
sum of squared z-scores (SSZ) 187
sum of squares (ANOVA table) 61, *66*, *70*, *77*, 82–91
summarising data, quick guide to 3
symbols, frequently used 216–18
systematic sampling 194–5, 199

t-test *see* Student's t-test
tails, choosing number of 32–3
tally charts 9
test sample/portion defined 190
test statistics 32, 33, 34
   Cochran test 53

Dixon outlier tests *51*
F-test 43, *44*, *45*
Grubbs outlier tests 51–2
one-sample *t*-test 35
paired *t*-test 41
two-sample *t*-test *37*, 38
treatment defined 114
triangular distribution 22
trimmed mean 54
trueness defined 150
true value 23, 31, 150, 152, 161
two-dimensional sampling 199–201
two-sample *t*-test 37–40
two-tailed test 32–3, 35, 37, 41, 43, 44
Type I error 32, 114, 122
Type II error 114, 122

uncertainty *see* measurement uncertainty; prediction uncertainty

unequal variance *t*-test 38, 40
uniform distribution 22

V-mask 177–8
validation *see* method validation
variance 27
  analysis of *see* ANOVA
  components (ANOVA) 61, 63, 64–5, 67, 77, 78, 79
  equal variance *t*-test 38
  F-test comparison 42–5
  unequal variance *t*-test 38, 40
variation, coefficient of 28
visual inspection of data 7–15
  and linear regression 93–4

W pattern sampling 199
working range (of a method) 156–7

z-score 184–5, *186*, 187–8

Lightning Source UK Ltd.
Milton Keynes UK
UKHW030642040220
358126UK00005B/131